Observation and Experiment in the Natural and Social Sciences

BOSTON STUDIES IN THE PHILOSOPHY OF SCIENCE

Editors

ROBERT S. COHEN, *Boston University*
JÜRGEN RENN, *Max-Planck-Institute for the History of Science*
KOSTAS GAVROGLU, *University of Athens*

Editorial Advisory Board

THOMAS F. GLICK, *Boston University*
ADOLF GRÜNBAUM, *University of Pittsburgh*
SYLVAN S. SCHWEBER, *Brandeis University*
JOHN J. STACHEL, *Boston University*
MARX W. WARTOFSKY†, *(Editor 1960–1997)*

VOLUME 232

OBSERVATION AND EXPERIMENT IN THE NATURAL AND SOCIAL SCIENCES

Edited by

MARIA CARLA GALAVOTTI

University of Bologna, Italy

KLUWER ACADEMIC PUBLISHERS

DORDRECHT / BOSTON / LONDON

A C.I.P. Catalogue record for this book is available from the Library of Congress.

ISBN 1-4020-1251-9

Published by Kluwer Academic Publishers,
P.O. Box 17, 3300 AA Dordrecht, The Netherlands.

Sold and distributed in North, Central and South America
by Kluwer Academic Publishers,
101 Philip Drive, Norwell, MA 02061, U.S.A.

In all other countries, sold and distributed
by Kluwer Academic Publishers,
P.O. Box 322, 3300 AH Dordrecht, The Netherlands.

Printed on acid-free paper

All Rights Reserved
© 2003 Kluwer Academic Publishers
No part of this work may be reproduced, stored in a retrieval system, or transmitted
in any form or by any means, electronic, mechanical, photocopying, microfilming, recording
or otherwise, without written permission from the Publisher, with the exception
of any material supplied specifically for the purpose of being entered
and executed on a computer system, for exclusive use by the purchaser of the work.

Printed in the Netherlands.

TABLE OF CONTENTS

MARIA CARLA GALAVOTTI / Foreword	vii
PATRICK SUPPES / From Theory to Experiment and Back Again	1
PAOLO LEGRENZI / Naïve Probability	43
LÀSZLÓ SZABÓ / From Theory to Experiments and Back Again ... and Back Again ... Comments on Patrick Suppes	57
REINHARD SELTEN / Emergence and Future of Experimental Economics	63
WENCESLAO GONZALEZ / Rationality in Experimental Economics: An Analysis of Reinhard Selten's Approach	71
ROBERTO SCAZZIERI / Experiments, Heuristics and Social Diversity: A Comment on Reinhard Selten	85
GERD GIGERENZER / Where Do New Ideas Come From? A Heuristics of Discovery in the Cognitive Sciences	99
DAVID PAPINEAU / Comments on Gerd Gigerenzer	141
JEANNE PEIJNENBURG / On the Concept of Discovery. Comments on Gerd Gigerenzer	153
URSULA KLEIN / Styles of Experimentation	159
ARISTIDES BALTAS / On French Concepts and Objects. Comments on Ursula Klein	187
DONALD GILLIES / Some Comments on "Styles of Experimentation" by Ursula Klein	199
GÜROL IRZIK / Improving "Styles of Experimentation". A Comment on Ursula Klein	203

DAVID ATKINSON / Experiments and Thought Experiments in
    Natural Science   209

DANIEL ANDLER / The Advantage of Theft over Honest Toil.
    Comments on David Atkinson   227

MIKLÓS RÉDEI / Thinking About Thought Experiments in
    Physics. Comment on "Experiments and Thought Experiments
    in Natural Science"   237

MICHAEL STÖLTZNER / The Dynamics of Thought Experiments.
    A Comment on David Atkinson   243

GIORA HON / An Attempt at a Philosophy of Experiment   259

RAFFAELLA CAMPANER / An Attempt at a Philosophy of
    Experimental Error. A Comment on Giora Hon   285

GEREON WOLTERS / O Happy Error. A Comment on Giora Hon   295

COLIN HOWSON / Bayesian Evidence   301

IGOR DOUVEN / On Bayesian Logic. Comments on Colin Howson   321

PAOLO GARBOLINO / On Bayesian Induction (and Pythagoric
    Arithmetic)   327

ILKKA NIINILUOTO / Probability and Logic. Comments on
    Colin Howson   333

Index of Names   339

FOREWORD

According to a long tradition in philosophy of science, a clear cut distinction can be traced between a context of discovery and a context of justification. This tradition dates back to the birth of the discipline in connection with the Circles of Vienna and Berlin, in the twenties and thirties of last century. Convicted that only the context of justification is pertinent to philosophy of science, logical empiricists identified its goal with the "rational reconstruction" of scientific knowledge, taken as the clarification of the logical structure of science, through an analysis of its language and methods. Stressing justification as the proper field of application of philosophy of science, logical empiricists intended to leave discovery out of its remit. The context of discovery was then discarded from philosophy of science and left to sociology, psychology and history.

The distinction between context of discovery and context of justification goes hand in hand with the tenet that the theoretical side of science can – and should – be kept separate from its observational and experimental components. Further, the final, abstract formulation of theories should be analysed apart from the process behind it, resulting from a tangle of context-dependent factors. This conviction is reflected by the distinction between theoretical and observational sentences underpinning the Hempelian view of theories as nets, whose knots represent theoretical terms, floating on the plane of observation, to which it is anchored by rules of interpretation. This view assigns to philosophy of science the task of clarifying the relationship between theoretical and observational terms, while taking the plane of observation as "given".

The view of theories upheld by logical empiricists, together with the distinction between context of discovery and context of justification, has been criticized in many ways, and has gradually been abandoned in favour of a more flexible viewpoint according to which theory and observation are not separate but strictly intertwined. Once it was admitted that the context of justification forms a continuum with the context of discovery, interest in the latter spread rapidly, and a whole array of new problems has been addressed in this connection. Observation and experimentation have become an important field of inquiry. The present volume is meant as a contribution to the ongoing debate on this topic.

The first paper is by Patrick Suppes, a forerunner of a constructive kind of epistemology that regards scientific theories as sets of models, ranging from empirical models describing experimental data, to mathematical models characterizing the abstract core of the theory. According to this perspective, scientific knowledge has to be analysed with reference to the logical as well as empirical structures characterizing its models, and philosophy of science is concerned as much with formal logic and set theory, as with probability and statistical inference. In his paper, Suppes argues that theory and experiment are engaged in a continuous interplay, substantiating this claim by two examples, taken respectively from the study of the brain and quantum physics.

Reinhard Selten focuses on experimental economics, a field to which he has contributed pioneering work. The beginnings of this discipline, which is now considered an important branch of economics, are recollected, together with its first developments in Germany and the USA. Selten then describes the standards of experimental economics, to be met by acceptable research in the discipline. Fully convinced of the utility of experimenting in economics, Selten predicts a growth surge in the future, as a complement of field research. Selten identifies the specific contribution of experimental economics with its capacity to test major assumptions of economic behaviour and to come up with facts that can lead to theories of limited application, which will hopefully converge towards a comprehensive theory of economic behaviour.

The paper by Gerd Gigerenzer claims the distinction between discovery and justification is "artificial", and argues for a heuristics grounded on the idea that the methods of justification adopted by scientists provide tools for the discovery of new theories. Inferential statistics and the digital computer are the examples discussed by Gigerenzer to illustrate his heuristics "from tools to theories", which originates in connection with cognitive science, but can be extended to other fields as well. Gigerenzer's analysis calls attention to the role of scientists' practice within scientific investigation and theory construction, a role widely overlooked by epistemologists.

Ursula Klein deals with experimentation in chemistry, through a careful examination of two case studies: the study of plant tissues by means of chemical operations in the eighteenth and early nineteenth centuries, and organic chemistry in Europe from the late 1820s onward. The discussion of these examples leads Klein to a distinction between "experimental analysis" and "experimental culture", meant to shed light on the multifarious and dynamic relationship between theory and experiment.

The paper by David Atkinson addresses the issue of thought experiments and their usefulness for the investigation of nature. Atkinson's analysis of three quite different cases, namely Galileo's, the EPR argument, and string theory, suggests insightful considerations on the relationship between real and thought experiments.

Giora Hon focuses on error, in the conviction that the theory of experiment can profit from a careful consideration and classification of the possible errors that can occur within an experimental setting. Such an analysis of empirical knowledge from the "negative" perspective of possible faults deepens its roots in the philosophy of Francis Bacon, to whom Hon refers to propose a slightly modified classification of experimental errors.

Colin Howson's paper addresses the assessment of probability evaluations in the face of empirical evidence. According to Bayesianism, the process of conditioning probability judgments upon the available evidence is subject to a consistency requirement, which a long tradition interprets as a rationality constraint having a direct bearing upon behaviour. Unlike this tradition, Howson puts forward a purely logical approach to personal probability, quite apart from the notions of utility or preference, usually associated with Bayesianism.

On the whole, the papers collected here span a wide range of problems related to experimentation, and the comments following them broaden the field even further, while adding considerably to the significance of the ensemble. Most of the main papers and comments ground epistemological considerations on historical remarks or case studies, according to a typical tendency of the European approach to epistemology, to combine ideas belonging to the history and the philosophy of science into a unique perspective.

In the year 2000, the European Science Foundation launched a Scientific Network on "Historical and contemporary perspectives of philosophy of science in Europe", intended to strengthen the European tradition and consolidate distinctive European perspectives in the field. During its three years of activity, the Network's coordinating committee, including Maria Carla Galavotti (Italy, chairperson), Aristides Baltas (Greece), Donald Gillies (United Kingdom), Theo Kuipers (Holland), Ilkka Niiniluoto (Finland), Michel Paty (France), Miklos Redei (Hungary), Friedrich Stadler (Austria) and Gereon Wolters (Germany), agreed to organize three workshops centred on the major themes under discussion within the discipline. This book collects the papers presented at the first workshop, held at the Bertinoro Conference Centre of the University of Bologna, from 30th September to 2nd October, 2001. The topic of the workshop: "Observation and experiment in the natural and social sciences" was chosen not only in view of its centrality in the ongoing debate in philosophy of science, but also as an ideal starting point for an appraisal of recent research in the field. To encourage a fruitful exchange of ideas between philosophers and scientists, both the main speakers and commentators were picked from among researchers with different backgrounds, such as physics, psychology, economics, chemistry, as well as philosophy of science.

The workshop was funded by the European Science Foundation, and received support from the "Federigo Enriques" Interdisciplinary Research Centre for Epistemology and History of Science and the Department of Philosophy of the University of Bologna. The publication of the proceedings benefited from a generous grant of the Alma Mater Studiorum – University of Bologna. As local organizer of the conference and editor of the proceedings, I wish to thank Raffaella Campaner for her help, and for extensive editorial work in preparing this volume for publication. On behalf of the other members of the coordinating committee of the ESF Scientific Network, I express deep gratitude to William Shea, President of the Standing Committee for the Humanities of the European Science Foundation, for supporting our work in so many ways.

MARIA CARLA GALAVOTTI

Department of Philosophy,
University of Bologna,
Bologna, Italy

PATRICK SUPPES

FROM THEORY TO EXPERIMENT AND BACK AGAIN

In this article I consider two substantive examples of the way in which there is continuing interaction in science between theory and experiment. The picture of theory often presented by philosophers of science is too austere, abstract and self-contained. In particular, the picture of theory that is painted is much too removed from the shock effects of new experiments. Perhaps even more to the point, in many parts of science the actual formulation of theory is much driven by the latest experiments.

The first example comes from scientific research I am currently doing on language and the brain. I begin by describing the work in broad terms. I then present the response of new experiments and new theoretical statistical analysis of the data to answer claims that the recognition rates for brain-wave representation of words and sentences is not significant, because of the large amount of information available. Here the use of the concept of an extreme statistic is used to answer this criticism in a detailed way. Discussion of this example will end with some brief remarks on how this use of more detailed statistical methods is now generating new experiments, and having an impact on the design of the experiments.

The second example deals with experiments and physical theory on the entanglement of particles, and the consequent nonlocality of standard quantum mechanics. After some general remarks on this area of research in quantum mechanics and its philosophical importance for our basic physical concepts, I turn to the theoretical work of Greenberg, Horne and Zeilinger and their proposed "GHZ-type" experiments.

First the purely theoretical result, formulated in probability-one terms, is stated. Then the question is asked, how can such probability-one theoretical results be tested, given the inevitable inefficiencies of particle detectors.

This prompts a new theoretical effort to derive inequalities, like those of Bell for other experiments, to deal with GHZ-type experiments. What comes out of the analysis is that better experimental results should be achievable with very careful design and use of current photon detectors. But the proof of this is rather detailed and relies on theory in critical ways at several points. These examples are but current illustrations, but the lesson is meant to be universal. The continual interaction between theory and experiment occurs in nearly every developed branch of science.

LANGUAGE AND THE BRAIN[1]

Some historical background

Aristotle said that the distinguishing feature of man as an animal is that he is a rational animal, but, in more biological and psychological terms, it is that of being a talking animal. Language is, in ways that we have not yet fully explored, the most distinguishing mark of man as an animal. Its processing is centered, above all, in the brain, not just for the production of speech, but for the intentional formation of what is to be said or for the comprehension of what has been heard or read. So it is the brain's processing of language that is the focus of this section. I begin with a historical sketch of the discovery of electrical activity in the brain.

An early reference to electricity being generated by muscles or nerves of animals comes from a study by Francesco Redi (1671), who describes in this way an experiment he conducted in 1666: "It appeared to me as if the painful action of the *torpedine* (electric ray) was located in these two sickle-shaped bodies, or muscles, more than in any other part." Redi's work was done in Florence under the Medici's. These electrical observations were fragmentary and undeveloped. But the idea of electrical activity in the muscles or nerves of various animals became current throughout the eighteenth century (Whittaker 1951, Galvani 1791). Yet it was more than 100 years after Redi before the decisive step was taken in Bologna by Luigi Galvani. He describes his first steps in the following manner:

The course of the work has progressed in the following way. I dissected a frog and prepared it ... Having in mind other things, I placed the frog on the same table as an electric machine. When one of my assistants by chance lightly applied the point of a scalpel to the inner crural nerves of the frog, suddenly all the muscles of the limbs were seen so to contract that they appeared to have fallen into violent tonic convulsions. Another assistant who was present when we were performing electrical experiments thought he observed that this phenomenon occurred when a spark was discharged from the conductor of the electrical machine. Marvelling at this, he immediately brought the unusual phenomenon to my attention when I was completely engrossed and contemplating other things. Hereupon I became extremely enthusiastic and eager to repeat the experiment so as to clarify the obscure phenomenon and make it known. I myself, therefore, applied the point of the scalpel first to one then to the other crural

---

[1] This section is taken from my forthcoming book *Representation and Invariance in Scientific Structures,* Stanford, CA: CSLI Publications.

nerve, while at the same time some one of the assistants produced a spark; the phenomenon repeated itself in precisely the same manner as before.

(Galvani 1791/1953, pp. 45–46)

Galvani's work of 1791 was vigorously criticized by the well-known Italian physicist Alessandro Volta (1745–1827), who was born in Como and was a professor of physics at the University of Pavia. Here are his words of criticism, excerpted from a letter by Volta to Tiberius Cavallo, read at the Royal Society of London:

The name of animal electricity is by no means proper, in the sense intended by Galvani, and by others; namely, that the electric fluid becomes unbalanced in the animal organs, and by their own proper force, by some particular action of the vital powers. No, this is a mere artificial electricity, induced by an external cause, that is, excited originally in a manner hitherto unknown, by the connexion of metals with any kind of wet substance. And the animal organs, the nerves and the muscles, are merely passive, though easily thrown into action whenever, by being in the circuit of the electric current, produced in the manner already mentioned, they are attacked and stimulated by it, particularly the nerves.

(Volta 1793/1918, pp. 203–208)

Galvani was able to meet these criticisms directly and in 1794 published anonymously a response containing the detailed account of an experiment on muscular contraction without the use of metals (Galvani 1794). The original and important nature of Galvani's work came to be recognized throughout Europe. The prominent German physicist Emil Du Bois-Reymond (1848) summarized in the following way Galvani's contribution:

1. Animals have an electricity peculiar to themselves, which is called Animal Electricity.

2. The organs to which this animal electricity has the greatest affinity, and in which it is distributed, are the nerves, and the most important organ of its secretion is the brain.

3. The inner substance of the nerve is specialized for conducting electricity, while the outer oily layer prevents its dispersal, and permits its accumulation.

4. The receivers of the animal electricity are the muscles, and they are like a Leyden jar, negative on the outside and positive on the inside.

5. The mechanism of motion consists in the discharge of the muscular fluid from the inside of the muscle via the nerve to the outside, and this discharge of the muscular Leyden jar furnishes an electrical stimulus to the irritable muscle fibres, which therefore contract.

(Du Bois-Reymond 1848/1936, p. 159)

A next event of importance was the demonstration by Carlo Matteucci (1844) that electrical currents originate in muscle tissue. It was, however, almost 100 years after Galvani, that Richard Caton (1875) of Liverpool detected electrical activity in an exposed rabbit brain, using the Thomson (Lord Kelvin) reflecting telegraphic galvanometer. In 1890, Adolf Beck of Poland detected regular electrical patterns in the cerebral cortex of dogs and rabbits. Beginning at the end of the nineteenth century Villem Einthoven, a Dutch physician and physiologist, developed a new electrocardiograph machine, based on his previous invention of what is called the string galvanometer, which was similar to the device developed to measure telegraphic signals coming across transatlantic cables. Using Einthoven's string galvanometer, significant because of its sensitivity, in 1914, Napoleon Cyblusky and S. Jelenska Macieszyna, of the University of Cracow in Poland, recorded a dog's epileptic seizures. Beginning about 1910, Hans Berger in Jena, Germany began an extensive series of studies that detected electrical activity through intact skulls. This had the great significance of being applicable to humans. His observations were published in 1929, but little recognized. Recognition came, however, when his findings were confirmed by Edward Douglas Adrian and B. H. C. Matthews of the University of Cambridge, who demonstrated Berger's findings at the Physiological Society in Cambridge in 1934, and the International Congress of Psychology in 1937. In the late 1930s and the early 1940s research on electrical activity in brains, or what we now call electroencephalography (EEG), moved primarily to North America—W. G. Lennox and Erna and F. A. Gibbs at the Harvard Medical School, H. H. Jasper and Donald Linsley at Brown University, and Wilder Penfield at McGill University. One of the first English-language reports to verify Berger's work was by Jasper and Carmichael (1935). Nearly at the same time, Gibbs et al. (1935) began using the first ink-writing telegraphic recorder for EEG in the United States, built by Garceau and Davis (1935). By the 1950s, EEG was widely used clinically, especially for the study of epilepsy, and for a variety of research on the nature of the electrical activity in the brain. This is not the place to summarize in any serious detail the work by a wide variety of scientists from 1950 to the present, but an excellent review of EEG, that is, of electrical activity, pertinent especially to cognition, is to be found in Rugg and Coles (1995).

Observing the brain's activity

The four main current methods of observing the brain are easy to describe. The first is the classical electroencephalographic (EEG) observations already mentioned, which, and this is important, have a time resolution of at least one millisecond. The second is the modern observation of the magnetic field rather than the electric field, which goes under the title of magnetoencephalography (MEG). This also has the same time resolution of approximately one millisecond. The third is positron emission tomography (PET), which has been widely used in the last several decades and is good for observing location, in some cases, of brain activity, but has a time resolution of only one second. Finally, the most popular current method is functional magnetic resonance imaging (fMRI), which does an excellent job of observing absorption of energy in well-localized places in the brain, but unfortunately, also has a time resolution of no less than a second.

Although many excellent things can be learned from PET and fMRI, they are not really useful if one wants to identify brain waves representing words or sentences, for the processing, although slow by modern computer standards, is much too fast to be able to accomplish anything with the time resolution of observation no better than one second. The typical word, for example, whether listened to or read, will be processed in not more than 4 or 5 hundred milliseconds, and often faster. My own serious interest, focused on the way the brain processes language, began from the stimulus I received by hearing a brilliant lecture in 1996 on MEG by Sam Williamson, a physicist who has been prominent from the beginning in the development of MEG. I was skeptical about what he said, but the more I thought about it, the more I realized it would be interesting and important to try using MEG to recognize the processing of individual words. This idea suggested a program of brain-wave recognition, as recorded by MEG, similar in spirit to speech recognition. I was familiar with the long history of speech recognition from the 1940s to the present, and I thought maybe the same intense analytical effort could yield something like corresponding results. So, in 1996, assisted especially by Zhong-Lin Lu, who had just taken a Ph.D. with Sam Williamson and Lloyd Kaufman at New York University, we conducted an MEG experiment at the Scripps Institute of Research in San Diego, California. When we proceeded to analyze the results of the first experiment, the problem of recognizing which one of seven words was being processed on the basis of either having heard the word or having read it on a computer screen, we were not able to get very good recognition results from the MEG recordings. Fortunately, it was a practice at the Scripps MEG facility, which is technically very much more expensive and complicated to run than standard EEG equipment, to also record the standard 20 EEG sensors used for many years. We proceeded to analyze the EEG data as well, and here we had much better recognition results (Suppes, Lu and Han 1997).

In the standard EEG system, widely used throughout the world for observing electrical activity in the brain, sensors to record the electrical activity are arranged in what are commonly called the 10-20 system, as shown in Figure 1, with the location on the surface of the skull of the head shown in the

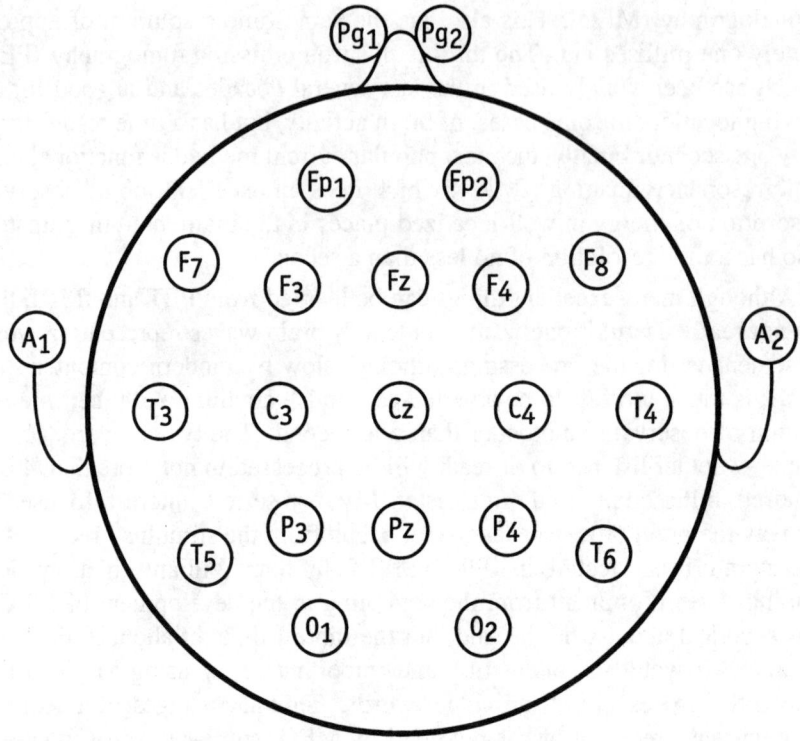

Figure 1: The 10-20 system of EEG sensors.

approximate form of a circle, with ears left and right and eyes at the top in the figure. The first letters of the initials used in the locations correspond to references to the location in the part of the brain, for example, F for frontal, C for center, T for temporal, P for parietal, O for occipital. Second, you will note that the odd-numbered sensors are located on the skull on top of the left hemisphere and the even-numbered sensors are located over the right hemisphere, with three sensors located approximately along the center line. There are more opinions than deeply articulated and established facts about what takes place in the left hemisphere or in the right hemisphere, possibly both, in the processing of language. My own view is that there is probably more duality than has been generally recognized, but I will not try to make an empirical defense of that view in the present context, although I have published data supporting duality (Suppes, Han and Lu 1998, Table 2).

Figure 2 shows a typical trial, in which the subject was given a visual, i.e., printed, sentence, one word at a time, on a computer screen. The trial

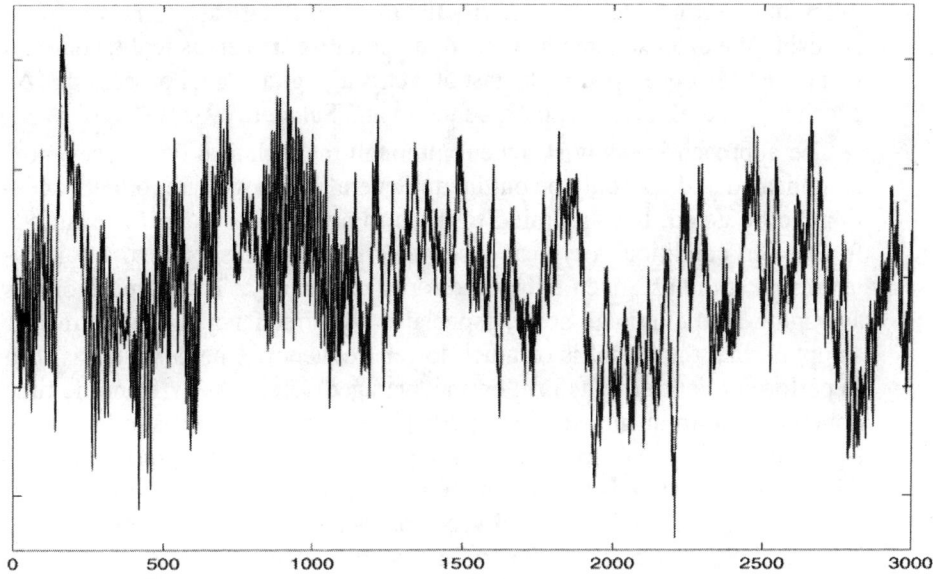

Figure 2: Unanalyzed EEG data from one trial and one sensor.

lasted for over 3000 milliseconds, with an observation of an amplitude of the observed wave plotted on the y-coordinate in microvolts every millisecond. Given that so much data are observed in a little over three seconds, just from one sensor, out of 20, it is easy to see that EEG recordings of language activity are rich in data and, in fact, we might almost say, swamped by data. It is not difficult to run an experiment with several subjects, each with a number of hours of recording, and have at the end between five and ten gigabytes of data of the kind to be seen in Figure 2. This means that the problem of trying to find waves corresponding to particular words and sentences is not going to be a simple matter. It is very different from behavioral experiments dealing with language, in which the observed responses of subjects are easily recorded in a few megabytes and then analyzed without anything like the same amount of computation.

Methods of data analysis

There is no royal road to finding words and sentences in the kind of data just described, so I will describe here the approach that I and my colleagues have used with some success in the past few years. The basic approach is very much taken from digital signal processing, but the application is a very different one from what is ordinarily the focus of electrical engineers or others using the extensive mathematical, quantitative and statistical techniques that have been developed in the last 50 years in digital signal processing. An excellent general reference is Oppenheim and Schafer (1975).

The approach is easily sketched, although the technical details are more complicated and will only be outlined. Generally, but with important exceptions to be noted, the first thing to do is to average the data, for example, for a given condition, a typical case being the brain response to a particular verbal stimulus, given either auditorily or visually. The purpose of this averaging is to eliminate noise, especially high-frequency noise, on the assumption that the signal is of much lower frequency. The next step is then to perform a Fourier transform on the averaged data, passing from the time domain to the frequency domain, perhaps the most characteristic feature of signal processing analysis. The third step, which can be done simultaneously with the second step, is to filter in the frequency domain, to reduce still further the bandwidth of the signal used for identifying the word or sentence being processed. An alternative, which we have explored rather thoroughly in some work, is to select in the frequency domain the frequencies with the highest energies, as measured by the absolute value of their amplitudes, and then to superpose the sine functions in the time domain. (Actually, in "selecting a frequency ω_i with amplitude A_i", we are in fact selecting a sine wave $A_i \sin(\omega_i t + \varphi_i)$, where φ_i is its phase. More on this on the next page.) In either case, by filtering or superposition, we get a much simpler signal as we pass by an inverse Fourier transform back to the time domain.

Speaking now of filtering, and ignoring for the moment the superposition approach, we go back to the time domain with a bandpass filter fixed by two parameters, the low frequency L and the high frequency H of the filter. We now select two more parameters that are of importance. Namely, what should we take to be the beginning s and the ending e of the signals for the words or sentences in a given experimental condition. As in other brain matters, it is not at a glance obvious from the recorded brain wave when the representation of a particular word or sentence begins or ends in the continually active electrical waves that are observed. When a quadruple (L, H, s, e) is selected, we then use that quadruple of parameters to make a classification of the brain waves of the words or sentences that are the stimuli in a given experimental condition.

Our aim is to optimize or maximize the number of brain waves correctly classified. We keep varying the quadruple of parameters until we get what seems to be the best result that can be found. We are doing this in a four-dimensional grid of parameters, but I show in Figure 3 the optimal surface for two parameters of the filter, although here what is used as a measure on the ordinate or y-axis is the difference between the high-frequency and low-frequency filter rather than the high filter itself. So we have on the abscissa the low filter measured in Hz and on the ordinate the difference W, which is the width in Hz of the bandpass filter. The smoothness of the surface shown in Figure 3 is characteristic of what we always observe and would be expected, of observations of an electrical field outside the skull, the place where we are observing them. (We are of course very fortunate that the electrical fields are strong enough to be observed without complete contamination by noise.) This isocontour map is for the first experiment with 48 geographic sentences discussed in more detail below, but the map shows clearly that the best recognition rate was 43 of the 48 sentences (approximately 90%). As the parameters of the bandpass filter are changed, the contour map shows definitely lower recognition rates.

Fourier analysis of EEG data

In the background of the Fourier analysis is the standard theory of the Fourier integral, but in practice our data are finite. The finite impulse data that we are interested in observing usually last for no more than a few seconds. For example, the sentences studied will ordinarily last not more than three or four seconds when spoken at a natural rate. To analyze frequencies with given amplitudes and phases, we use the discrete Fourier transform. As indicated, our goal is to find the frequencies that contain the signal and eliminate the noise. (The artifacts generated by eye blinks or other such events are discussed at the end of this section.)

Let N be the number of observations, equally spaced in time, usually one millisecond apart. We then represent the finite sequence of observations $x(n), 0 \leq n \leq N - 1$ by Fourier series coefficients $\tilde{X}(k)$ as a periodic sequence of period N, so we have the dual pair

$$\tilde{X}(k) = \sum_{n=0}^{N-1} \tilde{x}(n) e^{-i(\frac{2\pi}{N})kn} \tag{1}$$

$$\tilde{x}(n) = \frac{1}{N} \sum_{k=0}^{N-1} \tilde{X}(k) e^{i(\frac{2\pi}{N})kn} \tag{2}$$

I first note the following:

1. The periodic sines and cosines are represented by the standard exponential terms.

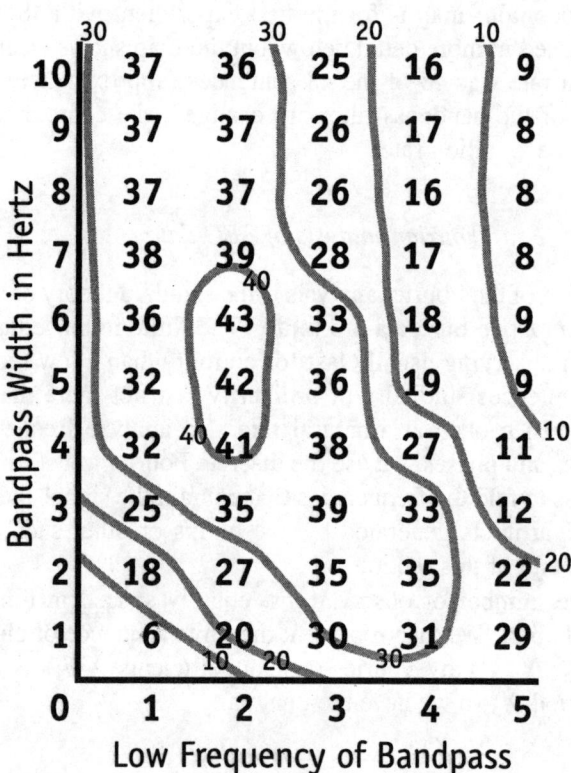

Figure 3: Typical contour map of recognition rate surface for bandpass-filter parameters L and W.

2. $\tilde{x}(n) = x(n) = \tilde{x}(n + kN)$, the tilde shows periodicity of length N and is for duality of time and frequency.

3. The kn part of the exponent gives us distinct exponentials, and thus sine and cosine terms for integer submultiples of the period N. This way we get in the representation frequencies that are an integer multiple of $\frac{2\pi}{N}$.

4. Using the periodicity N gets us duality between the time and frequency domains.

The properties of the two equations (1) and (2) are:

1. Linearity: If $\tilde{x}_1(n)$ and $\tilde{x}_2(n)$ have period N, so does

$$\tilde{x}_3(n) = a\tilde{x}_1(n) + b\tilde{x}_2(n)$$

and

$$\tilde{X}_3(k) = a\tilde{X}_1(k) + b\tilde{X}_2(k).$$

2. Invariance under shift of a sequence

$$n \to n + m$$

3. Various symmetry properties, e.g., $|\tilde{X}(k)| = |\tilde{X}(-k)|$.

4. Convolution of $\tilde{x}_1$ and $\tilde{x}_2$ of period N has period N:

$$\tilde{x}_3(n) = \sum_{m=0}^{N-1} \tilde{x}_1(m)\tilde{x}_2(n-m).$$

Of importance is the efficient fast discrete Fourier transform, an algorithm due to Cooley and Tukey (1965) and others, a variant of which was used in the computations reported below.

Filters. The principle of filter construction is simple. Details are not. A bandpass filter, e.g., 1-20 Hz simply "filters all the frequencies below 1 Hz and above 20 Hz." There are many developments in the electrical engineering literature on the theory and art of designing filters, which it is not possible to survey here. The important point is always to design a filter with some criterion of optimality.

If the signal is known, then the engineering objective is to optimize its transmission. Our problem, as already mentioned, is that, in our experiments, the signal carrying the word or sentence in question is unknown. So our solution is to optimize the filter to predict the correct classification. The parameters we used have been discussed above. In addition we often make a smoothing correction around the edges of the filter by using a 4th-order Butterworth filter, although in the work reported here, something simpler would serve the purpose just about as well.

Three experimental results

I turn now to three of the most important results we have obtained so far.

Invariance between subjects

In the first experiment, we presented 48 sentences about the geography of Europe to 9 subjects. The subjects were asked to judge the truth or falsity of the sentences, and while they were either listening to or reading the sentences displayed one word at a time on a computer screen, we made the typical EEG recordings. The semantic task was simple, but because the sentences were separated by only four seconds, the task of judging their truth or falsity was not trivial. Typical sentences were of the form *The capital of Italy is not Paris,* and *Warsaw is not the largest city in Austria.* Taking now the data from five subjects to form prototypes of the 48 sentences, by averaging the data from the five subjects, and taking the other four subjects to form corresponding averaged test samples of each sentence, we applied the Fourier methods described above and found an optimal bandpass filter from a predictive standpoint. (The data are for the visual condition of reading the sentences, one displayed word at a time.) We were able to recognize correctly 90% of the test samples, using as a criterion for selection a classical least-squares fit between a test sample and each of the 48 prototypes, after filtering (Suppes, Han, Epelboim and Lu 1999a).

Let $x_i(n), 0 \leq n \leq N - 1$, be the ith prototype (in the time domain), and $y_j(n), 0 \leq n \leq N - 1$, the jth test sample. Then the sum of squared differences is S_{ij}, where

$$S_{ij} = \sum_{n=0}^{N-1} (x_i(n) - y_j(n))^2.$$

The test sample $y_j(n)$ is correctly classified if

$$S_{jj} = \min_i S_{ij},$$

with the minimum being unique.

The surprising invariance result is that the data for prototypes and for test samples came from different subjects. There was no overlap in the two groups. Theoretically this is an efficient aspect of any much used communication system. My brain-wave representation of words and sentences is much like yours, so it is easy to understand you. But it is a theoretical point that needs strong empirical support to have it accepted. Another angle of comment is that the electric activity in the cortex is more invariant across subjects performing the same task than is the detailed anatomical geometry of their brains. I return to this invariance between subjects a little later, when I respond to some skeptical comments.

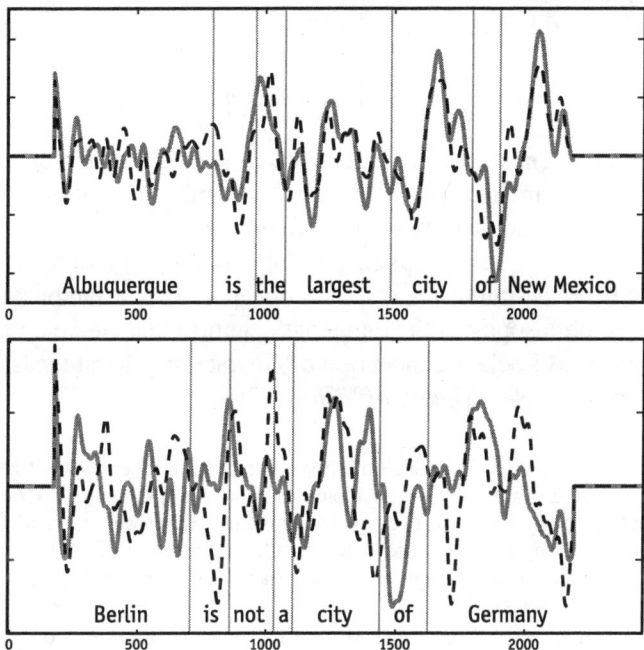

Figure 4: Prototypes (grey lines) and test samples (dashed black lines) generated by the best fitting sentence (upper panel) and worst fitting sentence (lower panel) correctly classified for subject S32. Time measurements after the onset of the sentence are shown in milliseconds on the abscissa.

One hundred sentences

I now turn to a second more recent experiment in which subjects were visually presented 100 different geography sentences (Suppes, Wong, et al., to appear). I concentrate here only on the remarkable result of correct recognition of 93 of the 100 sentences for one subject (S32). Using the methods described, the best recognition rate achieved for a single subject (S32) was 93%, i.e., 93 of the 100 test samples. These results were achieved with $L = 1.25$ Hz, $W = 21.25$ Hz, $s = 180$ ms after onset of the visual presentation of the first word of each sentence, and $e = 2200$ ms, marking the ending of the recordings used for the least-squares criterion of fit. The best bipolar sensor was C4-T6. In Figure 4 we show at the top the best and at the bottom the worst fit, as measured by the least-squares criterion, for the 93 sentences correctly recognized. The sum of squares for the worst was more than three times that for the best.

Invariance between visual images and their names

The third experiment showed that the visual images generated on a computer screen, of a familiar shape, such as a circle or triangle, were very similar to the brain images generated by the corresponding word (Suppes, Han, Epelboim and Lu 1999b). This surprising result very much reinforced a classical solution of how the mind has general concepts. It is a famous episode in the history of philosophy in the eighteenth century that Berkeley and Hume strongly criticized Locke's conception of abstract or general ideas. Berkeley has this to say in *A New Theory of Vision* (1709/1901):

> It is indeed a tenet, as well of the modern as the ancient philosophers, that all general truths are concerning universal abstract ideas; without which, we are told, there could be no science, no demonstration of any general proposition in geometry. But it were no hard matter, did I think it necessary to my present purpose, to shew that propositions and demonstrations in geometry might be universal, though they who make them never think of abstract general ideas of triangles or circles.
>
> After reiterated efforts and pangs of thought to apprehend the general idea of a triangle, I have found it altogether incomprehensible. And surely, if any one were able to let that idea into my mind, it must be the author of the *Essay concerning Human Understanding*: he, who has so far distinguished himself from the generality of writers, by the clearness and significancy of what he says. Let us therefore see how this celebrated author describes the general or which is the same thing, the abstract idea of a triangle. "It must be", says he, "neither oblique nor rectangle, neither equilateral, equicrural, nor scalenum; but all and none of these at once. In effect it is somewhat imperfect that cannot exist; an idea, wherein some parts of several different and inconsistent ideas are put together." (*Essay on Human Understanding*, B. iv. ch. 7. s. 9.) This is the idea which he thinks needful for the enlargement of knowledge, which is the subject of mathematical demonstration, and without which we could never come to know any general proposition concerning triangles. Sure I am, if this be the case, it is impossible for me to attain to know even the first elements of geometry: since I have not the faculty to frame in my mind such an idea as is here described.
>
> (Berkeley, pp. 188–189.)

Hume, in a brilliant exposition and extension of Berkeley's ideas, in the early pages of *A Treatise of Human Nature,* (1739/1951) phrased the matter beautifully in the opening paragraph of Section VII, entitled *Of Abstract Ideas*:

> A very material question has been started concerning *abstract* or *general* ideas, *whether they be general or particular in the mind's conception of them.* A great philosopher has disputed the receiv'd opinion in this particular, and has asserted, that all general ideas are nothing but particular ones, annexed to a certain term, which gives them a more extensive signification, and makes them recall upon occasion other individuals, which are similar to them. As I look upon this to be one of the greatest and most valuable discoveries that has been made of late years in the republic of letters, I shall here endeavour to confirm it by some arguments, which I hope will put it beyond all doubt and controversy. (Hume, *Treatise*, p. 17.)

Although not discussed by Berkeley and Hume, we also confirmed that the same is true of simple patches of color. In other words, a patch of red and the word "red" generate similar brain images in the auditory part of the cortex.

The specific significant results were these. By averaging over subjects as well as trials, we created prototypes from brain waves evoked by stimuli consisting of simple visual images and test samples from brain waves evoked by auditory or visual words naming the visual images. We correctly recognized from 60% to 75% of the test-sample brain waves. Our general conclusion was that simple shapes and simple patches of color generate brain waves surprisingly similar to those generated by their verbal names. This conclusion, taken together with extensive psychological studies of auditory and visual memory, support the solution conjectured by Berkeley and Hume. The brain, or, if you prefer, the mind, associates individual visual images of triangles, e.g., to the word *triangle*. It is such an associative network that is the likely procedural replacement for the mistaken attempt by Locke to introduce abstract ideas.

Comparisons of averaged and filtered brain waves generated by visual images and spoken names of the images are shown in Figure 5. Time after the onset of the stimulus (visual image or word) is shown in milliseconds on the abscissa. In the upper panel the solid curved line is the prototype brain wave generated by the color blue displayed as a blank computer screen with a blue background. The dotted curved line is the test-sample brain wave generated by the spoken word *blue*. In the lower panel are the prototype brain wave (solid line), generated by display of a triangle on the screen, and the test-sample brain wave (dotted line), generated by the spoken word *triangle*. In neither case is the match perfect, for even when the same stimulus is repeated, the filtered brain waves do not match exactly, since the brain's

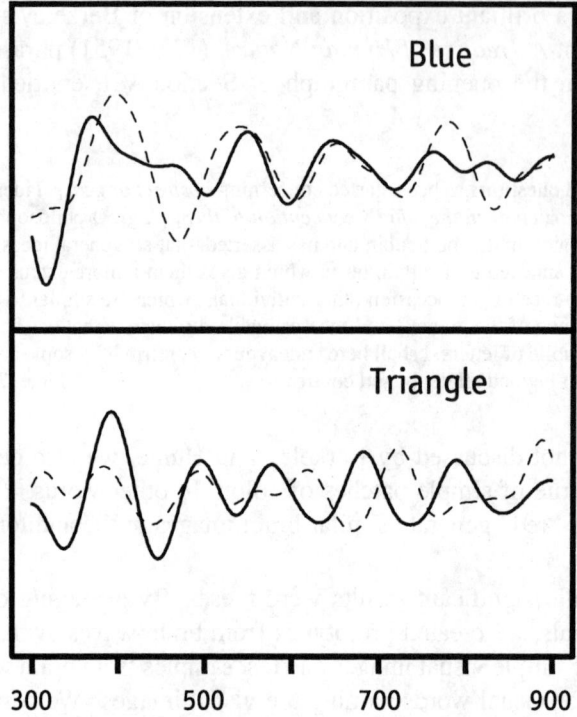

Figure 5: Comparison of filtered brain waves generated by visual images (solid curves) with those generated by the spoken names (dotted curves) of the images.

electric activity continually changes from moment to moment in numerous ways. But, all the same, there are many invariances necessary for human communication, and even at this early stage we can identify some of them.

Criticisms of results and response

I first sketch the general nature of the criticisms. In many brain imaging experiments the data are very rich and complex. Consequently, a complicated procedure may also be used to find, for given conditions, an optimal value. The search for this optimal value, which here is the best correct recognition rate, is analogous to computing an extreme statistic for a given probability distribution. The basis of the analogy is that the search corresponds to possibly many repetitions of a null-hypothesis experiment. These repetitions require computation of the appropriate extreme statistic. Moreover, if several parameters are estimated in finding such an optimal value, the significance of the value found may be challenged. The basis of such a challenge is the claim that for rich data and several estimated parameters, even a random assignment of the meaningful labels in the experiment may still produce a pretty good predictive result with a large enough number of chance repetitions.

Extreme statistics

We meet this criticism in two ways. The first is to derive the extreme-statistic distribution under the null hypothesis of a binomial distribution arising from such random assignments of labels. We compute how many standard deviations the scientifically interesting predictive result is from the mean of the extreme-statistic distribution. Physicists usually accept at least three or four standard deviations as a significant result. Other scientists and statisticians usually prefer a p value expressing the probability of the observed result under the null hypothesis. Here we report both measures.

The second approach is meant to answer those who are skeptical that the null hypothesis of a binomial distribution with a single parameter for the chance probability of a correct classification will adequately characterize the structure of the data even after a random permutation of the labels. To respond to such possible skeptics, we also compute a recognition rate for a sample of 50 random permutations of labels. We then fit a beta distribution to each such sample for a given experimental condition to compare with the corresponding extreme statistic distribution arising from the null hypothesis.

We first derive the extreme statistic under the null hypothesis.

Let p = probability of a success, a correct classification in our case, on a single trial, and $q = 1 - p$. Let $\mathbf{X}$ be the random variable whose value is the number k of successes in n independent trials. The probability of at least k successes is:

$$P(\mathbf{X} \geq k) = \sum_{j=k}^{n} P(\mathbf{X} = j) = \sum_{j=k}^{n} \binom{n}{j} p^j q^{n-j}. \qquad (3)$$

Now we repeat the experiment governed by a binomial distribution. So we have r independent repetitions of the n independent trials. For r repetitions ($r = 21,000$ in the 100-sentences experiment), the random variable representing the extreme statistic is

$$\mathbf{Y} = \max(\mathbf{X}_1, \ldots, \mathbf{X}_r). \qquad (4)$$

Let $P(\mathbf{Y} \geq k)$ be the probability that $\mathbf{Y}$ is at least k in at least one of the r

repetitions, the extreme statistic of interest. Then clearly

$$P(\mathbf{Y} \geq k) = 1 - P(\mathbf{X} < k)^r, \tag{5}$$

$$= 1 - \left[\sum_{j=0}^{k-1} \binom{n}{j} p^j q^{n-j}\right]^r.$$

We also need the theoretical density distribution of $\mathbf{Y}$, to compare to various empirical results later. This is easy to compute from (3).

$$P(\mathbf{Y} = k) = P(\mathbf{Y} \geq k) - P(\mathbf{Y} \geq k+1), \tag{6}$$
$$= P(\mathbf{X} < k+1)^r - P(\mathbf{X} < k)^r,$$
$$= \left[\sum_{j=1}^{k} \binom{n}{j} p^j q^{n-j}\right]^r - \left[\sum_{j=1}^{k-1} \binom{n}{j} p^j q^{n-j}\right]^r.$$

From (4) we can compute the mean and standard deviation of the extreme statistic $\mathbf{Y}$.

Beta distribution fitted to empirical sample

Second, we report results for the beta distribution on $(0,1)$ fitted to the empirical sample of extreme statistics. The density $f(x)$ of the beta distribution is:

$$f(x) = \begin{cases} \frac{\Gamma(a+b)}{\Gamma(a)\Gamma(b)} x^{a-1}(1-x)^{b-1}, & a, b, > 0, \ 0 < x < 1, \\ 0 & \text{otherwise,} \end{cases} \tag{7}$$

where $\Gamma(a)$ is the gamma function. If $\mathbf{Z}$ is a random variable with a beta distribution, then its mean and variance are given as simple functions of the parameters a and b.

$$\mu_{\mathbf{Z}} = E(\mathbf{Z}) = \frac{a}{a+b}, \tag{8}$$

$$\sigma_{\mathbf{Z}}^2 = \mathrm{Var}(\mathbf{Z}) = \frac{ab}{(a+b)^2(a+b+1)}. \tag{9}$$

The probability that the random variable $\mathbf{Z}$ has a value equal to or greater than $\frac{k}{n}$ is:

$$P(\mathbf{Z} \geq \frac{k}{n}) = \frac{\Gamma(a+b)}{\Gamma(a)\Gamma(b)} \int_{\frac{k}{n}}^{1} x^{a-1}(1-x)^{b-1} dx. \tag{10}$$

The computation of $P(\mathbf{Z} \geq \frac{k}{n})$ is difficult for the extreme tail of the distribution. In some cases we use a mathematically rigorous upper bound that is not the best possible, but easy to compute, namely, just the area of the rectangle with height $f(\frac{k}{n})$ containing the tail of the distribution to the right of $f(\frac{k}{n})$:

$$P(\mathbf{Z} \geq \frac{k}{n}) \leq f(\frac{k}{n})(1 - \frac{k}{n}), \tag{11}$$

where $f(\frac{k}{n})$ is defined by (5).

Computation of extreme statistics

I begin with the second experiment using 100 sentences. As a check on the null hypothesis, we constructed an empirical distribution of the extreme statistic by sampling 50 random permutations. Several points are to be noted.

1. A permutation of the 100 sentence "labels" is randomly drawn from the population of 100! possible permutations, and the sentence test samples are relabeled using this permutation.

2. Exactly the same grid of parameters (L, W, s, e) is now run for each bipolar pair of sensors, as for the correct labeling on the data of subject S32, to obtain, by Fourier analysis, filtering and selection of temporal intervals (s, e), a best rate of recognition or classification for the random label assignment. For the 100-sentences experiment, the number of points on the grid tested for each random permutation is $7 \times 10 = 70$ for $L \times W$, $5 \times 4 = 20$ for $s \times e$ and 15 for the number of sensors, so the number of repetitions r, from the standpoint of the null hypothesis, is $70 \times 20 \times 15 = 21,000$.

3. This random sampling of label permutations is repeated, and the recognition results computed, until a sample of 50 permutations has been drawn.

In Figure 6 I show the cumulative computation of the mean m and standard deviation s for the sample of 50 label permutations for the data of subject S32. For the full sample of 50 the mean $m = 6.04$ and the standard deviation $s = 0.77$. In Figure 7 I show: (i) the frequency distribution of the null-hypothesis extreme statistic $\mathbf{Y}$ with $n = 100, p = 0.01$ and $r = 21,000$, (ii) the empirical histogram of the maximum number of successes obtained for the 50 sample points with $r = 21,000$, and (iii) the fitted beta distribution as well. From Figure 7 it is visually obvious that the correct classification of more than 80 of the 100 sentences for S32 is not compatible with either

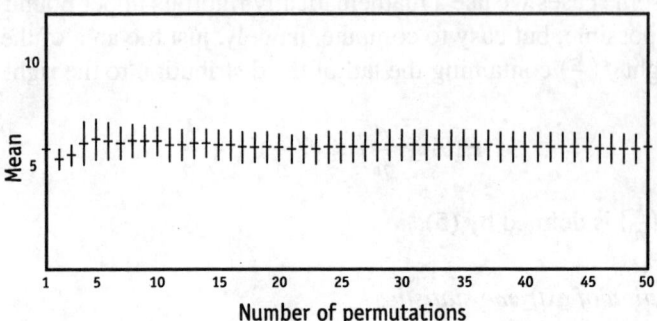

Figure 6: Cumulative mean and standard deviation of the recognition rate of the sample of random permutations.

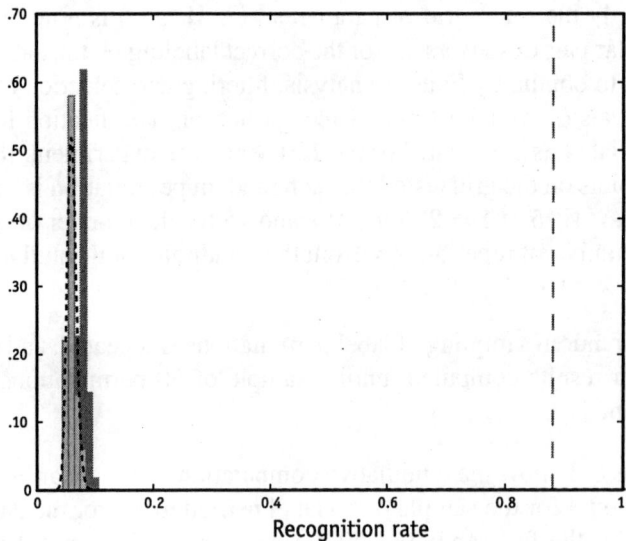

Figure 7: The frequency distribution (dark area) of the null-hypothesis extreme statistic **Y**, the histogram of the sample of 50 random permutations (lightly shaded areas), the fitted beta distribution (dotted line), and on the right (dashed vertical line) the recognition rate for S32.

the distribution of the extreme statistic **Y** or the estimated beta distribution for the sample grid computations based on 50 random permutations of the labels. The fact that the beta distribution fits slightly better than the distribution of **Y** is not surprising, since no free parameters were estimated for the latter. A finer search, with much larger r, yielding the higher result of 93 out of 100, is discussed in the next paragraph.

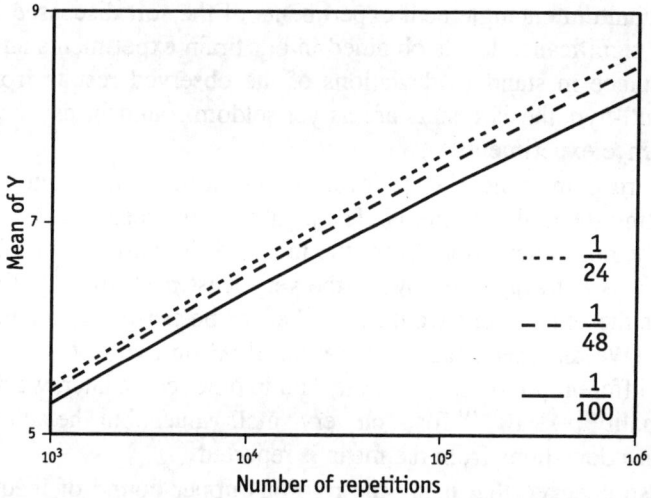

Figure 8: Mean of the null-hypothesis extreme statistic **Y** as a function of the number r of repetitions.

What is perhaps surprising is that the mean $\mu = 6.95$ of the null-hypothesis extreme statistic **Y** is slightly larger than the mean $m = 6.04$ of the empirical sample distribution. Three points are worth noting. First, the standard deviation $s = 0.77$ of the empirical sample is larger than the standard deviation $s = 0.67$ of the extreme statistic **Y**. I comment on this difference below. Second, I show in Figure 8 the rate of growth of the recognition rate for the null-hypothesis extreme statistic **Y**, for $n = 100$ and $p = 0.01$, and some other values of p and r used later, as r is increased by one or more orders of magnitude. As can be seen, under the null hypothesis the correct-recognition growth rate is slow. As an important example, we refined by extensive search the grid for the data of S32. We did not use a complete grid, but refined and extended only in promising directions. Extended comparably in all directions, the order of magnitude of r would be 10^7, i.e., 10,000,000 repetitions. So we computed the null-hypothesis distribution of **Y** for this large value of r, which is much larger than any actual computation we made. Even for this large grid, the mean of the extreme statistic **Y** for $r = 10^7$ only moved to $\mu = 9.60$. With the standard deviation now reduced to 0.62, the number of standard deviation units of the distance between 93 and 9.60 is 134.5, larger than before.

In reflecting on these results, it is important to keep in mind that physicists are usually very happy with a separation of 6 or 7 standard deviations between the classical or null-hypothesis prediction and new observed results, e.g., in quantum entanglement experiments of the sort discussed in Section 7.2. The significance levels obtained in our brain experiments and the very large distance in standard deviations of the observed results from the expected null-hypothesis results are as yet seldom found in psychological or neuroscience experiments.

The third point concerns the level of significance, or p value, we report for rejecting the null hypothesis. The p value of the result of $k = 93$, which is 134.5 standard deviations from the mean of the null-hypothesis extreme statistic **Y** is extravagantly low, at the very least $p < 10^{-100}$. Every other aspect of the experiment would have had to be perfect to support such a p value. (We did check the computation of 93 on two different computers running different programs.) So here, and in other cases later, we report only the inequality $p < 10^{-10}$ for such very small values, but the actual number of standard deviations from the mean is reported.

We also checked that using the rigorous upper bound of inequality (9), $P(\mathbf{X} \geq 93)$, computed for the fitted beta distribution, is also on the order of $p < 10^{-100}$. This is further support for the view that the p value inequality used later, namely, $p < 10^{-10}$, is highly conservative.

Analysis of earlier studies

I have emphasized the gain in predictive results from averaging across subjects as well as trials. The best result of the second experiment of 93% for one individual subject prompted us to review the best individual results in earlier experiments. In each experiment we have performed, the analysis of at least one individual subject's brain waves yielded a correct classification greater than 90%, with the exception of the 48-sentence experiment, mentioned already, which was 77%. (In Suppes, Han, Epelboim and Lu (1999a), this 77% was reported as 79%, because a finer grid was used.) Results for the best subject in the various experiments are summarized in Table 1.

With one exception, the p values shown in Table 1 are highly significant, by most standards of experimental work, extravagantly so. The exception is for the visual-image experiment in which 8 simple visual images were presented as stimuli. For four of the subjects, as shown in Table 1, we were

Table 1: *Exceptional Recognition Rates*[a]

Experiment	Subj.	Number of Successes	Chance Prob.	% Cor.	Repet. r	Statistic Y				Significance		p value
						μ	m	σ	s	#σ	# s	
7 visual words[1]	S1	32 of 35	1 of 7	91	2925	13.40		0.92		20.2		$<10^{-10}$
7 auditory words[1]	S3	34 of 35	1 of 7	97	3600	13.55		0.91		22.5		$<10^{-10}$
12 sentences[2]	S8	56 of 60	1 of 12	93	60,480	16.10		0.90		44.3		$<10^{-10}$
24 visual sent.[3]	S18	24 of 24	1 of 24	100	30,800	6.83	5.64	0.63	0.79	27.3	23.2	$<10^{-10}$
48 visual sent.[3]	S26	38 of 48	1 of 48	79	30,800	7.04	6.20	0.63	0.94	49.1	33.8	$<10^{-10}$
8 visual images[4]	4 Ss	8 of 8	1 of 8	100	95,550	6.28	4.92	0.46	0.69	3.7	4.5	$<.01$
100 visual sent.	S32	88 of 100	1 of 100	88	21,000	6.95	6.04	0.67	0.77	121.0	106.4	$<10^{-10}$
100 visual sent.	S32	93 of 100	1 of 100	93	10^7	9.60		0.62		134.5		$<10^{-10}$

1. Suppes, Lu, and Han (1997); 2. Suppes, Han and Lu (1998)
3. Suppes, Han, Epelboim and Lu (1999a); 4. Suppes, Han, Epelboim and Lu (1999b)

[a] The first column lists the experiment, with the last two entries being for the 100-sentence one. The subjects, listed in the second column, are numbered continuously from the experiments first reported in Suppes, Lu and Han (1997). The third column shows the maximum number of test samples successfully recognized out of the total presented. The fourth column shows the chance probability of a correct classification, which is simply 1 divided by the number of prototypes. The fifth column records the percent correct, as computed from the third column. The sixth columns shows the number r of repetitions used in the particular experiment to compute the extreme statistic. The number r is also the number of repetitions originally used in the grid for the initial search with correct labels. The seventh column records the mean μ of the null-hypothesis extreme statistic Y. The eighth column records the mean of the empirical samples of extreme statistics for the experiments for which we made this computation. The ninth column shows the standard deviation σ of the extreme statistic Y, and the tenth column the corresponding standard deviation s of the empirical samples. The eleventh column records the number $\frac{k-\mu}{\sigma}$, which is the number of standard deviations that the number k of successes recorded in column three is from the mean μ of the null-hypothesis distribution of the extreme statistic Y, and the twelfth column the corresponding number for the empirical sample. The thirteenth column shows a conservative bound for the p value of the observed number k of successes, with respect to the distribution of the extreme statistic Y, as given by equation (6). In the case of the four subjects in the visual-image experiment, m and s are the average for the four. The superscript on the description of each experiment is the reference to the published study. (The EEG sensor or bipolar pair of sensors and the optimal filter for each subject, except the four subjects of Suppes, Han, Epelboim and Lu (1999b), were as follows: S1:T6, 1-10 Hz; S3:T3, 3-11 Hz; S8:C4-T6, 2.5-9 Hz; S18:P4-T6, 0.5-10 Hz; S26:C4-C6, 1-15 Hz; S32:C4-T6, 1.25-22.5 Hz. The optimal parameters were often not unique.)

able to classify all 8 brain waves correctly, but this perfect result of 100 percent was significant only at the level of $p < 0.01$ for the null hypothesis, because with enough repetitions the best guesses under the null hypothesis do pretty well also, with $\mu = 6.28$.

The lesson for experimental design of this last point is obvious. If the data are massive and complex, as in the brain experiments described, and extensive search for optimal parameters is required, then the probability p of a correct response under the null hypothesis should be small. Figure 8 graphically makes the point. When p is small the number of repetitions can be very large, without affecting very much the mean of $\mathbf{Y}$, the extreme statistic of r repetitions. As can be seen also, from Table 1, when $p = 0.01$, the binomial parameter of the 100-sentences experiment, even 10,000,000 repetitions under the null hypothesis of 100 trials, increases $E(\mathbf{Y})$ only slightly to 9.60. To put the argument dramatically, at the rate of 1 second per trial, it would take more time than the present estimated age of the universe to have enough repetitions to obtain $E(\mathbf{Y}) \geq 93$.

I say in the preceding paragraph that p should be small, but that is too simple. The other way out, used in the first two experimental conditions of Table 1, 7 visual words and 7 auditory words, is to increase the number of test samples. In those two conditions, $p = \frac{1}{7}$, but the number of test samples was 35, and, as can be seen from the table, the null hypothesis was rejected at a level better than 10^{-10}. Reanalysis of the data from the visual-image experiment with $p = \frac{1}{8}$, in a similar approach by increasing the number of test samples from 8 to 24 yielded some better levels of rejection of the null hypothesis. The details are reported below.

More skeptical questions

As in all regimes of detailed experimentation, there is no sharp point after which further experiments need not be conducted, because all relevant questions have been answered. Galison (1987) made a detailed study of several important research programs of experimentation in physics. It seems likely that the main aspects of his analysis apply to many other areas of science.

Amidst the varied tests and arguments of any experimental enterprise, experimentalists must decide, implicitly or explicitly, that their conclusions stand *ceteris paribus:* all other factors being equal. And they must do so despite the fact that the end of an experimental demonstration is not, and cannot be, based purely on a closed set of procedures. ... Certain manipulations of apparatus, calculations, assumptions, and arguments give confidence to the experimentalist: what are they? ... When do experimentalists stake their claim on the reality of an effect? When do they assert that the counter's pulse or the spike in a graph is more than an artifact of the apparatus or environment? In short: How do experiments end?

(Galison 1987, pp. 3–4.)

In the context of the present brain experiments, the question is not really when do they end, but when do the computations on the experimental data come to an end? I examine three more different, but typical skeptical questions that are asked about new research with strong statistical support for its validity.

Other pairs in the first experiment with 48 sentences

Some skeptics commented to us that we were just lucky in the particular 2-element partition of the subjects we analyzed. So, we ran all 510 2-element partitions of the 9 subjects, with the same optimal values as in Table 1, without trying all points on the grid (Suppes, Wong, et al., to appear). In Figure 9 we show the histogram of these 510 partitions. The level of significance of the results is $p < 10^{-10}$ for all but 4 of the 510 possibilities, and one of these 4 has $p < 10^{-7}$. The best result is 46 out of 48, which holds for several partitions. So the brain-wave invariance between subjects argued for in the earlier study is robustly supported by the present more thorough statistical analysis. Another view of the same data is shown in Figure 10, where the number of subjects in the prototype of each 2-element partition is plotted on the abscissa and on the ordinate is shown the mean number of correct classifications of the 48 sentences for each type of prototype. Surprisingly, the mean results are good ($p < 10^{-10}$) when the prototype has only 1 subject or all but 1 subject, i.e., 8 subjects. The evidence is pretty convincing that our original choice of a partition was not just some happy accident.

Test of a timing hypothesis for the experiment with 100 sentences

In discussing with colleagues the high recognition rate of 93% obtained in the second experiment reported above, and the earlier results summarized in Table 1, several persons skeptically suggested that perhaps our recognition rates are just coming from the different timing of the visual presentation of words in different sentences. Sentences were presented one word at a time in the center of the computer screen, with the onset time of each visual word the same as the onset time of the corresponding auditory presentation of the sentence. Visually displaying one word at a time avoids many troublesome eye movements that can disrupt the brain waves, and it has also been shown to be an effective fast way to read for detailed content (Rubin and Turano 1992). The duration of each visual word of a sentence also matched the auditory duration within a few milliseconds.

To test this timing idea, which is supported by the presence of an evoked response potential at the onset of most visual words, we used a recognition

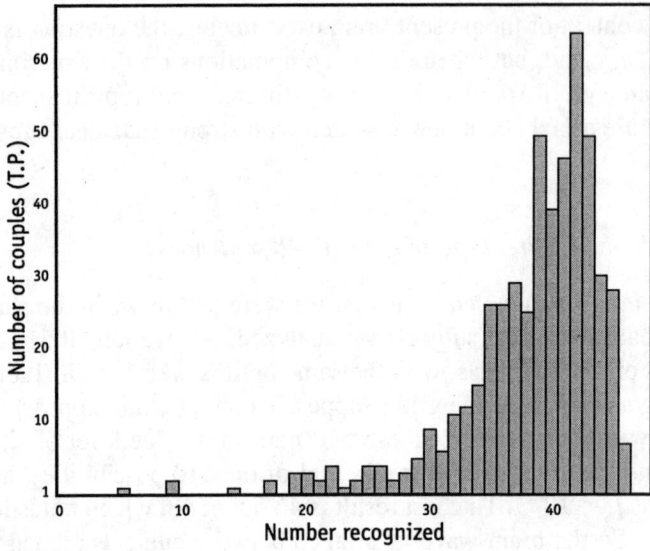

Figure 9: Histogram of the correct recognition rates for the 510 2-element partitions of the 9 subjects in the 48-sentences experiment.

model that depended only on an initial segment of the brain-wave response to each word in a sentence (Suppes, Wong, et al., to appear). The model replaces the two parameters s and e for the temporal interval by two different parameters. The first is α, which is the estimated time lag between the onset of each word in every sentence and the beginning of the corresponding brain wave in the cortex. The second is β, which is the proportion of the displayed length of each word, starting from its onset, used in the prototype for recognition after the delay time α for the signal to reach the cortex. Because of the variable length of words and sentences, we normalized the least squares computation by dividing by the number of observations used. If only timing, and not the full representation of the word, matters in recognition, then only a small portion of the initial segment of a word is needed, essentially the initial segment containing the onset-evoked response potential. On the other hand, if the full representation of the word is used in successful recognition, in terms of our least squares criterion, then the larger β is, the better for recognition. To adjust β to the temporal length of each word displayed, we expressed β as a decimal multiple of the temporal display length of word i of each sentence. The best predictive result was for $\alpha = 200$ ms and $\beta = 1.25$, with a recognition rate of 92%. The recognition rate as a function of $0.125 \leq \beta \leq 2.00$ is shown in Figure 11. The rate of correct recognition in-

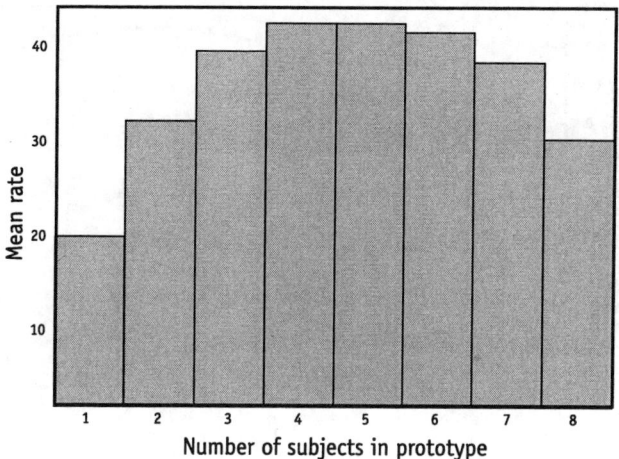

Figure 10: Histogram of mean number of the correct recognition rate for the 510 2-element partitions, indexed from 1-8 for the number of subjects in the prototype. So, e.g., 1 on the abscissa corresponds to all the 2-element partitions having exactly one subject used for the prototype.

creases monotonically with β up to $\beta = 1.25$ and then declines slowly after $\beta = 1.50$. These results support two conclusions. First, timing is important. The recognition rate of 45% for $\beta = 0.125$ is much greater than a chance outcome. But, second, the more complete the brain-wave representation of the words in a sentence, the better for recognition purposes.

Censoring data in the visual-image experiment

One kind of skeptical question that keeps the computations going is about artifacts. Perhaps the remarkable levels of statistical significance are due to some artifacts in the data. Now there is a long history of the problems of artifacts in EEG research. A main source is eye blinks and saccadic or pursuit eye movements, another is ambient current in the environment, mainly due to the 60 Hz oscillation of the standard alternating current in any building in which experiments are ordinarily conducted, and still another source is in the instrumentation for observing and recording the electric brain waves. This list is by no means exhaustive. There is a large literature on the subject from several different angles, but it would be too much to survey it here.

Given that the extreme statistics of the random permutations had a mean close to the low mean of the null hypothesis for the second experiment, it is extremely unlikely that any artifacts could account for the correct classifica-

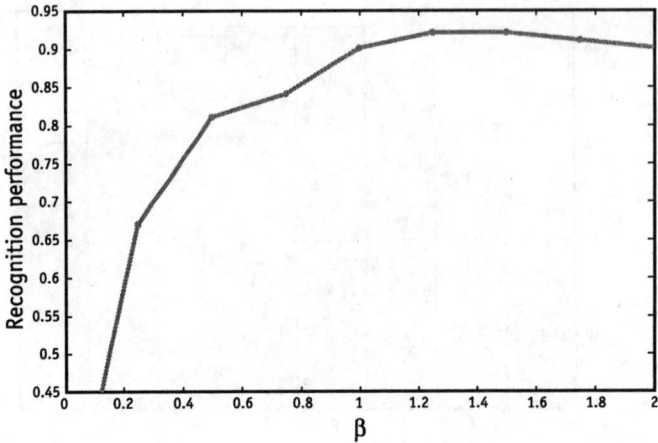

Figure 11: Classification results using different initial segments of brain-wave data after onset of each word in a sentence. Segment i begins after α ms following onset of word i in a sentence and segment i ends after $\beta \cdot l_i$ ms, where l_i is the length in ms of the visual display of word i.

tion of 93% in the second experiment, or, in fact, in any of the others with $p < 10^{-10}$. But artifact removal remains an important topic, not so much to meet ill-informed skeptical questions, but to improve the classification results, as is the case for an example that follows.

I restrict myself to describing how we used a rather familiar statistical approach, rather than any visual inspection of the recorded data for eye blinks or other artifacts. In our larger experiments with more than 40,000 trials it is impractical to try to use traditional observational methods to detect artifacts. The approach was to censor the data, but to introduce a free parameter to optimize the censoring—optimize in the sense already described of maximizing the correct recognition rate.

Let $\mathbf{X}_{ik}$ = observation i on trial k, and let ω be the number of trials averaged to create a prototype or test sample. Then

$$m_i = \overline{\mathbf{X}}_i = \frac{1}{\omega} \sum_{k=1}^{\omega} \mathbf{X}_{ik}.$$

In similar fashion, we compute the variance s_i^2. Let α be the free parameter for censoring such that if

$$|\mathbf{X}_{ik} - m_i| > \alpha s_i$$

eliminate obervation i of trial k from the data being averaged. The computational task is to find the $\hat{\alpha}$ that optimizes classification. In the example reported here, we ran a one-dimensional grid of 20 values of s_i to approximate the best value $\hat{\alpha}$.

The experiment for which the extreme statistics were not highly significant was the visual-image one already described, the one that was relevant to the eighteenth-century controversy about abstract ideas. The data, as I said earlier, supported in a direct way the skeptical views of Berkeley and Hume, but the statistical support was not very strong. So we reanalyzed the data, creating 24 rather than 8 test samples, and we also ran the experiment with four monolingual Chinese (Mandarin) speakers to confirm the verbal part, with auditory or visual words, in Chinese as well as English. The details are reported in Suppes, Wong, et al. (to appear).

Table 2 shows the significant results for cross-modal classification. The first two conditions are for the original experiment using English. For the feminine-auditory-voice representing brain waves (AWF) as prototypes and the visual-image brain waves as test samples, 15 of the 24 test samples were correctly classified after censoring, an improvement from 11 of 24 without censoring, for a resulting significance level of $p < 0.016$. When the roles of prototype and test sample were reversed the results of censoring were better, 16 of the 24 test samples correctly classified after censoring, with $p < 0.001$. Note that the significance levels here are conservative, based on the complete grid search equal to $r = 1,470,000$, the number of repetitions under the extreme-statistic null hypothesis.

In the case of the Chinese, the best results were in the comparison of the auditory and visual presentation of the eight words, with the best result being for the visual Chinese words (VW) as prototypes and the auditory Chinese words (AW) as test samples, in the censored case, 17 of 24 correctly classified, with p approximately 0.0001, and with, as before, the number of repetitions r, under the null hypothesis, greater than a million. So we end by strengthening the case for Berkeley and Hume, without claiming the evidence is as yet completely decisive.

An appropriate stopping point for this analysis is to emphasize that censoring does not guarantee improvement in classification. Most of the other results in Table 1 showed little improvement from censoring, but then for all of the experiments reported there, except for the visual-image one, the results were highly significant without censoring.

As in many areas of science, so with EEG recordings, statistical and experimental methods for removing artifacts and other anomalies in data constitute a large subject with a complicated literature. I have only reported a common statistical approach here, but I am happy to end with this one exam-

| | R | Y | | Significance | |
Exper.		μ	σ	#σ	p-value
English					
AWF/VI	11	11.57	.692	0.8	≈ .500
censor	15	13.01	.642	3.1	< .016
VI/AWF	11	11.57	.692	0.8	≈ .500
censor	16	13.01	.642	4.7	< .001
Chinese					
AW/VW	9	11.42	.690		
censor	16	12.86	.662	4.7	< .001
VW/AW	13	11.42	.690	2.3	≈ .05
censor	17	12.86	.662	6.3	≈ .0001

Table 2: Cross-modal results in visual-image experiment with censored data.

ple of a typical method of "cleaning up" data. It is such censored data that should be used to form a representation suitable for serving as a test of some theory or, as is often the case, some congerie of theoretical ideas.

QUANTUM MECHANICAL ENTANGLEMENT

The literature on hidden variables in quantum mechanics is now enormous. This section covers mainly the part dealing with probabilistic representation theorems for hidden variables, even when the hidden variables may be deterministic. Fortunately, this body of results can be understood without an extensive knowledge of quantum mechanics, which is not developed *ab initio* here. Many of the results given are taken from joint work with Acacio de Barros and Gary Oas (Suppes, de Barros and Oas 1998; de Barros and Suppes 2000).

First, I state, and sketch the proof, of the fundamental theorem that there is a factoring hidden variable for a finite set of finite or continuous observables, i.e., random variables in the language of probability theory, if and only if the observables have a joint probability distribution. The physically important aspect of this theorem is that under very general conditions the existence of a hidden variable can be reduced completely to the relationship between the observables alone, namely, the problem of determining whether or not they have a joint probability distribution compatible with the given data, e.g., means, variances and correlations of the observables.

I emphasize that although most of the literature is restricted to no more than second-order moments such as covariances and correlations, there is no necessity to make such a restriction. It is in fact violated in the third-order or fourth-order moments that arise in the well-known Greenberger, Horne and Zeilinger (1989) three- and four-particle configurations providing new Gedanken experiments on hidden variables, which are discussed later.

Factorization

In the literature on hidden variables, the principle of factorization is sometimes baptized as a principle of locality. The terminology is not really critical, but the meaning is. We have in mind a quite general principle for random variables, continuous or discrete, which is the following. Let $\mathbf{X}_1, \ldots, \mathbf{X}_n$ be random variables, then a necessary and sufficient condition that there is a random variable λ, which is intended to be the hidden variable, such that $\mathbf{X}_1 \ldots, \mathbf{X}_n$ are conditionally independent given λ, is that there exists a joint probability distribution of $\mathbf{X}_1, \ldots, \mathbf{X}_n$, without consideration of λ. This is the general fundamental theorem relating hidden variables and joint probability distributions of observable random variables.

THEOREM 1. (Suppes and Zanotti 1981, Holland and Rosenbaum 1986) *Let n random variables $X_1, \ldots, X_n$, finite or continous, be given. Then there exists a hidden variable λ such that there is a joint probability distribution F of $(\mathbf{X}_1, \ldots, \mathbf{X}_n, \lambda)$ with the properties*

(i) $F(x_1, \ldots, x_n \mid \lambda) = P(\mathbf{X}_1 \leq x_1, \ldots, \mathbf{X}_n \leq x_n \mid \boldsymbol{\lambda} = \lambda)$

(ii) *Conditional independence holds, i.e., for all $x_1, \ldots, x_n, \lambda$,*

$$F(x_1, \ldots, x_n \mid \lambda) = \prod_{j=1}^{n} F_j(x_j \mid \lambda),$$

if and only if there is a joint probability distribution of $\mathbf{X}_1, \ldots, \mathbf{X}_n$. Moreover, λ may be constructed so as to be deterministic, i.e., the conditional variance given λ of each $\mathbf{X}_i$ is zero.

To be completely explicit in the notation

$$F_j(x_j \mid \lambda) = P(\mathbf{X}_j \leq x_j \mid \boldsymbol{\lambda} = \lambda). \tag{12}$$

Idea of the proof. Consider three ± 1 random variables $\mathbf{X}$, $\mathbf{Y}$ and $\mathbf{Z}$. There are 8 possible joint outcomes $(\pm 1, \pm 1, \pm 1)$. Let p_{ijk} be the probability of outcome (i, j, k). Assign this probability to the value λ_{ijk} of the hidden variable λ we construct. Then the probability of the quadruple (i, j, k, λ_{ijk}) is just p_{ijk} and the conditional probabilities are deterministic, i.e.,

$$P(\mathbf{X} = i, \mathbf{Y} = j, \mathbf{Z} = k \mid \lambda_{ijk}) = 1,$$

and factorization is immediate, i.e.,

$$P(\mathbf{X} = i, \mathbf{Y} = j, \mathbf{Z} = k \mid \lambda_{ijk}) = \\ P(\mathbf{X} = i \mid \lambda_{ijk}) P(\mathbf{Y} = j \mid \lambda_{ijk}) P(\mathbf{Z} = k \mid \lambda_{ijk}).$$

Extending this line of argument to the general case proves the joint probability distribution of the observables is sufficient for existence of the factoring hidden variable. From the formulation of Theorem 1 necessity is obvious, since the joint distribution of $(\mathbf{X}_1, \ldots, \mathbf{X}_n)$ is a marginal distribution of the larger distribution $(\mathbf{X}_1 \ldots, \mathbf{X}_n, \lambda)$.

It is apparent that the construction of λ is purely mathematical. It has in itself no physical content. In fact, the proof itself is very simple. All the real mathematical difficulties are to be found in giving scientifically interesting criteria for observables to have a joint probability distribution.

Locality

The next systematic concept to discuss is locality. What John Bell meant by locality is made clear in the following quotation from his well-known 1964 paper (Bell 1964).

> It is the requirement of locality, or more precisely that the result of a measurement on one system be unaffected by operations on a distant system with which it has interacted in the past, that creates the essential difficulty. ... The vital assumption is that the result B for particle 2 does not depend on the setting **a**, of the magnet for particle 1, nor A on **b**.

To make the locality hypothesis explicit, we need to use additional concepts. For each random variable $\mathbf{X}_i$, we introduce a vector M_i of parameters for the local apparatus (in space-time) used to measure the values of random variable $\mathbf{X}_i$.

DEFINITION 1. (LOCALITY CONDITION I)

$$E(\mathbf{X}_i^k \mid M_i, M_j, \lambda) = E(\mathbf{X}_i^k \mid M_i, \lambda),$$

where $k = 1, 2$, corresponding to the first two moments of $\mathbf{X}_i$, $i \neq j$, and $1 \leq i, j \leq n$.

Note that we consider only M_j on the supposition that in a given experimental run, only the correlation of $\mathbf{X}_i$ with $\mathbf{X}_j$ is being studied. Extension to more variables is obvious. In many experiments the direction of the measuring apparatus is the most important parameter that is a component of M_i.

DEFINITION 2. (LOCALITY CONDITION II) *The distribution of λ is independent of the parameter values M_i and M_j, i.e., for all functions g for which the expectation $E(g(\lambda))$ and $E(g(\lambda)|M_i, M_j)$ are finite,*

$$E(g(\lambda)) = E(g(\lambda)|M_i, M_j).$$

Here we follow Suppes and Zanotti (1976). In terms of Theorem 3, locality in the sense of Condition I is required to satisfy the hypothesis of a fixed mean and variance for each $\mathbf{X}_i$. If experimental observation of $\mathbf{X}_i$ when coupled with $\mathbf{X}_j$ were different from what was observed when coupled with $\mathbf{X}_{j'}$, then the hypothesis of constant means and variances would be violated. The restriction of Locality Condition II must be satisfied in the construction of λ and it is easy to check that it is.

These remarks are summarized in Theorem 2.

THEOREM 2. *Let n random variables $\mathbf{X}_1, \ldots, \mathbf{X}_n$ be given satisfying the hypothesis of Theorem 3. Let M_i be the vector of local parameters for measuring $\mathbf{X}_i$, and let each $\mathbf{X}_i$ satisfy Locality Condition I. Then there is a hidden variable λ satisfying Locality Condition II and the Second-Order Factorization Condition if there is a joint probability distribution of $\mathbf{X}_1, \ldots, \mathbf{X}_n$.*

The next theorem states two conditions equivalent to an inequality condition given in Suppes and Zanotti (1981) for three random variables having just two values.

THEOREM 3. *Let three random variables $\mathbf{X}, \mathbf{Y}$ and $\mathbf{Z}$ be given with values ± 1 satisfying the symmetry condition $E(\mathbf{X}) = E(\mathbf{Y}) = E(\mathbf{Z}) = 0$ and with covariances $E(\mathbf{XY}), E(\mathbf{YZ})$ and $E(\mathbf{XZ})$ given. Then the following three conditions are equivalent.*

(i) *There is a hidden variable λ satisfying Locality Condition II and equation (a) of the Second-Order Factorization Condition holds.*

(ii) *There is a joint probability distribution of the random variables $\mathbf{X}, \mathbf{Y}$, and $\mathbf{Z}$ compatible with the given means and covariances.*

(iii) *The random variables $\mathbf{X}, \mathbf{Y}$ and $\mathbf{Z}$ satisfy the following inequalities.*

$$-1 \leq E(\mathbf{XY}) + E(\mathbf{YZ}) + E(\mathbf{XZ})$$
$$\leq 1 + 2\mathrm{Min}(E(\mathbf{XY}), E(\mathbf{YZ}), E(\mathbf{XZ})).$$

There are several remarks to be made about this theorem, especially the inequalities given in (iii). A first point is how do these inequalities relate to Bell's well-known inequality (Bell 1964).

$$1 + E(\mathbf{YZ}) \geq | E(\mathbf{XY}) - E(\mathbf{XZ}) |. \tag{13}$$

Bell's inequality is in fact neither necessary nor sufficient for the existence of a joint probability distribution of the random variables $\mathbf{X}, \mathbf{Y}$ and $\mathbf{Z}$ with values ± 1 and expectations equal to zero.

The next well-known theorem states two conditions equivalent to Bell's Inequalities for random variables with just two values. This form of the inequalities is due to Clauser et al. (1969), referred to as CHSH. The equivalence of (ii) and (iii) is due to Fine (1982).

THEOREM 4. (BELL'S INEQUALITIES) *Let n random variables be given satisfying the locality hypothesis of Theorem 4. Let $n = 4$, the number of random variables, let each $\mathbf{X}_i$ be discrete with values ± 1, let the symmetry condition $E(\mathbf{X}_i) = 0$, $i = 1, \ldots, 4$ be satisfied, let $\mathbf{X}_1 = \mathbf{A}$, $\mathbf{X}_2 = \mathbf{A}'$, $\mathbf{X}_3 = \mathbf{B}$, $\mathbf{X}_4 = \mathbf{B}'$, with the covariances $E(\mathbf{AB})$, $E(\mathbf{AB}')$, $E(\mathbf{A}'\mathbf{B})$ and $E(\mathbf{A}'\mathbf{B}')$ given. Then the following three conditions are equivalent.*

(i) *There is a hidden variable λ satisfying Locality Condition II and equation (a) of the Second-Order Factorization Condition holds.*

(ii) *There is a joint probability distribution of the random variables $\mathbf{A}$, $\mathbf{A}'$, $\mathbf{B}$ and $\mathbf{B}'$ compatible with the given means and covariances.*

(iii) *The random variables $\mathbf{A}$, $\mathbf{A}'$, $\mathbf{B}$ and $\mathbf{B}'$ satisfy Bell's inequalities in the CHSH form*

$$-2 \leq E(\mathbf{AB}) + E(\mathbf{AB}') + E(\mathbf{A}'\mathbf{B}) - E(\mathbf{A}'\mathbf{B}') \leq 2$$

$$-2 \leq E(\mathbf{AB}) + E(\mathbf{AB}') - E(\mathbf{A}'\mathbf{B}) + E(\mathbf{A}'\mathbf{B}') \leq 2$$

$$-2 \leq E(\mathbf{AB}) - E(\mathbf{AB}') + E(\mathbf{A}'\mathbf{B}) + E(\mathbf{A}'\mathbf{B}') \leq 2$$

$$-2 \leq -E(\mathbf{AB}) + E(\mathbf{AB}') + E(\mathbf{A}'\mathbf{B}) + E(\mathbf{A}'\mathbf{B}') \leq 2.$$

GHZ-type Experiments

Changing the focus, I now first consider GHZ-type experiments. Good references are Greenberger, Horne and Zeilinger (1989), the more extended discussion in Greenberger, Horne, Shimony and Zeilinger (1990) and Mermin (1990).

I follow the quantum-mechanical argument given in Mermin (1990). We start with the three-particle entangled state

$$|\psi\rangle = \frac{1}{\sqrt{2}}(|+\rangle_1|+\rangle_2|+\rangle_3 + |-\rangle_1|-\rangle_2|-\rangle_3), \tag{14}$$

This state is an eigenstate of the following spin operators:

$$\hat{\mathbf{A}} = \hat{\sigma}_{1x}\hat{\sigma}_{2y}\hat{\sigma}_{3y}, \quad \hat{\mathbf{B}} = \hat{\sigma}_{1y}\hat{\sigma}_{2x}\hat{\sigma}_{3y}, \tag{15}$$

$$\hat{\mathbf{C}} = \hat{\sigma}_{1y}\hat{\sigma}_{2y}\hat{\sigma}_{3x}, \quad \hat{\mathbf{D}} = \hat{\sigma}_{1x}\hat{\sigma}_{2x}\hat{\sigma}_{3x}. \tag{16}$$

If we compute quantum mechanically the expected values for the correlations above, we obtain at once that $E_Q(\hat{\mathbf{A}}) = E_Q(\hat{\mathbf{B}}) = E_Q(\hat{\mathbf{C}}) = 1$ and $E_Q(\hat{\mathbf{D}}) = -1$. (To exhibit all the details of this setup is too lengthy to include here, but the argument is elementary and standard, in the context of quantum mechanics.)

Now we note that

$$E_Q(\mathbf{ABC}) = (s_{1x}s_{2y}s_{3y})(s_{1y}s_{2x}s_{3y})(s_{1y}s_{2y}s_{3x}) \tag{17}$$

$$= s_{1x}s_{2x}s_{3x}(s_{1y}^2 s_{2y}^2 s_{3y}^2), \tag{18}$$

but since the s_{ij} can only be 1 or -1, we obtain at once that

$$\hat{s}_{1y}^2 = s_{2y}^2 = s_{3y}^2 = 1, and \tag{19}$$

$$E_Q(\mathbf{ABC}) = s_{1x}s_{2x}s_{3x} = E_Q(\hat{\mathbf{D}}) = -1. \tag{20}$$

In stark contrast, we have the following elementary theorem of classical probability.

THEOREM 5. *Let* $\mathbf{A}, \mathbf{B},$ *and* $\mathbf{C}$ *be* ± 1 *random variables having a joint probability distribution such that* $E(\mathbf{A}) = E(\mathbf{B}) = E(\mathbf{C}) = 1$. *Then* $E(\mathbf{ABC}) = 1$.

Proof. Since $E(A) = 1$, $P(\bar{a}) = P(\bar{a}bc) = P(\bar{a}b\bar{c}) = P(\bar{a}\bar{b}c) = P(\overline{abc}) = 0$, where $P(\overline{abc}) = P(A = -1, B = 1, C = 1)$, etc. By similar argument for $E(B)$ and $E(C)$, we are left with $P(abc) = 1$, which implies at once the desired result.

So, rather than inequalities, we have a flat contradiction. Classically

$$E(\mathbf{ABC}) = 1.$$

but, as shown above, quantum mechanically

$$E_Q(\mathbf{ABC}) = -1$$

Of course, using now also Theorem 1 we infer at once that there can be no factoring hidden variable for the quantum mechanical case.

This striking characteristic of GHZ's theoretical predictions, however, has a major problem. How can one verify experimentally predictions based on probability-one statements, since experimentally one cannot in the relevant experiments obtain events perfectly correlated? Fortunately, the correlations present in the GHZ state are so strong that even if we allow for experimental errors, the non-existence of a joint distribution can still be verified, as we show in the following theorem and its corollary.

THEOREM 6. (deBarros and Suppes 2000) *If* $\mathbf{A}$, $\mathbf{B}$, *and* $\mathbf{C}$ *are three* ± 1 *random variables, a joint probability distribution exists for the given expectations* $E(\mathbf{A})$, $E(\mathbf{B})$, $E(\mathbf{C})$, *and* $E(\mathbf{ABC})$ *if and only if the following inequalities are satisfied:*

$$-2 \leq E(\mathbf{A}) + E(\mathbf{B}) + E(\mathbf{C}) - E(\mathbf{ABC}) \leq 2, \quad (21)$$

$$-2 \leq E(\mathbf{A}) + E(\mathbf{B}) - E(\mathbf{C}) + E(\mathbf{ABC}) \leq 2, \quad (22)$$

$$-2 \leq E(\mathbf{A}) - E(\mathbf{B}) + E(\mathbf{C}) + E(\mathbf{ABC}) \leq 2, \quad (23)$$

$$-2 \leq -E(\mathbf{A}) + E(\mathbf{B}) + E(\mathbf{C}) + E(\mathbf{ABC}) \leq 2. \quad (24)$$

Proof. First we prove necessity. Let us assume that there is a joint probability distribution consisting of the eight atoms abc, $ab\bar{c}$, $a\bar{b}c$, ...$\bar{a}\bar{b}\bar{c}$. Then,

$$E(\mathbf{A}) = P(a) - P(\bar{a}),$$

where

$$P(a) = P(abc) + P(a\bar{b}c) + P(ab\bar{c}) + P(a\bar{b}\bar{c}),$$

and

$$P(\bar{a}) = P(\bar{a}bc) + P(\bar{a}\bar{b}c) + P(\bar{a}b\bar{c}) + P(\bar{a}\bar{b}\bar{c}).$$

Similar equations hold for $E(\mathbf{B})$ and $E(\mathbf{C})$. For $E(\mathbf{ABC})$ we obtain

$$\begin{aligned} E(\mathbf{ABC}) &= P(\mathbf{ABC} = 1) - P(\mathbf{ABC} = -1) \\ &= P(abc) + P(a\bar{b}\bar{c}) + + P(\bar{a}b\bar{c}) + P(\bar{a}\bar{b}c) \\ &\quad -[P(a\bar{b}c) + P(ab\bar{c}) + P(\bar{a}bc) + P(\bar{a}\bar{b}\bar{c})]. \end{aligned}$$

Corresponding to the first inequality above, we now sum over the probability expressions for the expectations

$$F = E(\mathbf{A}) + E(\mathbf{B}) + E(\mathbf{C}) - E(\mathbf{ABC}),$$

and obtain the expression

$$\begin{aligned} F &= 2[P(abc) + P(\bar{a}bc) + P(a\bar{b}c) + P(ab\bar{c})] \\ &\quad -2[P(\bar{a}\bar{b}\bar{c}) + P(\bar{a}\bar{b}c) + P(\bar{a}b\bar{c}) + P(a\bar{b}\bar{c})], \end{aligned}$$

and since all the probabilities are nonnegative and sum to ≤ 1, we infer at once inequality (21). The derivation of the other three inequalities is very similar.

To prove the converse, i.e., that these inequalities imply the existence of a joint probability distribution, is slightly more complicated. We restrict ourselves to the symmetric case

$$P(a) = P(b) = P(c) = p,$$

$$P(\mathbf{ABC} = 1) = q$$

and thus

$$E(\mathbf{A}) = E(\mathbf{B}) = E(\mathbf{C}) = 2p - 1,$$

$$E(\mathbf{ABC}) = 2q - 1.$$

In this case, (21) can be written as

$$0 \leq 3p - q \leq 2,$$

while the other three inequalities yield just $0 \leq p + q \leq 2$. Let

$$x = P(\bar{a}bc) = P(a\bar{b}c) = P(ab\bar{c}),$$

$$y = P(\bar{a}\bar{b}c) = P(\bar{a}b\bar{c}) = P(a\bar{b}\bar{c}),$$

$$z = P(abc),$$

and

$$w = P(\bar{a}\bar{b}\bar{c}).$$

It is easy to show that on the boundary $3p = q$ defined by the inequalities the values $x = 0$, $y = q/3$, $z = 0$, $w = 1 - q$ define a possible joint probability distribution, since $3x + 3y + z + w = 1$. On the other boundary, $3p = q + 2$ so a possible joint distribution is $x = (1-q)/3$, $y = 0$, $z = q$, $w = 0$. Then, for any values of q and p within the boundaries of the inequality we can take a linear combination of these distributions with weights $(3p-q)/2$ and $1 - (3p-q)/2$, chosen such that the weighed probabilities add to one, and obtain the joint probability distribution:

$$\mathbf{x} = \left(1 - \frac{3p-q}{2}\right)\frac{1-q}{3},$$

$$y = \left(\frac{3p-q}{2}\right)\frac{q}{3},$$

$$z = \left(1 - \frac{3p-q}{2}\right)q,$$

$$w = \frac{3p-q}{2}(1-q),$$

which proves that if the inequalities are satisfied a joint probability distribution exists, and therefore a noncontextual hidden variable as well, thus completing the proof. The generalization to the asymmetric case is tedious but straightforward.

As a consequence of the inequalities above, the correlations present in the GHZ state can be so strong that even if we allow for experimental errors, the non-existence of a joint distribution can still be verified (deBarros and Suppes 2000), as is shown in the following.

COROLLARY 1. *Let* $\mathbf{A}$, $\mathbf{B}$, *and* $\mathbf{C}$ *be three* ± 1 *random variables such that*
(i) $E(\mathbf{A}) = E(\mathbf{B}) = E(\mathbf{C}) \geq 1 - \epsilon$,
(ii) $E(\mathbf{ABC}) \leq -1 + \epsilon$,
where ϵ represents a decrease of the observed GHZ correlations due to experimental errors. Then, there cannot exist a joint probability distribution of $\mathbf{A}$, $\mathbf{B}$, *and* $\mathbf{C}$ *if*

$$\epsilon < \frac{1}{2}. \tag{25}$$

Proof. To see this, let us compute the value of F define above. We obtain at once that
$$F = 3(1 - \epsilon) - (-1 + \epsilon).$$
But the observed correlations are only compatible with a noncontextual hidden variable theory if $F \leq 2$, hence $\epsilon < \frac{1}{2}$. Then, there cannot exist a joint probability distribution of $\mathbf{A}$, $\mathbf{B}$, and $\mathbf{C}$ satisfying (i) and (ii) if

$$\epsilon < \frac{1}{2}. \tag{26}$$

From the inequality obtained above, it is clear that any experiment that obtains GHZ-type correlations stronger than 0.5 cannot have a joint probability distribution. For example, the recent experiment made at Innsbruck (Bouwmeester et al. 1999) with three-photon entangled states supports the quantum mechanical result that no noncontextual hidden variable exists that explains their correlations. Thus, with this reformulation of the GHZ theorem it is possible to use strong, yet imperfect, experimental correlations to prove that a noncontextual hidden-variable theory is incompatible with the experimental results.

On the other hand, as is shown in de Barros and Suppes (2000), the mean result of the Innsbruck experiment is not far from the classical regime. The distance is slightly less than two standard deviations from the classical boundary, so a more refined experiment, with mean results further from the boundary, would be desirable as a next step, and should be possible without any major technological changes in the experimental instruments.

Ventura Hall, Stanford University
Stanford, California

REFERENCES

de Barros, J. A. and Suppes, P. 2000. "Inequalities for Dealing with Detector Inefficiencies in Greenberger-Horne-Zeilinger-Type Experiments". *Physical Review Letters* 84: 793–797.

Bell, J. S. 1964. "On the Einstein-Podolsky-Rosen Paradox". *Physics* 1: 195–200.

Berkeley, G. 1901. "An Essay Towards a New Theory of Vision". In A. C. Fraser (ed.), *Berkeley's Complete Works*. London: Oxford University Press, vol. 1, pp. 93–210. First published in 1709.

Bouwmeester, D., Pan, J. W., Daniell, M., Weinfurter, H. and A., Z. 1999. "Observation of Three-Photon Greenberger-Horne-Zeilinger Entanglement". *Physical Review Letters* 82: 13–45.

Caton, R. 1875. "The Electric Currents of the Brain". *British Medical Journal* 2: 278.

Clauser, J. F., Horne, J. F., Shimony, A. and Holt, R. A. 1969. "Proposed Experiment to Test Local Hidden-Variable Theories". *Physical Review Letters* 23: 880–884.

Cooley, J. W. and Tukey, J. W. 1965. "An Algorithm for the Machine Computation of Complex Fourier Series". *Math. Computation* 19: 297–301.

Du Bois-Reymond, E. 1848. *Untersuchungen über Thierische Elektricität*. Berlin: Verlang von G. Reiner. Passage quoted translated by Hebbel E. Hoff, "Galvani and the Pre-Galvanic Electrophysiologists". *Annals of Science* 1 (1936), 157–172.

Fine, A. 1982. "Hidden Variables, Joint Probability, and the Bell Inequalities". *Physical Review Letters* 48: 291–295.

Galison, P. 1987. *How Experiments End*. Chicago: University of Chicago Press.

Galvani, L. 1791. "De Viribus Electricitatis in Motu Musculari". *De Bononiensi Scientiarum et Artium Instituto atque Academia, Comm.* 7: 363–418. Translated by Margaret Glover Foley, as *Luigi Galvani, Commentary on the Effects of Electricity on Muscular Motion*. Notes and introduction by I. Bernard Cohen. Norwalk, Connecticut, USA: Burndy Library, 1953.

Galvani, L. 1794. *Dell'uso, e dell'attivitá, dell'arco conduttore nelle contrazioni dei muscoli*. Bologna: A. S. Tommaso d'Aquino. Published anonymously.

Greenberger, D. M., Horne, M. A., Shimony, A. and Zeilinger, A. 1990. "Bell's Theorem without Inequalities". *American Journal of Physics* 58: 1131–1143.

Greenberger, D. M., Horne, M. A. and Zeilinger, A. 1989. "Going Beyond Bell's Theorem". In M. Kafatos (ed.), *Bell's Theorem, Quantum Theory, and Conceptions of the Universe*, Dordrecht: Kluwer Academic Press.

Holland, P. W. and Rosenbaum, P. R. 1986. "Conditional Association and Unidimensionality in Monotone Latent Trait Models". *Annals of Statistics* 14: 1523–1543.

Hume, D. 1739. *A Treatise of Human Nature*. London: John Noon. Quotations taken from L. A. Selby-Bigge's edition (1951), Oxford University Press, London.

Matteucci, C. 1844. *Traité des Phenomènes Electrophysiologiques des Animaux*. Paris: Fortin Masson.

Mermin, N. D. 1990. "Quantum Mysteries Revisited". *American Journal of Physics* 58(8): 731–734.

Oppenheim, A. V. and Schafer, R. W. 1975. *Digital Signal Processing*. Englewood Cliffs, NJ: Prentice-Hall.

Redi, F. 1671. *Esperienze intorno a diverse cose naturali, e particolarmente a quelle, che ci son portate dall'Indie*. Florence: Piero Matini. First published in 1671.

Rubin, G. S. and Turano, K. 1992. "Reading without Saccadic Eye Movements". *Vision Research* 32(5): 895–902.

Rugg, M. D. and Coles, M. G. 1995. *Electrophysiology of Mind: Event-Related Brain Potentials and Cognition*. New York: Oxford University Press.

Suppes, P., de Barros, J. A. and Oas, G. 1998a. "A Collection of Probabilistic Hidden-Variable Theorems and Counterexamples". In R. Pratesi and L. Ronchi (eds.), *Waves, information and foundations of physics. Conference proceedings, Vol. 60*, Bologna: Società Italiana Di Fisica, pp. 267–291.

Suppes, P., Han, B., Epelboim, J. and Lu, Z. L. 1999a. "Invariance Between Subjects of Brain Wave Representations of Language". *Proceedings of the United States National Academy of Sciences* 96: 12953–12958.

Suppes, P., Han, B., Epelboim, J. and Lu, Z. L. 1999b. "Invariance of Brain-wave Representations of Simple Visual Images and Their Names". *Proceedings of the United States National Academy of Sciences* 96: 14658–14663.

Suppes, P., Han, B. and Lu, Z. L. 1998b. "Brain-wave Recognition of Sentences". *Proceedings of the National Academy of Sciences* 95: 15861–15866.

Suppes, P., Lu, Z. L. and Han, B. 1997. "Brain-wave Representations of Words". *Proceedings of the United States National Academy of Sciences* 94: 14965–14969.

Suppes, P., Wong, D. K., Perreau-Guimaraes, M., Uy, E. T. and Yang, W. Forthcoming. "High Statistical Recognition Rates for Some Persons' Brain-wave Representations of Sentences".

Suppes, P. and Zanotti, M. 1976. "Necessary and Sufficient Conditions for Existence of a Unique Measure Strictly Agreeing with a Qualitative Probability Ordering". *Journal of Philosophical Logic* 5: 431–438.

Suppes, P. and Zanotti, M. 1981. "When Are Probabilistic Explanations Possible?" *Synthese* 48: 191–199.

Volta, A. 1793/1918. "Letter to Tiberius Cavallo, 22 May 1793". In *Le opere di Alessandro Volta*. Milan: Ulrico Hoepli, vol. 1, pp. 203–208.

Whittaker, E. T. 1951. *The History of the Theories of Aether and Electricity*, vol. 1. London: Nelson.

PAOLO LEGRENZI

NAÏVE PROBABILITY

The theory of mental models and extensional probability

Suppose that someone tells you: "If the director is in the office, then her secretary is in the office too". You start to think about the different possibilities compatible with the conditional. You think of the possibility of the director in the office, and so her secretary is in the office too. You think about what happens if the director is not in the office: in one possibility, the secretary is in the office; in another possibility, the secretary is not in the office, either. You have envisaged the three possibilities that are compatible with the truth of the conditional assertion, which we summarize as follows, using "¬" to denote negation:

Director in office	Secretary in office
¬ Director in office	Secretary in office
¬ Director in office	¬ Secretary in office

Following philosophers and logicians, we refer to such possibilities as the "extensions" of the conditional assertion, i.e., possibilities to which it refers. And when individuals infer probabilities by considering the extensions of assertions, we shall say that they are reasoning *extensionally*.

You can tackle the same problem in a different way. You know that directors are unlikely to spend as much time in the office as their secretaries. This stereotype may have occurred to you as you were thinking about the problem, and you might have based your inference on it. When you think in this way, you do not consider the extensions of assertions, but rather you use some index – some evidence or knowledge – to infer a probability. We use "non-extensional" as an umbrella term to cover the many ways in which people can arrive at probabilities without thinking about extensions. Of course, you might think about a problem both extensionally and non-extensionally.

Given a problem about a set of events, you can consider its partition, that is, the exhaustive set of possible conjunctions of individual events. In the problem about the director and the secretary, there are four such possibilities, which comprise this "partition" for the problem:

Director in office	Secretary in office
Director in office	¬ Secretary in office
¬ Director in office	Secretary in office
¬ Director in office	¬ Secretary in office

Once you know the probabilities for each possibility in a partition, you know everything that is to be known from a probabilistic standpoint. So let us introduce some probabilities, which for convenience we state as chances out of a hundred:

		Chances
Director in office	Secretary in office	50
Director in office	¬ Secretary in office	0
¬ Director in office	Secretary in office	30
¬ Director in office	¬ Secretary in office	20

You can now deduce the probability of any assertion about the domain, including conditional probabilities, such as:

> The probability that the director is not in the office given that the secretary is in the office: 30/80

The mental model theory postulates that each mental model represents a possibility, and that its structure and content capture what is common to the different ways in which the possibility might occur. For example, when individuals understand that either the director or else the secretary is in the office, but not both, they construct two mental models to represent the two possibilities:

director

secretary

where each line represents an alternative model, "director" denotes a model of the director in the office, and "secretary" denotes a model of the secretary in the office. Likewise, a conjunction, such as:

The director is in the office and the secretary is in the office

has only a single mental model:

> director   secretary

Granted that individuals construct mental models to represent the possibilities described in assertions, they can reason by formulating a conclusion that holds in their mental models, and they can test its *validity* by checking whether it holds in all possible models of the discourse. They can establish the invalidity of a conclusion by finding a counterexample, i.e., a model of the discourse in which the conclusion is false.

The theory makes a fundamental assumption, which is known as the principle of *truth*:

> Individuals represent assertions by constructing sets of mental models in which, first, each model represents a true possibility, and, second, the clauses in the assertions, affirmative or negative, are represented in a mental model only if they are true in the possibility.

Consider an exclusive disjunction in which only one of the two clauses is true:

The director is not in the office or else the secretary is in the office.

The mental models of the disjunction represent only the two true possibilities, and within them, they represent only the two clauses in the disjunction when they are true within a possibility:

> ¬ director
>
>   secretary

The first model represents the possibility that the director is not in the office, but it does not represent explicitly that it is false that the secretary is in the office. The second model represents the possibility that the secretary is in the office, but it does not represent explicitly that it is false that the director is not in the office (i.e. the director *is* in the office).

The mental models of conditionals are simple. For a conditional, such as:

If the director is in the office then the secretary is in the office

the mental models represent explicitly only the possibility in which the two clauses are true, whereas the possibilities in which the antecedent clause (the director is in the office) is false are represented by a wholly implicit model (shown here as an ellipsis):

      director   secretary
         . . .

A mental footnote on the implicit model stipulates that the antecedent is false in the possibilities that this model represents. If individuals retain this footnote, they can construct fully explicit models:

      director   secretary
  ¬ director   secretary
  ¬ director   ¬ secretary

Table 1 summarizes the mental models and the fully explicit models for four major sentential connectives.

Table 1: The mental models and the fully explicit models for four sentential connectives: "¬" denotes negation and ". . ." denotes a wholly implicit model. Each line represents a model of a possibility.

CONNECTIVE	MENTAL MODELS		FULLY EXPLICIT MODELS	
A and B:	A	B	A	B
A or else B:	A		A	¬B
		B	¬A	B
A or B, or both:	A		A	¬B
		B	¬A	B
	A	B	A	B
If A then B:	A	B	A	B
	...	¬	A	B
			¬A	¬B

All the principal predictions of the model theory follow from the previous account. But, to explain probabilistic reasoning, it is necessary to make some additional assumptions (Johnson-Laird, Legrenzi, Girotto, Legrenzi, and Caverni 1999). An important assumption is as follows:

1. The *equiprobability* principle: Each model represents an equiprobable possibility unless individuals have beliefs to the contrary, in which case they will assign different probabilities to the models representing the different possibilities.

The equiprobability principle works closely with:

2. The *proportionality* principle: Granted equiprobability, the probability of an event, A, depends on the proportion of models in which the event occurs, i.e., $p(A) = n_A/n$, where n_A is the number of models containing A, and n is the number of models.

Proportionality predicts that a description, such as: The director or her secretary, or both of them, are in the office, is compatible with three possibilities, which will each be assigned a probability of 1/3. An analogous principle applies to numerical probabilities:

3. The *numerical* principle: If assertions refer to numerical probabilities, then their models can be tagged with the appropriate numerical values, and an unknown probability can be calculated by subtracting the sum of the remaining known probabilities from the overall probability of all the possibilities in the partition.

The procedure is still extensional, but it generalizes to any sort of numerical values, including frequencies and probabilities expressed as fractions, decimals, or percentages.

How do naïve individuals infer conditional probabilities, and in particular posterior probabilities? According to the model theory, they rely on a simple procedure:

4. The *subset* principle: Granted equiprobability, a conditional probability, $p(A \mid B)$, depends on the subset of B that is A, and the proportionality of A to B yields the numerical value of the conditional probability.

Experimental test of the theory

Experimental tests have corroborated the model theory of extensional reasoning about probabilities. In one of these studies, the participants were given assertions about the contents of a box, and their task was to estimate the probabilities of various other assertions (Johnson-Laird et al. 1999). For example, given an exclusive disjunction, such as:

> There is a box in which there is a either a yellow card or else a brown card, but not both

individuals should construct models of the alternative possible contents of the box:

> Yellow-card
>
> Brown-card

and so they should infer probabilities of 50% for the following two assertions:

> There is at least a yellow card in the box.
>
> There is a yellow card in the box and there is not a brown card in the box.

The experiment examined three sorts of initial assertions:

> Exclusive disjunctions: Either A or else B, but not both.
>
> Inclusive disjunction: A or B, or both.
>
> Conditionals: If A then B.

Table 1 shows the mental models for each sort of these assertions, and they predict the probabilities that reasoners should assign to various categorical assertions presented after the initial assertions:

> At least A.
>
> A and B.
>
> A and not B.
>
> Neither A nor B.

In the case of an inclusive disjunction, for instance, reasoners should assign a probability of 67% to *at least A*, and a probability of 33% to *A and B*. Because participants should construct certain models only when they are asked questions about the corresponding possibilities, particularly in the case of conditionals, they are likely to overestimate probabilities so that they sum of their estimates of the different propositions in a partition should be greater than 100%.

Table 2: The predictions and the numbers of participants who made inferences within ± 5% of the predictions (n = 22) in Experiment 1 from Johnson-Laird et al. (1999).

	TYPES OF JUDGMENT			
A or B, not both	p(A)	p(A&B)	p(A&notB)	p(neither A nor B)
Predictions:	50	0	50	0
No. of participants within ± 5%	16	19	18	16
A or B, or both				
Predictions:	67	33	33	0
No. of participants within ± 5%	9	15	14	20
If A then B				
Predictions:	50	50	0	50
No. of participants within ± 5%	14	12	22	12

Table 2 presents the results of this study. The student participants estimated probabilities that tended to be within ±5% of the predicted values. For every initial assertion, their estimates were closer to the prediction than one would expect by chance (Sign tests varied from p < .0005 to < than 1 in four million). As the theory predicts, some participants appeared to forget the implicit model of the conditional, and thus four of them inferred a 100% probability for *at least A*, and eight of them inferred a 100% probability for *A and B*. The model theory also predicts that participants should tend to infer higher probabilities for *A and B* than for *neither A nor B*; both are possible for conditional and biconditional interpretations, but only the former corresponds to an initially explicit model. This difference was reliable. The inferences for *B and not-A*, which are not shown in Table 2, reflect the interpretation of the conditional: 12 participants inferred a probability of 0% (the biconditional interpretation), four participants inferred a probability of 50% (the conditional interpretation), and the remaining six participants inferred some other probability. In either case, we have inferences for all four possibilities in the partition, and they ought to sum to 100%. A biconditional has fewer explicit models than a conditional, and those participants who made the biconditional interpretation tended to infer probabilities that summed correctly to 100%, whereas those participants who made a conditional interpretation tended to infer probabilities that summed to more than 100%. This difference was reliable. Reasoners failed to bring to mind all the models of the conditional, and so they overestimated the probability of the model that corresponds to the event for which they are trying to infer a probability (*cf.* the "subadditivity" predicted by Tversky and Koehler 1994, theory of non-extensional reasoning).

The results supported the model theory. The participants appeared to infer probabilities by constructing models of the premises using the equiprobability principle, and assessing the proportion of models in which the events occur. Experts tend to baulk at the questions in our experiment or else to describe the range of possible probabilities. In contrast, naïve individuals such as those in our experiment have intuitions about probability based on equiprobable possibilities. We now consider a prediction that is unique to the model theory.

Systematic biases in extensional reasoning

The theory predicts the occurrence of systematic biases in extensional reasoning, because models represent what is true, not what is false. As readers will recall, this *principle of truth* applies at two levels: individuals construct models that make explicit only true possibilities, and these models make explicit only those clauses in premises that are true. It is important to bear in mind that what is omitted concerns falsity, not negation. A negative sentence can be true, in which case it will be represented in a mental model. For certain assertions, the failure to represent what is false should produce biased extensional inferences. The mental models of such assertions yield partitions that differ from the partitions corresponding to the fully explicit models of the assertions. The theory predicts that biased inferences should occur because they are based on mental models rather than fully explicit models. Other assertions, however, have mental models that yield partitions corresponding to the fully explicit models of the assertions. The theory predicts that inferences from these assertions should be unbiased. A computer program implementing the construction of mental models and fully explicit models searched systematically for both sorts of assertions in the vast space of possible assertions.

Consider the following example:

> The director at least is in the office, or else both the secretary and the chauffeur are in the office, but the three of them are not all in the office.
> What is the probability that the director and the chauffeur are in the office?

The mental models of the assertion are as follows

Director

Secretary   Chauffeur

where "director" denotes a model of the director in the office, and so on. Reasoners should therefore infer that the probability of the director and the chauffeur both being in the office is 0%. The fully explicit models of the assertion, however, take into account that when it is true that the director is in the office, there are three distinct ways in which it can be false that both the secretary and the chauffeur are in the office:

Director	Secretary	¬ Chauffeur
Director	¬ Secretary	Chauffeur
Director	¬ Secretary	¬ Chauffeur
¬ Director	Secretary	Chauffeur

It follows that the probability of the director and the chauffeur being in the office is not 0%. If each possibility is equiprobable, then the probability is 25%.

In general, an unbiased inference is one that applies the equiprobability principle to the alternatives corresponding to fully explicit models, which are the correct representation of the possibilities compatible with assertions. The following control problem should elicit an unbiased inference, because its mental models yield the same partition as its fully explicit models:

> The director at least is in the office and either the secretary or else the chauffeur is in the office, but the three of them are not all in the office.
>
> What is the probability that the director and the chauffeur are in the office?

The assertion has the mental models:

| Director | Secretary | |
| Director | | Chauffeur |

and so reasoners should respond, 50%. The fully explicit models of the assertion are as follows:

| Director | Secretary | ¬ Chauffeur |
| Director | ¬ Secretary | Chauffeur |

They support the same inference, and so it is an unbiased estimate. We carried out an experiment that investigated a set of nine experimental problems and nine control problems. The results corroborated the model theory's predictions (see Johnson-Laird et al. 1999, Experiment 3).

Conditional probabilities lie on the boundary of naive reasoning ability. Consider, for instance, the following problem:

> The director has two secretaries: A and B. One of them is a woman. What's the probability that the other is a woman?

If you have the stereotype that secretaries are women, then you make a non-extensional inference of, say, 90%. But, if you suppress this stereotype, you can draw the extensional conclusion that the probability is about 1/2. You assume that there are two possibilities for the other secretary:

Woman

Man

and equiprobability yields the conclusion. In fact, the problem calls for an estimate of a conditional probability, p (one secretary is a woman | other secretary is a woman). The partition is therefore as follows

Secretary 1	Secretary 2
Woman	Woman
Woman	Man
Man	Woman
Man	Man

Because at least one secretary is a woman, we can eliminate the last of these four possibilities. Given that secretary 1 or 2 is a woman, it follows that the probability that the other secretary is a woman is 1/3. Readers will note that if the female secretary is identified in some way, e.g., she is secretary 1, then the probability that secretary 2 is a woman does indeed equal 1/2.

General knowledge is readily triggered by any materials to which it seems relevant. Consider one last time our example about the director:

If the director is in the office then the secretary is too.

Your knowledge of the typical hours that directors and secretaries work, as we mentioned, may yield an answer to the question of who is more likely to be in the office. Tversky and Kahneman (1983) have shown that knowledge may also lead to a "conjunction fallacy" in which individuals rate a conjunction as having a higher probability that one of its conjuncts. For instance, given a description of a woman called Linda, which stressed her independence of mind and other features typical of feminists, individuals rated the conjunction:

Linda is a feminist and a bank teller

as more probable than its constituent proposition:

> Linda is a bank teller.

They rated as most probable, however, its other constituent:

> Linda is a feminist.

This pattern of ratings violates the general principle that a proposition has a probability greater than, or equal to, the probability of any conjunction in which it occurs.

Studies of the fallacy have shown that a conjunction is often rated as more probable than only *one* of its constituents. In a recent study of non-extensional probabilistic reasoning, however, we have established a stronger version of the fallacy (Legrenzi, Girotto, Legrenzi, and Johnson-Laird 2000). The key to this phenomenon is the nature of causal explanations. We presented the participants with a series of logically inconsistent assertions, such as:

> If a person pulls the trigger then the pistol will fire. Someone has pulled the trigger but the pistol did not fire. Why not?

The task was to rank order the probabilities of a series of putative explanations. One of the explanations was a causal chain consisting of a *cause* and an *effect*, where the effect in turn accounted for the inconsistency:

> A prudent person had unloaded the pistol and there were no bullets in the chamber.

This explanation was rated as having a higher probability than either the statement of the cause alone or the statement of the effect alone. The underlying mechanism in our view is the modulation of models of assertions by models in general knowledge – a mechanism that we have implemented in a computer program. Knowledge enables individuals to infer an effect from its cause, but it is harder to infer a cause from its effect, because effects may have other causes. Hence, the theory predicts the trend (for a similar account that anticipates our own, see Tversky and Kahneman 1983, p. 305). Such modulations can occur even in extensional reasoning, and the mixture of extensional and non-extensional processes is typical in daily life.

The model theory of probabilistic reasoning is based on a small number of simple principles. Reasoners make inferences from mental models representing what is true. By default, they assume that models represent equiprobable alternatives. They infer the probabilities of events from the proportions of models in which they hold. If the premises include numerical probabilities, reasoners tag their models with numerical probabilities, and use simple arithmetic to calculate probabilities. Problems that cannot be solved in these ways are probably beyond the competence of naive reasoners.

Acknowledgements

This research was supported in part by grants from MIUR (Italian Ministry of Universities and Research) and CNR.

Faculty of Architecture, IUAV
Venice, Italy

REFERENCES

Johnson-Laird, P. N., Legrenzi, P., Girotto, V., Legrenzi, M. and Caverni, J. P. 1999. "Naïve Probability: A Model Theory of Extensional Reasoning". *Psychological Review* 106: 62–88.

Legrenzi, M., Girotto, V., Legrenzi, P. and Johnson-Laird, P.N.F. Forthcoming. *Reasoning to Consistency: A Theory of Nonmonotonic Reasoning*.

Tversky, A. and Kahneman, D. 1983. "Extensional versus Intuitive Reasoning: The Conjunction Fallacy in Probability Judgment". *Psychological Review* 90: 293–315.

Tversky, A. and Koehler, D.K. 1994. "Support theory". *Psychological Review* 101: 547–567.

LÀSZLÓ E. SZABÓ

FROM THEORY TO EXPERIMENTS AND BACK AGAIN ... AND BACK AGAIN ... COMMENTS ON PATRICK SUPPES

"The picture of theory often presented by philosophers of science is too austere, abstract and self-contained" Professor Suppes writes. While, as it turns out from the two substantive examples considered in the paper, a closer analysis of the experimental details, the method of data processing and the most important features of the measuring equipments can be fruitful in understanding the basic concepts and the metaphysical conclusions drawn from the theoretical description of the experimental scenario.

Since my field of interest is closer to quantum mechanics, I would like to focus on Suppes' second example based on de Barros and Suppes (2000) general analysis of the realistic GHZ experiments, where experimental error reduces the perfect correlations of the ideal GHZ case. The following important question motivated their analysis: "How *can one verify experimentally predictions based on correlation-one statements, since experimentally one cannot obtain events perfectly correlated?*" De Barros and Suppes' analysis makes use of inequalities which are said to be "*both necessary and sufficient for the existence of a local hidden variable*" for the experimentally realizable GHZ correlations. In applying their analysis to the Innsbruck experiment, however, they only count events in which all the detectors fire. While necessary for the analysis of that experiment, they recognize that this selective procedure weakens the argument for the nonexistence of local hidden variables.

In Szabó and Fine (2002) we pointed out that their analysis does not rule out a whole class of local hidden variable models in which the detection inefficiency is not (only) the effect of the random errors in the detector equipment, but it is a more fundamental phenomenon, the manifestation of a predetermined hidden property of the particles. This conception of local hidden variables was first suggested in Fine's *prism model* (1982) and, arguably, goes back to Einstein.

Both, de Barros and Suppes' analysis and our polemics, confirm, however, Suppes' thesis about the continuing interaction in science between theory and experiment.

Theory => Experiment

De Barros and Suppes approach the problem in the following way. Without loss of generality, the space of hidden variable can be identified with $O = \{+, -\}^6$, the set of the $2^6 = 64$ different 6-tuples of possible combinations of the values of $\sigma_{1x}, \sigma_{1y}, \ldots \sigma_{3y}$. Then the GHZ contradiction amounts to the assertion that no probability measure over O reproduces the expectation values.

De Barros and Suppes demonstrate this by concentrating on the product observables *(A, B, C* and *ABC)* for which they derive a system of inequalities that play the same role for GHZ that the general form of the Bell inequalities do for EPR-Bohm type experiments; namely, they provide necessary and sufficient conditions for a certain class of local hidden vaxiable models. Their inequalities axe just

$$-2 \leq E(A) + E(B) + E(C) - E(ABC) \leq 2$$
$$-2 \leq E(A) + E(B) - E(C) + E(ABC) \leq 2$$
$$-2 \leq E(A) - E(B) + E(C) + E(ABC) \leq 2$$
$$-2 \leq E(A) + E(B) + E(C) + E(ABC) \leq 2$$

and clearly this is violated by

$$E(A) = E(B) = E(C) = 1$$
$$E(ABC) = -1$$

Experiment => Theory

In the realistic experiments, due to inefficiencies in the detectors or to dark photon detection, the observed correlations were reduced by some factor e; that is:

$$E(A) = E(B) = E(C) = 1 - \varepsilon$$
$$E(ABC) = -1 + \varepsilon$$

Theory => Experiment

Then, it follows immediately from the inequalities that, *"the observed correlations are only compatible with a local hidden variable theory"* if $\varepsilon > 1/2$. De Barros and Suppes (2000) translated this condition into the language of the dark-count rate and the detector efficiency.

Experiment => Theory

Estimating the realistic values of the dark-count rate and the detector efficiency, they found that the Innsbruck experiment is not compatible with a local hidden variable theory.

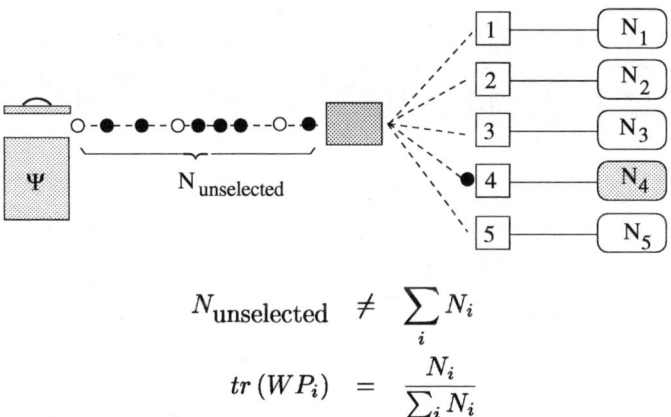

$$N_{\text{unselected}} \neq \sum_i N_i$$
$$tr(WP_i) = \frac{N_i}{\sum_i N_i}$$

Figure 1: *In a typical quantum measurement, quantum mechanical "probabilities" are equal to the relative frequencies taken on a* sub-ensemble of *objects producing any outcome.*

Theory => Experiment

As in the case of the Bell inequalities, however, the de Barros and Suppes derivation is based on the assumption that the variables $\sigma_{1x}, \sigma_{1y}, \ldots \sigma_{3y}$ are two valued (either +1 or -1).

Consider, however, a typical configuration of a quantum measurement shown in Figure 1. We have no information about the original ensemble of emitted particles. (quantum mechanical "probabilities" are equal to the relative frequencies taken on a *sub-ensemble of* objects producing any outcome (passing the analyzer).

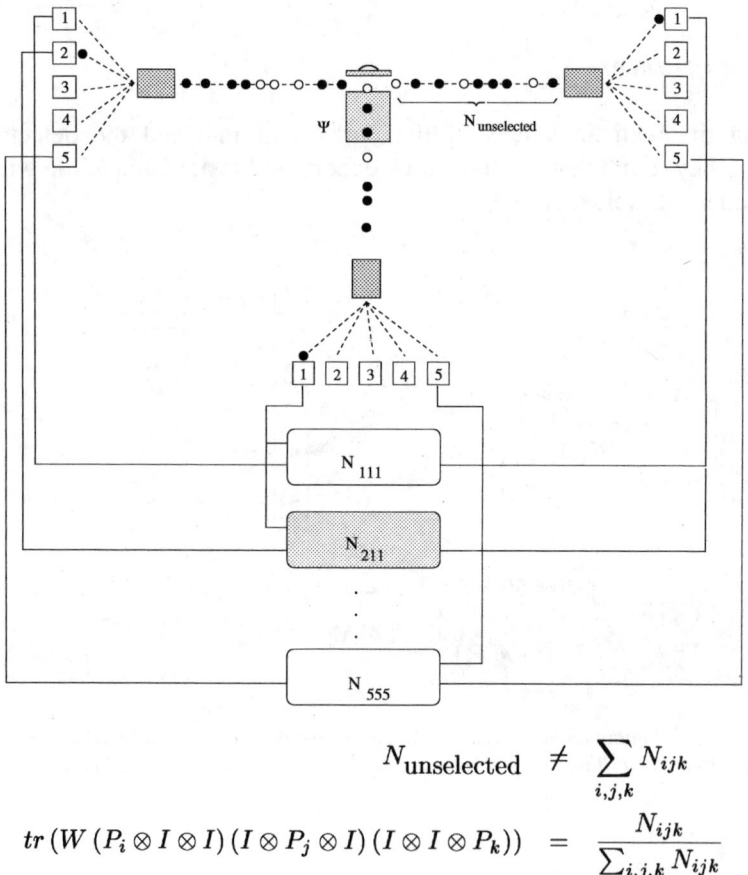

$$N_{\text{unselected}} \neq \sum_{i,j,k} N_{ijk}$$

$$tr\left(W\left(P_i \otimes I \otimes I\right)\left(I \otimes P_j \otimes I\right)\left(I \otimes I \otimes P_k\right)\right) = \frac{N_{ijk}}{\sum_{i,j,k} N_{ijk}}$$

Figure 2: *Quantum mechanical "probabilities" are experimentally identified with the relative frequencies calculated on a* sub-ensemble *of the complex systems that produce triple detection coincidences.*

In case when the conjunction of three properties are measured (Fig. 2), like the GHZ experiment, quantum mechanical "probabilities" are experimentally identified with the relative frequencies calculated on a *sub-ensemble* of the complex systems that produce triple detection coincidences.

Fine's prism model reflects the above experimental scenario. The variables can take on a third value, "D", corresponding to an inherent "no show" or defectiveness. Consequently, the space Λ of hidden variables is a subset of $\{+, -, D\}^6$. In Szabó and Fine (arXiv:quant-ph/000102 v4, 2001) we gave explicit prism models for a GHZ experiment with perfect detector efficiency and with zero dark-photon detection probability. Each element of Λ is a 6-tuple that corresponds to combinations like

$$\sigma_{1x}, \sigma_{1y}, \sigma_{2x}, \sigma_{2y}, \sigma_{3x}, \sigma_{3y} - (+ - D - + +)$$

which, for example, stands for the case when particle 1 is predetermined to produce the outcome +1 if x-measurement is performed, -1 if the setup is y in the measurement, particle 2 is x-defective, i.e., it gives no outcome if for an x-measurement, but produces an outcome -1 for y, particle 3 produces outcome +1 for both cases. Some of these combinations have probability zero, which rule out a large number of 6-tuples. One can show that we achieve the best efficiency if we take for Λ the subset, listed in Table 1, and simply omit all the others. Each atomic element has probability 1/48. Each GHZ event is represented as a subset $U \subseteq \Lambda$. For instance, $U_{x^+y^-y^-}$ stands for the triple outcome $x^+y^-y^-$ with probability

$$p\left(U_{x^+y^-y^-}\right) = p\left(\{\lambda_{32}, \lambda_{34}, \lambda_{37}, \lambda_{39}, \lambda_{41}, \lambda_{44}\}\right) = \frac{6}{48}$$

The probability of a triple detection for the measurement setups x, y, y:

$$p\left(U_{x \neq D\, y \neq D\, y \neq D}\right) = p\left(\{\underbrace{\lambda_1, \lambda_4, \lambda_5, \lambda_8, \lambda_9, \lambda_{10}, \lambda_{10}, \ldots \lambda_{44}, \lambda_{45}, \lambda_{48}}_{24}\}\right) = \frac{24}{48}$$

Quantum probabilities are reproduced as conditional probabilities:

$$p_{QM}\left(x^+y^-y^-\right) = \underbrace{\frac{1}{8}\left(1 + \sin\left(\frac{\pi}{2} + 0 + 0\right)\right)}_{\frac{1}{4}}$$

$$= p\left(U_{x^+y^-y^-} \middle| U_{x \neq D\, y \neq D\, y \neq D}\right) = \frac{\frac{6}{48}}{\frac{24}{48}} = \frac{1}{4}$$

etc. All quantum probabilities and the GHZ correlations are correctly reproduced in the model. The triple detection efficiency = ½!

Experiment => Theory

The question is what is the triple detection/emission ratio in the realistic GHZ experiments. Although the reported triple detection probability is very low ($\approx 10^{-4}$), this question is, actually, irrelevant in case of the Innsbruck experiment. The reason is that the preparation of GHZ entangled states is performed on selected sub-ensembles conditioned by the triple coincidence detections. Therefore, all of these experimental observations will be treated by our local hidden variable model.

Theoretical Physics Research Group of the Hungarian Academy of Sciences
Department of History and Philosophy of Science, Eötvös University
Budapest, Hungary

REFERENCES

De Barros, J. A. and Suppes, P. 2000. "Inequalities for Dealing with Detector Inefficiencies in GHZ-type Experiments". *Phys. Rev. Lett.* 84: 793.

Fine, A. 1982. "Some Local Models for Correlation Experiments". *Synthese* 50: 279–294.

Szabó, L. E. and Fine, A. 2002. "A Local Hidden Variable Theory for the GHZ Experiment". *Physics Letters* A.

REINHARD SELTEN

EMERGENCE AND FUTURE OF EXPERIMENTAL ECONOMICS*

1. Beginnings of experimental economics in Frankfurt

I would first like to relate something on the subject of this paper. Experimental research in economics[1] investigates the behavior of trial subjects in laboratory situations of interest for economics. In the laboratory an artificial economic reality is constructed, for example a market or an auction. Monetary payments linked to success provide for realistic economic incentives.

An important part of the earlier research in experimental economics took place in the Department of Economics at the University of Frankfurt am Main, where I studied and did research until my first professorship in Berlin, and which recently became one hundred years old. I am reporting on it as a direct witness. Heinz Sauermann and I published our article "Ein Oligopolexperiment" in the *Zeitschrift für die gesamte Staatswissenschaft* in 1959. Since then, at Heinz Sauermann's chair there has been a small group of young scientists working experimentally. To this changing group belonged, besides myself, Reinhard Tietz, Otwin Becker, Volker Häselbarth, Klaus G. Schuster, Claus C. Berg, Karl-Heinz Fischer, Manfred Reinfeld and others. Reinhard Tietz continued this work even after the retirement of Heinz Sauermann.

When we began with experimental economics in Frankfurt, this area did not yet exist. The area was only first developing in the sixties, approximately simultaneously in Frankfurt and in the USA. At the time, German economics was mainly preoccupied with catching up with the USA. However, we followed no foreign models and we did things not done anywhere else in the world.

Our experimental work was at first not taken seriously by many. My colleagues among the assistants[2] called me the "Dr. Mabuse of gamers". The [widely distributed] re-election posters of [the West German

Chancellor] Konrad Adenauer sporting the slogan "No Experiments" were a welcome occasion for jokes. Nevertheless our research also found its supporters. The German Research Association (*Deutsche Forschungsgemeinschaft*) supported experimental economics in Frankfurt for many years. This was thanks to favorable referees who repeatedly supported our requests for extensions and renewals.

Heinz Sauermann was editor of a book series entitled *Beiträge zur experimentellen Wirtschaftsforschung* (*Contributions to Experimental Economics*). The first three volumes appeared in 1967, 1970 and 1972. The first two volumes contained, with one exception, only work from Frankfurt. Heinz Sauermann organized the first International Conference on Experimental Economics in the world in 1971 in Kronberg in the Taunus district [Hessia, West Germany]. The proceedings of this conference constituted the third volume of the series.

The first three volumes also contained bibliographies on the area, which were prepared by Volker Häselbarth (1967), Claus C. Berg (1970) and Hans-Jürgen Weber (1972). There one finds only five publications from the years before 1950, 34 from the fifties, but already 245 from the sixties. The early publications predominantly arose in connection with experimental psychology. The emergence of experimental economics as a sub-discipline of economics took place in the sixties.

A point in time this emergence occurred cannot be determined precisely, however. The situation is analogous to finding the source of the River Ems, which I visited in the seventies. I there found a wet meadow, in which at some indefinable spot a small rivulet emerged. Later someone confined the source of the Ems in stone, thereby giving it an exact position in the landscape. A precise dating of the beginnings of experimental economics would be just as artificial.

2. Why experimental economics is so recent

Sometimes people ask me why experimental economics did not emerge earlier, and also how we in Frankfurt took up the idea of doing experiments. Wilhelm Wundt had already in 1875 founded the first laboratory for experimental psychology in Leipzig.

An important reason why experimentation had not entered the picture for economists was the generally unquestioned presupposition of a "homo oeconomicus". They believed they could derive economic behavior from rationality assumptions and therefore held experiments to be superfluous. To be sure, it was not always clear what was to be understood by rationality. Reflecting this, oligopoly theory, for example, had developed a confusing plethora of approaches. Game theory gave hope at first to the possibility of overcoming the confusion. But in the theory of cooperative games a similarly confusing plethora of solution concepts were soon developed. The confusion was not eliminated but merely lifted to a higher level.

In this historical situation, the first experiments on coalition games were undertaken by Kalish, Milnor, Nash and Nering in 1954. These experiments, as well as a report on earlier enterprise planning games [*Unternehmensplanspiele*] by Ricciardi in 1957 gave me the idea of doing experiments on oligopolies. While studying mathematics, I had heard some lectures by the Frankfurt Gestalt psychologist, Edwin Rausch, with whom I even took a seminar. Techniques of experimental psychology were thus familiar to me.

Professor Heinz Sauermann, for whom I worked starting in 1957, was convinced of the fruitfulness of experimentation. He expended much effort to support the area then emerging. Without him, experimental economics would not have existed in Germany.

3. Parallel developments in the U.S.A.

In parallel to what was going on in Germany, experimental economics was developing in the USA. Most influential were the books by Siegel and Fouraker in 1960, and Fouraker and Siegel in 1963. Vernon Smith achieved a breakthrough with his market experiments published in 1962. In the market organization of a double auction introduced by him into the literature, one obtains with identical repetitions rapid convergence to competitive equilibrium, even with relatively few market players.

Following Vernon Smith, people in the USA strove for a long time to extend the domain of the experimentally confirmed established theory as far as possible. This is a reasonable research program, only it soon runs into limitations.

We in Frankfurt took another path. The ideas of Herbert Simon on bounded rationality we found convincing, and we were further strengthened in our convictions by the outcomes of our experiments. Boundedly rational economic behavior cannot be invented in a scientist's easy chair, it must be discovered through experiments. For us, experiments were trips to an unknown continent. They served in the main not to test traditional theories, they rather had an exploratory character.

Influenced by experimental results, appreciation for the idea of bounded rationality had grown in the USA as well. Thus, in the eighties the distinction just described between German-language and American experimentation became weaker and weaker.

Many important discoveries have been made in experimental economics. We now know that economic motivation is not so self-centered as has usually been assumed in theory. Fairness and reciprocity are very important in behavior. The pioneering work of Güth, Schmittberger and Schwarze of 1982 on ultimatum bargaining gave rise to a flood of investigations which have enormously enriched our understanding of economic motivation. These experiments were done in Cologne. Later on, Güth attained a professorship in Frankfurt, where he fruitfully cooperated with Reinhard Tietz. In other areas great progress was achieved as well. Unfortunately I cannot go into them here, not even in outline.

4. Standards for experimental economics

Experimental economics has developed its own standards. By such standards we mean specifications for acceptable research work. Some of these standards have since become universally accepted.

First of all we mention the requirement of a full description of the experimental procedure. Journals demand that the instructions to subjects be reproduced in an appendix.

The second standard is the requirement of truthful instruction. Subjects must not be lied to. Not everything must be revealed, but what *is* instructed must be true.

Finally we have the standard of monetary payments dependent on success. Subjects must have a genuine material incentive to make careful decisions.

Such standards are not at all a matter of course in experimental psychology. Experimental setups are often not described precisely. Deception of subjects is frequent, and success-dependent monetary payments are rather exceptional. Perhaps experimental psychology must have other standards, for it investigates many behaviors other than economic ones.

The standard of full description is reasonable, for the reader must be able to follow what happened exactly. Details of experimental setups which seem unimportant at first can turn out to be important in retrospect.

The standard of truthful instruction is intended to establish credibility of instructions. Only in this way can it be assured that subjects take their tasks seriously.

The standard of success-dependent monetary payments is based on the interest of the economist in economic behavior driven by material incentives. In experimental psychology, completely different aims can of course be taken.

5. Forecasts for experimental economics

It is now time to take stock of future prospects for the area. I now hazard to make a few forecasts. Of course these are not intended as scientifically secured forecasts, just personal assessments. I am nevertheless pretty confident about those I make here.

Forecasts: In the course of the next twenty years, every superior economics department will

1. have at least one chair for experimental work,
2. be equipped with a computer-supported laboratory,
3. offer courses on experimental economics.

Naturally I realize that experimental economics is not now everywhere recognized as an indispensable part of economics. That will change, however.

I believe I can safely predict that experimental research will have a much greater influence on developments in economics in the future than is the case now. People will demand experimental evidence for theoretical claims much more than today, especially where economic behavior is concerned.

Experimental research has already now attained considerable importance in developing rules for auctioning radio frequencies (i.e. frequency bands for telecommunications, including mobile telephony and broadcasting). The German UMTS auction in 2000 yielded almost 100 billion German marks (about $60 billion). Experiments have also been used in strategic consultation of participants in the auctioning of radio frequencies. My colleagues and I have several times had the opportunity to gain experience with them.

We can count on much increased use in the future of experimental economics in solving problems involving development of institutions (e.g.

auctioning rules), as well as strategic consultation for corporate and governmental participants.

6. The future of experimental economics

Let me now indicate some of my hopes for future research in experimental economics. Experimental research of course cannot replace field research. The institutional environment of economic participants must be investigated in the real world. Once we are sure we have modeled such an environment, however, we can and must test or redevelop the behavioral assumptions of theory in the laboratory.

Empirical investigations lying a few decades in the past are often no longer taken seriously. We are easily inclined to believe that conclusions no longer hold because circumstances have changed. It is in fact not easy to estimate to what extent empirical results really depend on institutional factors or instead on the behavior of economic participants.

The objection of not being up to date implies a short "shelf life" [*kurze Verfallszeit* = quick expiration] for empirical investigations. Experiments are substantially more durable. They can be repeated even after one hundred years without worrying about significantly different outcomes. Economic behavior is presumably partially culture-dependent, but culture changes much more slowly than the economy. One can therefore conjecture that experimental research leads to a lasting body of facts. Such a body of facts can be steadily extended and can serve as a foundation for theory formation.

Patient experimental research will yield new behavioral theories of limited application. Some theories of this sort are already available, e.g. the entitlement adjustment theory (*Anspruchsausgleichstheorie*) of Tietz and Weber, which was developed in Frankfurt. (Its first publication is to be found in vol. 3 of the *Beiträge zur experimentellen Wirtschaftsforschung* of 1972.) I hope that one day a large number of such theories of limited application may grow together to form a comprehensive theory of boundedly rational economic behavior.

Institut für Gesellschafts- und Wirtschaftswissenschaften, Universität Bonn, Bonn, Germany.

NOTES

* Translated by Eckehart Köhler, University of Vienna.

1 The usual German term is *experimentelle Wirtschaftsforschung*, but "experimental economics", common in the English-language literature, will be used here. (Translator note)
2 Research and teaching positions assigned to chairs in Germany and Austria, comparable in status to assistant professors in the US, but without authority to offer courses independently – except for those who achieve the habilitation and become docents. (Translator note)

REFERENCES

Berg, C.C., Selten, R. 1970. "Ein Gerät zur akustischen Anzeige von Entscheidungspunkten in Oligopolexperimenten mit kontinuierlicher Zeit". In H. Sauermann (ed), 1970, pp. 162–221.

Güth, W., Schmittberger, R., Schwarze B. 1983. "An Experimental Analysis of Ultimatum Bargaining". *Journal of Economic Behavior and Organization* 3: 367–388.

Häselbarth, V. 1967. "Literaturhinweise zur experimentellen Wirtschaftsforschung". In H. Sauermann (ed.), 1967, pp. 267–276.

Kagel, J.H., Roth, A.E. (eds.) 1995. *The Handbook of Experimental Economics*. Princeton:Princeton University Press.

Kalish, G.K., Milnor, J.W., Nash, J.F. and Nering, F.D. 1954. "Some Experimental *n*-Person Games". In Thrall, R.M., Coombs, C.H., Davis, R.S. (eds.), *Decision Processes*. New York: John Wiley.

Ricciardi. F.M. 1957. *Top Management Decision Simulation. The AMA Approach*. New York.

Sauermann, H., Selten R. 1957. "Ein Oligopolexperiment". *Zeitschrift für die gesamte Staatswissenschaft* 115: 427–471. Reprinted in H. Sauermann (ed.), 1967, pp. 9–59.

Sauermann, H. 1967. *Beiträge zur experimentellen Wirtschaftsforschung*, I. Tübingen: J.C.B. Mohr.

Sauermann, H. 1970. *Beiträge zur experimentellen Wirtschaftsforschung*, II. Tübingen: J.C.B. Mohr.

Sauermann, H. 1972. *Beiträge zur experimentellen Wirtschaftsforschung*, III. Tübingen: J.C.B. Mohr.

Selten, R. 1978. "The Chain-Store Paradox". *Theory and Decision* 9: 127–159.

Siegel, S. and Fouraker, L.F. 1960. *Bargaining and Group Decision Making - Experiments in Monopoly*. New York - Toronto - London: McGraw-Hill.

Siegel, S. and Fouraker, L. E. 1963. *Bargaining Behavior* New York - Toronto - London: McGraw-Hill.

Simon, H. 1959. "Theories of Decision-Making in Economics and Behavioral Science". *American Economic Review* 49: 253–283.

Simon, H. 1972. "Theories of Bounded Rationality". In C.B. Radner and R. Radner (eds.), *Decision and Organization.* Amsterdam: North-Holland Publishing Co. Reprinted in H. Simon, *Models of Bounded Rationality*, II. Cambridge, MA: MIT Press, 1982.

Smith, V. 1962. "An Experimental Study of Competitive Market Behavior". *Journal of Political Economy* 70: 111–137.

Tietz, R. and Weber, H. J. 1972. "On the Nature of the Bargaining Process in the KRESKO Game". In H. Sauermann (ed.), 1972, pp. 305–334.

Weber, H.-J. 1972. "Bibliography". In H. Sauermann (ed.), 1972, pp. 690–699.

WENCESLAO J. GONZALEZ

RATIONALITY IN EXPERIMENTAL ECONOMICS: AN ANALYSIS OF REINHARD SELTEN'S APPROACH

Experimental economics is a branch of economics which has received increasing attention since the mid-1980's. Its first, informal, precedent has been set by Alvin Roth as early as Daniel Bernoulli[i], but it is only since the second half of the twentieth century that it has been clearly developed[ii]. In fact, it is in the period between 1975 and 1985 when experimental economics undergoes the transformation from "a seldom encountered curiosity to a well-established part of economic literature" (Roth 1987, p. 147). The process was consolidated around 1985, when the *Journal of Economic Literature* initiated a separate bibliographical category for "Experimental Economic Methods"[iii].

Among its most influential specialists is Professor Reinhard Selten, who – as it is well known – received along with John Nash and John Harsanyi the Nobel Prize in economics for his work on game theory. Selten's relation with experimental economics appears to have its first expression in 1959, when he publishes with Heinz Sauermann the paper *Ein Oligopolexperiment*[iv]. A few years later, in 1962, both of them – Sauermann and Selten – show their approval of Herbert A. Simon's concept of "bounded rationality", which they use to develop an "aspiration adaptation theory of the firm"[v]. Thereafter, the bounded rationality approach regarding human behavior and the methodological perspective of experimental economics have been closely related in Selten's intellectual trajectory.

During all this period – four decades – of special interest in bounded rationality, Selten has made important contributions to economics. Frequently, his papers include criticisms of mainstream economics, especially of the principle of subjective expected utility maximization. His publications are usually critical of assumptions of mainstream game theory[vi], which is deeply imbued with instrumental rationality. In fact, he links one of his most famous contributions to game theory – the chain store paradox – to the need for a bounded rationality supported by experimental evidence. He considers that the attempts to save the behavioral relevance of full rationality miss the point (*cf.* Selten 1990, p. 651).

When Selten develops his economic approach, he presents area theories, such as the theory of equal division payoff bounds, which are based on a limited rationality (*cf.* Selten 1987, pp. 42–98). Furthermore, he offers us a series of phenomena which confirm experimentally the existence of a bounded rationality (*cf.* Selten 1998a, pp. 413–436). In this paper, after the presentation of his approach to experimental economics, the emphasis will be on two aspects: the existence of three kinds of rationality (epistemic, practical, and evaluative) and his conception of economic rationality as *bounded*.

1. Selten's approach to experimental economics

Epistemologically, Selten stresses the importance of *empirical knowledge* over theoretical knowledge, a position which is more in tune with an empiricist framework than with a rationalist one. His views, furthermore, differ from the claims of critical rationalism insofar that he is dissatisfied with a negative role of experience and that he highlights the need for experience understood in positive terms. In this regard, he maintains that "we know that Bayesian decision theory is not a realistic description of human economic behavior. There is ample evidence for this, but we cannot be satisfied with negative knowledge – knowledge about what human behavior fails to be. We need more positive knowledge on the structure of human behavior. We need theories of bounded rationality, supported by experimental evidence, which can be used in economic modelling as an alternative to exaggerated rationality assumptions"[vii].

Selten's recommendation against the attempts to derive human behavior from a few general principles – either psychological or biological[viii] – is the gaining of empirical knowledge. In addition, he is critical of unrealistic principles, thus opposing a view held by influential mainstream economists, and he does not accept criticism of the use of *ad hoc* assumptions insofar as they are empirically supported[ix]. He maintains that successful explanations of experimental phenomena should be built up along the primacy of empirical knowledge. That knowledge reveals diversity: "experiments show that human behavior is *ad hoc*. Different principles are applied to different decision tasks. Case distinctions determine which principles are used where" (Selten 1991a, p. 19). Moreover, against the dominant position in favor of full rationality, he affirms that the "attempts to save the rationalistic view of economic man by minor modifications have no chance of succeeding" (Selten 1993, p. 135).

Methodologically, Selten seems to be sympathetic towards the research on induction[x]. His approach to experimental economics tends to identify some empirical regularities based on experimental data and thereafter to construe a formal theory to explain them, instead of beginning with a formal theory which is submitted to test in the laboratory. This kind of methodological approach is different from other methodological possibilities frequent among experimental economists, of which there are basically three: 1) experiments designed for testing and modifying formal economic theories; 2) experiments designed to collect data on interesting phenomena and relevant institutions, in the hope of detecting unanticipated regularities; and 3) experiments associated with having a direct impact in the realm of policy-making (*cf.* Roth 1986, pp. 245–246).

The fourth possibility, which can be found in Selten's papers, stems from the dissatisfaction of a present theory in the light of the data and the need for an alternative theory based directly on observed behavior. The experimental results are used to identify some empirical regularities. This evidence may suggest theoretical considerations which eventually can lead to the construction of a formal theory. This theory is ordinarily of a limited range, because usually experimental results support only theories of limited range, whereas an empirical-based general theory appears as a task of the future (*cf.* Selten 1998a, p. 414).

This kind of methodological approach, which can be seen in Selten's theory of equal division payoff bounds[xi], is different from the methodological case of theory-oriented experiments insofar as the starting point is different. In one case – the first methodological view – the research starts with a body of formal theory and then proceeds to develop a set of experiments which allow some conclusions to be drawn about the theory, whereas in the other case – the fourth methodological view – the research starts with a body of data from experimental games, which leads to a theory (*cf.* Roth 1986, pp. 266–267). The theory can take "the form of a hypothetical reasoning process which looks at the players in order of their strength" (Selten 1982/1988, p. 301). (According to Selten, "typically, game-theoretic solution concepts are based on definitions that describe the proposed solution by inner properties ... The theory of equal division payoff bounds has a different character. The payoff bounds are not characterized by inner properties. They are constructively obtained by straightforward commonsense arguments based on easily recognizable features of the strategic situation" (Selten 1987, p. 78)).

Underlying Selten's methodological approach on experimental economics is a rejection of key methodological views of mainstream economics: "the success of the theory of equal division payoff bounds confirms the methodological point of view that the limited rationality of human decision behavior must be taken seriously. It is futile to insist on explanations in terms of subjectively expected utility maximization. The optimization approach fails to do justice to the structure of human decision processes" (Selten 1987, p. 95). This methodological recognition of the need for a bounded rationality approach to human decision making in the economic activity seems to me very relevant.

Nevertheless, we still have certain methodological problems in experimental economics, mainly in the sphere of *methodological limitations*: how much of what is obtained in the economic laboratory can be applied directly to the complex situation of economic activity within the real world? It is not a minor problem, because – as Selten himself recognizes – "also field data are important, but they are more difficult to obtain and harder to interpret" (Selten 1998a, p. 414)[xii]. This aspect can have repercussions in two ways: on the one hand, in the characterization of *economic activity* as such (i.e., in giving the real features of human decision making in ordinary circumstances, instead of in an artificial environment); and, on the other hand, in the analysis of economic activity as *interconnected* with other human activities in a changeable historical setting, because economic activity is *de facto* connected with other human activities, and in a context which is also historical[xiii].

2. Epistemic rationality, practical rationality and evaluative rationality

Within economics as a whole, the concept of "rationality" plays a key role, and experimental economics is not an exception. *Rationality* is a notion that is closely linked to "choice" and "decision making". In this regard, it is customary in economics to present rationality in twofold terms: on the one hand, *normative rationality*, which points to what one should do in order to attain some specific aim and, on the other, *descriptive rationality*, which is used to reflect human endeavours in order to explain them or to predict them. Both aspects – normative and descriptive – assume that human behavior – the conduct of *homo economicus* – is goal oriented, and the emphasis is usually put on the relation from means to ends. Thus, a rational economic choice appears frequently as a selection of *adequate means* to attain *given ends*.

If the ends have more weight in the characterization of rationality and we understand that "rationality consists in the intelligent pursuit of appropriate ends" (Rescher 1988, p. 1), then the notion of *rationality* becomes wider than in the dominant tendency in mainstream economics. In fact, there are three different *dimensions of rationality* regarding choice and decision making: a) epistemic or cognitive rationality, b) practical rationality, and c) evaluative rationality.

From Kant on, the philosophical tradition "sees three major contexts of choice, those of *belief*, of accepting or endorsing thesis or claims, of *action*, of what overt acts to perform, and of *evaluation*, of what to value or disvalue. These represent the spheres of cognitive, practical, and evaluative reason, respectively" (Rescher 1988, pp. 2–3). Thus, not all deliberative reasoning – including the economic one – is means-end reasoning: there are three kinds of rationality according to the objects of rational deliberation: *epistemic rationality*, which deals with what is possible to believe or accept in the realm of knowledge; *practical rationality*, which decides regarding actions; and *evaluative rationality*, which judges what to prefer or prize (it assesses values, goals or ends).

According to a characteristic analysis of rational choice (especially of a single-agent) within the mainstream tendency, there is first an attribution of *practical rationality* (the optimatility of one's action is assumed, given one's desires and beliefs: if agent *a* desires *d* and believes that action *r* will secure *d*, the agent is practically rational in choosing *r*); and there is a second attribution of *epistemic rationality* to the actor (where rationality is then an attribute of belief, and consists in recognizing its correctness, given the evidence at the actor's disposal)[xiv]. But ordinarily there is no mention at all of *evaluative rationality*: the ends are given – they are not evaluated – and a rational agent is instrumentally rational, i.e., he or she should make practical decisions on means to attain the given ends. This is also the case in Herbert Simon's conception of rationality[xv], in spite of his being clearly critical of the outlook on rationality of mainstream economics.

Although they share a common interest in bounded rationality, I think that Selten's views on economic rationality go beyond Simon's instrumental rationality insofar as Selten accepts the presence of *evaluative rationality* in addition to practical rationality and epistemic rationality. Even though he does not use that terminology in his papers, it seems to me that these three dimensions of rationality underlie what he calls "three stages of reasoning", which he finds in the boundedly rational strategy construction: i) superficial analysis, ii) goal formation, and iii) policy formation (*cf.* Selten 1990, p. 656).

In this differentiation, the "superficial analysis" is when there is an easily accessible information, and the examination is qualitative rather than quantitative. Here the presence of *epistemic rationality* is undeniable. The "goal formation" seems to have implicit (or even explicit) the use of an evaluative rationality: when some concept of "fairness" intervenes (be it in terms of equal profits, profits proportional to Cournot profits, ...) in order to determine the quantities for players which can be called "an ideal point" (a cooperative goal), then a *rational evaluation* of the aim is made. The "policy formation" looks at the means to reach the end: it is necessary to determine a way in which the goal (cooperative at the ideal point) can be achieved. This case is a characteristic use of *practical rationality*.

Recently, Selten has offered us a good example of *evaluative rationality* through an experimental solidarity game (*cf.* Selten and Ockenfels 1998, pp. 517–539), because the other-directed motivations of the players can include reasoning about the *ends* themselves. On the one hand, solidarity aims at a reciprocal relationship, but it is a more subtle relation than giving after one has received. Solidarity is different from reciprocity insofar as the gifts made are not reciprocated. And, on the other hand, the subjects have to decide how much, in the case of their winning, they are willing to give to a loser, when he or she is the only one in the group, or to each one of the losers, when these are two. What Selten and Ockenfels have found is quite different from utility maximization: the players have "a decision process which first fixes the total amount to be sacrificed for solidarity and then distributes it (up to rounding) among the losers regardless of their number" (Selten and Ockenfels 1998, p. 525). The decision process which deliberates on the aim is based on the value of solidarity and it is different from a practical rationality of an instrumental kind.

3. Bounded rationality in the context of experimental economics

On the one hand, it is clear that Selten insists on the deficiencies of the dominant view on economic rationality: "experiments have shown again and again that the way in which human beings choose among alternatives is not adequately described by the theory of subjectively expected utility maximization" (Selten 1987, p. 43). Intransitivities in choice behavior and the phenomenon of preference reversal are mentioned as examples of the difficulties of the economic principle of utility maximization.

Yet, on the other hand, it should be emphasized that he complains about the present situation of experimental economics in this issue. He considers that "relatively few experimentalists contribute to the development of a theory of boundedly rational economic behavior. Too many experimentalists are in search for a confirmation of orthodox theory and go to great length in explaining away deviations which cannot be overlooked.

Nevertheless the necessity for a radical reconstruction of microeconomic theory becomes more and more visible" (Selten 1990, p. 650).

Bounded rationality in Selten can be understood in two different ways because he distinguishes between cognitive bounds and motivational bounds. *Cognitive bounds* are the limits related to the human capability to think and to compute; whereas *motivational bounds* are the failures to behave according to one's rational insights (*cf.* Selten 1993, pp. 132–133). The first kind – cognitive bounds – has been stressed by Simon from his very first papers on bounded rationality to his latest writings on this issue[xvi]. The second kind – motivational bounds – appears more clearly in Selten[xvii]. He discovered these rational limits along with the chain store paradox: it is not a lack of epistemic power but rather a failure to behave according to the rational insights. For him, "many phenomena of everyday life can be understood as caused by motivational bounds of rationality. Somebody who is convinced that it would be best for him to stop smoking may nevertheless find himself unable to do this" (Selten 1993, p. 133).

Between Simon's approach ("empirically grounded economic reason"[xviii]) and Selten's conception (experimentally limited reason) there is a difference on bounded rationality: the former has accentuated the cognitive limitations (mainly as a limitation of computational capacity), whereas the latter offers a broader panorama insofar as *motivational limitations* are explicitly added to cognitive limitations. Moreover, Selten considers that motivation is the mental process which acts as the driving force of human economic behavior (*cf.* Selten 1998a, p. 414), and he emphasizes the second kind of rational limits – motivational bounds –[xix]. He maintains that "the motivational limits of rationality are due to the separation of cognition and decision. The problem is known in philosophy under the name of 'acrasia' or 'weakness of the will'. A person may know very well what action is best for him and yet may find himself unable to take it" (Selten 1990, p. 651).

Motivational limits are, then, experimentally based (the experimental research in recent years has dealt with questions of motivation, such as the influence of reciprocity, which affect rational decision making), and they are also analyzed philosophically (both in philosophy of mind and in ethics). Once their existence is accepted there is a problem: how to understand the role "motivation" in economic behavior. In this regard, Selten explicitly recognizes: "I do not claim to be in possession of a valid theory of human motivation" (Selten 1994, p. 43). And his position includes an important assertion: "unfortunately we have no clear understanding of the interaction of different motivational forces. This is a serious difficulty for the development of a comprehensive theory of bounded rationality" (Selten 2001, p. 32).

Usually in philosophy, motivation is seen as something originally extrinsic to human will and which moves it towards a chosen end[xx], whereas in Selten it seems to be primarily intrinsic[xxi]: "the human motivational system determines the goal pursued by boundedly rational decision making" (Selten 2001, p. 32). His remarks on motivation seem to suggest the idea of a human factor which is not in principle similar to "substantive rationality" – something more or less established in *homo economicus* – but is rather regarded as a *kind of process* related to "procedural rationality" (i.e., open to variability and in accordance with particular circumstances). Motivation appears as a process which includes human deficiencies: "motivation is concerned with what behavior aims at and how a multitude of fears and desires combine to determine human action" (Selten 1998a, p. 414). In his view, "motivation and bounded rationality are not completely separable" (Selten 1998a, p. 414).

Again, in comparison with previous economic views, I think it is an interesting improvement to maintain that cognitive bounds and motivational bounds are not disconnected. In other words, it seems to me that the limits of epistemic rationality and the limits of practical rationality are related in human economic activity, which is a process with a complex structure. In the example of reciprocity – I do unto you as you do unto me – Selten seems open to this nexus between a bounded rational *cognition* and a bounded rational *motivation*: "reciprocity means that there is a tendency to react with friendliness to friendly acts and with hostility to hostile acts. This requires an interpretation of acts of others as friendly, hostile, or neutral. Here boundedly rational cognition enters the picture. Whether an act is perceived as friendly, neutral, or hostile depends on boundedly rational reasoning process" (Selten 1998a, p. 415).

For Selten, the *theory of decision making* has three levels: 1) the routine level, when routine decisions arise spontaneously without any thinking; 2) the level of imagination, which derives decisions from selected scenarios – the imagined courses of future play of limited length–; and 3) the level of analysis, which requires abstract thinking (*cf.* Selten 1993, p. 132). These three levels suggest the idea of increasing room for *more complex situations* for the economic decision making than those situations that should be addressed by the relatively simple principle of bounded rationality based on cognitive bounds.

Regarding the issue of the relations between the *complexity* of possible cases (the recognition that different arguments can apply to different cases) and the idea of *simplicity* in economic decision being guided by bounded rationality, Selten points to a middle ground: "in theories of limited rationality one should not look for the simplicity of abstract principles of sweeping generality. A combination of complex case distinctions with very simple decision rules for every single case seems to be very typical for limited rationality decision making" (Selten 1987, p. 79).

Concerning the future, there is a constant thought in Selten's writings: the need for an empirically supported *general theory* of bounded rationality. In 1993 he foresaw it as an aim for the long run: "it will take decades of painful experimental research until an empirically defendable general theory of bounded rationality emerges" (Selten 1993, p. 118). In a more recent paper, dated in 2001, he insists on the necessity of that general theory: "a comprehensive, coherent theory of bounded rationality is not available. This is a task for the future. At the moment, we must be content with models of limited scope" (Selten 2001, p. 14). I think that bringing economic theory into line with empirical evidence is a project of great importance.

Faculty of Humanities, University of A Coruña
Ferrol, Spain

NOTES

[i] Bernoulli 1738, pp. 175–192, translated in Bernoulli 1954, pp. 23–36. *Cf.* Roth 1988, p. 974, reprinted in Roth 1993, p. 3.
[ii] Even though Volker Häselbarth in 1967 lists 20 publications before 1959, R. Selten stresses that "experimental economics as a field of economic research did not emerge before the 1960s" (Selten 1993, p. 118).
[iii] The same year – 1985 – the Fifth World Congress of the Econometric Society included a paper on experimental economics, *cf.* Roth 1986, p. 245.
[iv] *Cf.* Sauermann and Selten 1959, pp. 427–471; Sauermann 1967, pp. 9–59.
[v] *Cf.* Sauermann and Selten 1962, pp. 577–597. *Cf.* Selten 1990, pp. 649–658, especially, p. 649.
[vi] John Nash considers that the book *A General Theory of Equilibrium Selection in Games*, written by J. Harsanyi and R. Selten, "is very controversial" (Nash 1996, p. 182).
[vii] Selten 1991a, p. 21. "The application of Bayesian methods makes sense in special contexts. For example, a life insurance company may adopt a utility function for its total assets; subjective probabilities may be based on actuarial tables. However, a general use of Bayesian methods meets serious difficulties. Subjective probabilities and utilities are needed as inputs. Usually these inputs are not readily available" (Selten 1991a, p. 19).

[viii] "We have to gain empirical knowledge. We cannot derive human economic behavior from biological principles" (Selten 1991a, p. 9).

[ix] "It is better to make many empirically supported ad hoc assumptions, than to rely on a few unrealistic principles of great generality and elegance" (Selten 1991a, p. 19).

[x] *Cf.* Selten 1990, p. 656. He is specially interested in the book Holland, Holyoak, Nisbett, and Thagard 1986.

[xi] "As more data became available, I developed a new descriptive theory for zero-normalized supperadditive three-person games in characteristic function form. This theory, called 'equal division pay-off bounds' (1983, 1987), derives lower bounds for the players' aspiration levels based on simple computations involving various equal shares. The improved version of this theory (1987), in particular, has had a remarkable predictive success" (Selten 1993, p. 120). *Cf.* Selten 1982, pp. 255–275, reprinted in Selten 1988, pp. 301–311. *Cf.* Selten 1987, pp. 42–98; especially, pp. 64–80.

[xii] Some experimental economists are really cautious regarding their work: "we do not go to the basement (laboratory) with the idea of reproducing the world, or a major part of it; that is better done (and it is hoped will be done) through 'field' observation. We go to the laboratory to study, under *relatively* controlled conditions, our *representations* of the world – most particularly our representations of *markets*" (Smith, McCabe, and Rassenti 1991, p. 197).

[xiii] On the distinction *economic activity–economics as activity*, *cf.* Gonzalez 1994, pp. 205–246.

[xiv] *Cf.* Bicchieri 1992, pp. 155–188; especially, pp. 161–162.

[xv] "We see that reason is wholly instrumental. It cannot tell us where to go; at best it can tell us how to get there. It is a gun for hire that can be employed in the service of whatever goals we have, good or bad" (Simon 1983, pp. 7–8).

[xvi] *Cf.* Gonzalez 1997, pp. 205–232. Simon's main publications on bounded rationality in recent years are Simon, Egidi, Marris, and Viale 1992; Simon 1997; and Simon 2000, pp. 25–39.

[xvii] Simon mentions "motivational constraints" in his writings, but he uses this expression in the context of a criticism to the traditional organization theory, which conceives the human organism as a *simple machine*. Thus, in his book with James March, he considers "the unanticipated consequences of treating an organization as though it were composed of such machines. This does not mean that the 'classical' theory is totally wrong or needs to be totally displaced. It means that under certain circumstances, which we will try to specify, dealing with an organization as a simple mechanism produces outcomes unanticipated by the classical theory" (March and Simon 1993, p. 53).

[xviii] This is the subtitle of Simon's *Models of Bounded Rationality,* vol. 3.

[xix] Selten points out his own personal experience regarding his discovery of the chain store paradox, *cf.* Selten 1990, p. 651.

[xx] "Motives" are not "reasons": the *motives* are what serve to impulse towards an action, but they are not the same as the *reasons* in favour of the action, *cf.* Rescher 1999, pp. 88 and 91.

[xxi] Due to this feature, in the conversations held in Italy (mainly, on 2. 10. 2001), I proposed to Selten a change in favor of the expression "volitive bounds" and its correlate "volitional bounds", in order to emphasize the primarily internal character of these limits. This feature is not adequately grasped with the phrase "motivational bounds".

REFERENCES

Bernoulli, D. 1738. "Specimen theoriae novae de mensura sortis". *Comentarii Academiae Scientiarum Imperialis Petropolitanae*, v. 5, pp. 175–192. (English translation by L. Sommer: "Exposition of a New Theory on the Measurement of Risk". *Econometrica* 22: 23–36, 1954).

Bicchieri, C. 1987. "Rationality and Predictability in Economics". *British Journal for the Philosophy of Science* 38: 501–513.

Bicchieri, C. 1992. "Two Kinds of Rationality". In N. de Marchi (ed.), *Post-Popperian Methodology of Economics*. Boston: Kluwer, pp. 155–188.

Bicchieri, C. 1993. *Rationality and Coordination*. Cambridge: Cambridge University Press.

Gonzalez, W. J. 1994. "Economic Prediction and Human Activity. An Analysis of Prediction in Economics from Action Theory". *Epistemologia* 17: 205–246.

Gonzalez, W. J. 1997. "Rationality in Economics and Scientific Predictions: A Critical Reconstruction of Bounded Rationality and its Role in Economic Predictions". *Poznan Studies in the Philosophy of Sciences and the Humanities* 61: 205–232.

Gonzalez, W. J. 1998. "Prediction and Prescription in Economics: A Philosophical and Methodological Approach". *Theoria* 13: 321–345.

Gonzalez, W. J. 2001. "De la Ciencia de la Economía a la Economía de la Ciencia: Marco conceptual de la reflexión metodológica y axiológica". In A. Avila, W. J. Gonzalez, and G. Marques (eds.), *Ciencia económica y Economía de la Ciencia: reflexiones filosófico-metodológicas*. Madrid: FCE, pp. 11–37.

Harsanyi, J. C. and Selten, R. 1988. *A General Theory of Equilibrium Selection in Games*. Cambridge, MA: MIT Press.

Harsanyi, J. C. 1996. "The Work of John Nash in Game Theory". *Journal of Economic Theory* 69: 158–161.

Hey, J. D. and Loomes, G. (eds.) 1993. *Recent Developments in Experimental Economics*. Aldershot: Elgar.

Holland, J. H., Holyoak, K. J., Nisbett, R. E. and Thagard, P. R. 1986. *Induction: Processes of Inference, Learning, and Discovery*. Cambridge, MA: MIT Press.

March, J. G. and Simon, H. A. 1993. *Organizations*. 2nd ed., Cambridge, MA: Blackwell.

Nash, J. 1996. "The Work of John Nash in Game Theory". *Journal of Economic Theory* 69: 182–183.

Rescher, N. 1988. *Rationality. A Philosophical Inquiry into the Nature and the Rationale of Reason*. Oxford: Oxford University Press.

Rescher, N. 1999. *Razón y valores en la Era científico-tecnológica*. Barcelona: Paidós.

Roth, A. 1986. "Laboratory Experimentation in Economics". *Economics and Philosophy* 2: 245–273.

Roth, A. 1987. "Laboratory Experimentation in Economics, and its Relation to Economic Theory". In N. Rescher (ed.), *Scientific Inquiry in Philosophical Perspective*. Lanham: University Press of America, pp. 147–167.

Roth, A. 1988. "Laboratory Experimentation in Economics: A Methodological Overview". *The Economic Journal* 98: 974–1031. Reprinted in J. D. Hey, and G. Loomes (eds.), *Recent developments in experimental economics*, vol. I. Aldershot: Elgar, 1993, pp. 3–60.

Sauermann, H. and Selten, R. 1959. "Ein Oligopolexperiment". *Zeitschrift für die gesamte Staatswissenschaft* 115: 427–471. Reprinted in H. Sauermann (ed.), *Beiträge zur experimentellen Wirtschaftsforschung*. Tübingen: J. C. B. Mohr (Paul Siebeck), 1967, pp. 9–59.

Sauermann, H. and Selten, R. 1962. "Anspruchsanpassungstheorie der Unternehmung". *Zeitschrift für die gesamte Staatswissenschaft* 118: 577-597.

Selten, R. 1978. "The Chain Store Paradox". *Theory and Decision* 9: 127–159. Reprinted in R. Selten, *Models of Strategic Rationality*. Dordrecht: Kluwer, 1988, pp. 33–65.

Selten, R. 1979. "Limited Rationality and Structural Uncertainty". In H. Berghel, A. Hübner, and E. Köhler (eds.), *Wittgenstein, The Vienna Circle and Critical Rationalism*. Vienna: Hölder-Pichler-Tempsky, pp. 476–483.

Selten, R. 1982. "Equal Division Payoff Bounds for Three-Person Characteristic Function Experiments". In R. Tietz (ed.), *Aspiration Levels in Bargaining and Economic Decision Making*. Berlin: Springer, pp. 255–275. Reprinted in R. Selten, *Models of Strategic Rationality*. Dordrecht: Kluwer, 1988, pp. 301–311.

Selten, R. 1987. "Equity and Coalition Bargaining in Experimental Three-Person Games". In A. E. Roth (ed.), *Laboratory Experimentation in Economics – Six Points of View*. Cambridge: Cambridge University Press, pp. 42–98.

Selten, R. 1990. "Bounded Rationality". *Journal of Institutional and Theoretical Economics* 146: 649-658.

Selten, R. 1991a. "Evolution, Learning, and Economic Behavior", 1989 Nancy Schwartz Memorial Lecture. *Games and Economic Behavior* 3: 3–24.

Selten, R. 1991b. "Properties of a Measure of Predictive Success". *Mathematical Social Sciences* 21: 153–167.

Selten, R. 1993. "In Search of a Better Understanding of Economic Behavior". In A. Heertje (ed.), *Makers of Modern Economics*. London: Harvestern Wheatsheaf, pp. 115–139.

Selten, R. 1994. "New Challenges to the Rationality Assumption: Comment". *Journal of Institutional and Theoretical Economics* 150: 42–44.

Selten, R. 1998a. "Features of Experimentally Observed Bounded Rationality". *European Economic Review* 42: 413–436.

Selten, R. 1998b. "Game Theory, Experience, Rationality". In W. Leinfellner, and E. Köhler (eds.), *Game Theory, Experience, Rationality*. Dordrecht: Kluwer, pp. 9–34.

Selten, R. and Ockenfels, A. 1998. "An Experimental Solidarity Game". *Journal of Economic Behavior and Organization* 34: 517–539.

Selten, R. 2001. "What is Bounded Rationality?". In G. Gigerenzer, and R. Selten (eds.), *Bounded Rationality: The Adaptive Toolbox.* Cambridge, MA: MIT Press, pp. 13–36.

Sensat, J. 1998. "Game Theory and Rational Decision". *Erkenntnis* 47: 379–410.

Simon, H. A. 1983. *Reason in Human Affairs.* Stanford: Stanford University Press.

Simon, H. A., Egidi, M., Marris, R., and Viale, R. 1992. *Economics, Bounded Rationality, and Cognitive Revolution.* Brookfield, VT: Elgar.

Simon, H. A. 1997. *Models of Bounded Rationality,* vol. 3. Cambridge, MA: MIT Press.

Simon, H. A. 2000. "Bounded Rationality in Social Science: Today and Tomorrow". *Mind and Society* 1: 25–39.

Smith, V. L., McCabe, K. A. and Rassenti, S. J. 1991. "Lakatos and Experimental Economics". In N. de Marchi, and M. Blaug (eds.), *Appraising Economic Theories.* Aldershot: Elgar, pp. 197–227.

ROBERTO SCAZZIERI

EXPERIMENTS, HEURISTICS AND SOCIAL DIVERSITY: A COMMENT ON REINHARD SELTEN

1. Introductory remarks

Professor Reinhard Selten wrote that: "Behaviour cannot be invented in the armchair. It has to be observed. Therefore, the development of theories of bounded rationality needs an empirical basis. Laboratory experimentation is an important source of empirical evidence. Of course, also field data are important, but they are more difficult to obtain and harder to interpret" (Selten 1998, p. 414). This passage brings to the fore the special relationship between the research programme of what is commonly referred to as "experimental economics" and the consideration of bounded rationality.

The connection between bounded rationality and economic experiments is examined in Professor Selten's contribution to this workshop. What I intend to do in this comment is to tackle some of the traditional issues that have been considered when discussing social science experiments, and to suggest that recent developments in economic theory, as well as in cognitive analysis and epistemology, may be useful in assessing the scope and status of experimental economics. Section 2 of this paper examines the relationship between bounded rationality and the context-dependence of economic behaviour. Section 3 discusses the specific nature of economic experiments under conditions of context-dependence. This section highlights that experiments in economics (or, more generally, experiments in social science) ought to aim at the heuristic identification of *possible* patterns of interaction, rather than at the repeated "performance" of causal processes. Section 4 brings the paper to a close by suggesting a research agenda that would take experimental economics considerably beyond its original core. This section calls attention to the "local" character of economic experiments and examines some of its implications in fields where economic causal processes are more explicitly associated with the agents' ability to transform the space of events in which human actions take place.

2. Social diversity and context-dependence

The intellectual paradigm of bounded rationality has been associated since its very beginning with the idea that human decisions derive from local information and suitable focussing (see, for instance, Simon 1983). In particular, the bounded rationality approach has brought about a remarkable shift from "cold reason" to attention. This means that human rationality is assessed not so much in terms of the single-minded pursuit of a clearly identified objective, as in terms of justified procedures and multiple *foci* of attention[1]. The relationship between context-dependence and rational economic behaviour stems from the fact that a successful choice of strategy is primarily a successful cognitive process. The latter often presupposes the ability to make use of multiple cognitive devices, and to identify a variety of features of salience. In this case, a successful strategy is also a successful criterion by which the rational agent selects one out of many feasible cognitive programmes and focuses upon a particular cluster of salient features. The likelihood that any given context of interaction ends up in co-ordination rather than conflict critically depends on the cognitive skills of individuals (or social groups). This means that individuals or groups are more likely to identify a feasible pattern of co-ordination if they are able to use multiple cognitive devices and to interact along a variety of social dimensions.

The boundedness of reason is a virtue, rather than a shortcoming, if it is associated with the ability to recognise multiple features of salience and to follow a variety of inferential paths. In this case, "bounded reason" entails that only a finite (and sufficiently narrow) set of logical steps can be envisaged at any given time. But ability to envisage (and possibly to carry out) only a limited number of logical steps is not necessarily a disadvantage if it is coupled with the ability to identify suitable focal points, that is, *foci* of attention that "draw" individuals from one inferential basis to another[2].

The above perspective suggests an inductive approach to (economic) rationality and calls attention to the fact that individuals (and social groups) are likely to make social inferences on the basis of a *narrow* set of premises. Under these conditions, the effectiveness of interaction is associated with the ability to choose premises adequate to the problem situation and the inferential capacities of interacting agents.

To sum up, rationality is to a large extent state-contingent. This means that the body of indirect knowledge that appears to be "reasonably grounded" reflects a shifting constellation of premises, which may vary as a result of a change in the concentration of attention by individuals or social groups. It follows that the collection of reasonable inferences is likely to change as agents' attention shifts from one set of characteristics (and capabilities) to another (see also Scazzieri 2001a). Indirect knowledge appears to be ephemeral if considered from the point of view of shifting attention. However, any given concentration of attention is associated with a definite focus and a finite set of feasible logical steps. Bounded rationality entails that reason may be sharply focussed. The structure and timing of attention determines whether that focus is effective or not .

The above argument suggests that contexts may be critically relevant in "extracting" a suitable concentration of attention. Considering that any given situation results from the co-existence of multiple features of salience entails that contexts may be central to the discovery of suitable *foci*. The reason is that a *general structure* of co-existing features is likely to provide inadequate information if the *specific texture* of overlaps is unknown. In particular, knowledge that k different features are shared within a given social group across n time periods is not a sufficient basis for the discovery of relevant *foci* unless the pattern of overlaps relevant for each time period is found. To take up a point of view expressed long ago, any given context may be defined in terms of "transversal data of time, place, circumstances" (Valeriani 1804, p. xxv). This means that context is coincident with overlap structure, and that any relevant (effective) focus derives from salient features that the overlap structure has brought into view. In short, context may be seen as a device by which information concerning critical features may be extracted from a social universe in which the identity of individuals (or groups) is not given once for all[3]. A social universe in which individuals (or groups) A_i and A_j are respectively associated with the vectors of characteristics $\{c_1, c_2, c_3, c_4\}$ and $\{c_2, c_4, c_5, c_6\}$ is one in which different patterns of partial resemblance are possible. Similarity features c_2 and c_4 may generate alternative patterns of resemblance, suggest different *foci* of attention and could be associated with distinct focal points for social co-ordination. This point is closely associated with Albert Hirschman's analysis of "shifting involvements" (Hirschman 1982).

As argued by Amartya Sen, social identity is, to a certain extent, a matter of choice (Sen 1999, pp. 15–19). An interesting implication is that attention to social characteristics is also subject to shifting *foci*, and that co-ordination may take a variety of routes as a result of shifting emotions and constraints[4]. Context turns a bundle of virtual identities into a set of realised social opportunities. This suggests that pragmatics has a most important role in determining the structure and evolution of co-ordination in a diverse social universe. Co-ordination reflects a process of social heuristics, that is, a process by which social actors reveal themselves and are in turn "revealed" in the course of social interaction. The above setting is characterised by co-ordination dialogues in which relevant information about individuals or groups is extracted by means of contingent signals and reactions. It seems reasonable to conjecture that economic experiments could be useful tools by which the ephemeral structure of contingent interaction may be "fixed" and the associated pattern of co-ordination be studied as the temporary outcome of interaction between dynamic factors of variable speed and intensity[5].

3. Economic experiments and social heuristics

Economic experiments may be defined as laboratory situations in which "an artificial economic reality is constructed" and "situations of interest for economics" are investigated (Selten, this volume, p. 63). The above description calls forth a careful investigation of what type of laboratory situation is best suited to the generation of an artificial reality in which interesting economic issues can be analysed.

In the previous section I have argued that rationality is to a large extent state contingent, and that the manipulation of a social situation by means of experiment may provide important insights into the causal structure of co-ordination patterns that could be volatile and short-lived. This point of view suggests a critical role for economic experiments. At the same time, it also suggests that effective economic experiments may be quite different from the canonical (Baconian) pattern of controlled manipulation inspired by general principles. The reason is that the setting of initial boundary conditions is a necessary prerequisite for classical experiments (as noted by Aristides Baltas in his contribution to workshop discussion, a brick and an apple may not be sharply different from the point of view of physics; but they are obviously very different from the point of view of natural history). In the case of economics, boundary conditions are not always the same, so that it may be unclear when boundary conditions are unchanged and when they are not[6]. The context-dependence of (successful) economic interaction suggests that economic experiments belong to the domain of conjectures and (social) heuristics, rather than to the domain of the identification of uniform patterns (to be observed in spite of experimental errors). The bounded character of human rationality brings to the fore the fact that economic interaction is to a large extent context-dependent (see above). Successful co-ordination often reflects not so much agents' computational and inferential abilities, as their ability to detect possible patterns of interaction, and to concentrate attention in a highly selective way upon a suitable focal point.

Patrick Suppes has recently argued that "[i]t is not the heavy tomes of mathematics, philosophy and physics, or, now, of biology, that reflect characteristic activity of the brain. It is the laughing, dancing, and learning of children, as they listen to the shouts of each other and ignore the calls of their mothers, that we shall come to understand in a new and deeper mental way. Philosophers like to insist that intentionality is the essential mark of the mental, but an equally strong case can be made empirically that pride of place must be shared with the concept of attention, the speed and efficiency of its operation being equally characteristic of the mental and probably even more important for survival" (Suppes 1999, p. 34).

This set of arguments entails that successful co-ordination may critically depend upon agents' cognitive skills, and in particular upon the agents' ability to concentrate attention upon the right object at the right time. In general, skilled social actors are capable to relate with one another along a variety of social dimensions. This means that any given individual could find herself (himself) in a "co-ordination mode" with respect to other individuals that may vary depending upon the "similarity criterion" that is being considered. As argued in section 2 above, a social universe in which individuals (or groups) A_i and A_j are identified by multi-dimensional vectors of characteristics, is one in which multiple bridges between individuals (or groups) are possible. If we introduce the assumption of finite social variety, the likelihood of social co-ordination would be enhanced by the multiplicity of admissible features. The reason is that, as the description of individuals (or groups) takes on an increasing number of characteristics, it is more likely that some similarity feature will be shared by a significant set of social actors. This means that social co-ordination is, to a large extent, conjectural, and that effective co-ordination often presupposes the ability to undertake social experiments. In other words, the "co-ordination mode" is one in which individuals and groups are open to a variety of social linkages and are prepared to identify locally effective patterns of interaction by means of trial and error.

The relationship between bounded rationality, attention, and effective co-ordination in a diverse social universe points to the primarily *heuristic* character of economic experiments. The reason is that context-dependence could make specific interaction outcomes very difficult to identify (and to predict). Under these conditions, the attempt to reduce experimental errors is not always conducive to a deeper knowledge of causal relationships. Indeed, a "stabilised" experimental process could be one in which informational content is significantly reduced. We may argue that the conjectural character of social co-ordination makes Baconian experiments ineffective (it may be impossible to identify the structure of causal relationships *in spite* of experimental errors precisely because what appears to be an experimental error could be essential in order to discover the relevant causal factor(s) under specific conditions). Economic experiments are best conceived as contrived patterns of interaction by means of which the economist can explore (in a way, "artificially") the diversity of the social universe. As a result, the structure of experiments should be such that the implications of *multiple* modes of perception and reasoning can be investigated.

4. Chance, causation and local experiments

The above argument calls attention to the conjectural character of social co-ordination in a variety setting. It also emphasises that economic experiments are different from Baconian experiments: their primary goal should be to uncover the inherent diversity of the social universe, and to explore the feasibility of alternative patterns of co-ordination. Diversity is sometimes concealed, which entails that important co-ordination schemes are not readily perceived. A reasonable goal of economic experiments is to explore social variety from the point of view of "potential co-ordination". In this case, experiments would have to be arranged so as to "extract" information from social actors. This means that, in principle, the description of social actors should be open-ended, and that individuals or groups should be "left free" to determine particular features of their own identity in the course of social interaction.

The above argument suggests a non-reductionist view of economic experiments. In other words, the experiment would not assume a fixed configuration space, and would explore social co-ordination by taking the concept of "economic agent" as a non-equilibrium concept[7]. This means that a desirable structure for economic experiments may be one in which initial and boundary conditions are left free to change in the course of the experiment. Any given experiment can be seen as an "open field" of interaction, in which (for example) primary and secondary features of individuals (or groups) may be allowed to vary as their interaction unfolds. This situation is *prima facie* one in which the distinction between experiment and observation gets blurred. The reason is that a characterising feature of classical experiments (the setting up of initial and boundary conditions) is missing in this particular case. However, an "open field" experiment remains one in which manipulation is central (just as in classical experiments). This is because a purpose of the experiment would be to extract information about features of similarity and modes of co-ordination that may not be apparent to a *prima facie* observer (see above). The open-endedness of the experiment is a situation in which serendipity is encouraged. The latter was defined by Horace Walpole, from the title of the fairy-tale *The Three Princes of Serendip,* as the art of making happy discoveries by "accidents and sagacity" (Walpole 1754). The open-ended structure of the experiment enhances the likelihood of discovery but presupposes "sagacity" as manipulative ability *vis-à-vis* the changing set of experimental conditions. It is worth noting that, in this case, manipulation takes a character that sets it apart from the type of manipulation common in Baconian experiments. The reason is that a Baconian experiment is primarily an attempt to reach a "stabilised" process by means of experimentally induced repetitiveness. In particular, the Baconian procedure entails that repetitiveness be associated with the identification of initial and boundary conditions, which suitable manipulation has been able to achieve. The "open-ended" experiment may be strikingly different. In this case, manipulation is not aimed at the identification of boundary conditions, but at the discovery of emergent features. This means that repetitiveness is not always a virtue, and that the primary goal of manipulation could be to extract information about the inherent diversity of the process(es) under investigation.

The above argument calls attention to the relationship between experimental conditions and the structure of similarity. In a standard Baconian experiment (and in the Humean view of causality that may be associated with it) "experimental laws" are derived from the setting of initial and boundary conditions, and the repetitiveness of the experiment is supposed to show that the initial and boundary conditions have been successfully identified. In an "open-ended" experiment, on the other hand, experimental laws (in the Baconian sense) cannot be discovered. What can be discovered is the diversity of phenomena and the multiple routes that causal processes may take. This approach has interesting implications for economic analysis. In particular, experimental research could extract useful information about the *degree of variety* of co-ordination schemes and the way in which alternative co-ordination schemes may be associated with different sets of boundary conditions. The "open-ended" experiment is primarily a test of alternative boundary conditions and a search for co-ordination outcomes that may be hidden unless a suitable experimental environment is brought about. Co-ordination is often associated with the ability to identify a suitable focal point. This means that the discovery of mutually compatible goals, interests, and so on, often derives from the ability to concentrate one's attention in the midst of diversity[8].

It has been argued that there is a formal symmetry between the identification of a focal point in strategy choice, and the identification of a focal point when a theory about the world is to be selected (see Rubinstein and Zhou 1999, pp. 205–6). This symmetry may suggest that strategy choice is embedded in theory, and that theory choice is embedded in a pragmatic framework. In either case, attention seems to take precedence over the inferential side of cognition. As a result, both strategy choice and theory choice appear to be closely related with the identification of focal points. In strategy choice, focal points attract agents' attention and make convergent expectations to appear. In theory choice, focal points serve as means by which complex processes may be reduced to a manageable set of relationships. In either case, the ability to detect focal points is an essential cognitive feature of human beings. In a purely theoretical exercise, focal points reduce the complexity of the social (or natural) world and allow the formation of epistemic conjectures (these are conjectures by which a complex process is reduced to its first constituents). In a co-ordination exercise, focal points reduce the complexity of social interaction and allow individuals (or groups) to relate with one another on the basis of a narrow set of behavioural beliefs[9]. Theory choice and strategy choice are often intertwined when particular social contexts are considered. This means that any given mental representation of the environment often emerges as a problem space, and that any particular problem space presupposes a conceptual representation of events.

The relationship between theory choice and strategy choice is different in Baconian and "open-ended" experiments. In the former case, the pragmatic context brings about a sort of contrived uniformity by "fixing" a particular set of initial and boundary conditions. In the latter case, the pragmatic context is meant to explore a *universe* of possible boundary conditions. This means that the configuration space of the social universe cannot be presupposed when the structure of the experiment is determined. The principal aim of the "open-ended" experiment is to "let" boundary conditions to change, and *new structures* to emerge as the experiment unfolds. In this case, the design of empirical investigation is associated with the conjecture that structural change (rather than structural invariance) is to be expected. However, changes in initial and boundary conditions are not easily detectable. For example, under certain conditions, structural parameters may be altered with little influence upon the structure of empirical data (see Hendry 2000). The "open-ended" experiment is a contrived empirical setting in which changes in initial and boundary conditions may be artificially induced, so that the inherent diversity of the social (or natural) world can be better explored. In a nutshell, process stabilisation is an important aim of Baconian experiments, but *not* of "open-ended" experiments. The latter are primarily interested in the generation of novelty.

The distinction between Baconian and "open-ended" experiments calls attention to a deeper layer of investigation, which is associated with the formal structure of induction and the analysis of similarity and distance. In a Baconian experiment, manipulation reduces the epistemic significance of chance: a successful experiment is one in which a particular causal core has been effectively identified, and unexpected occurrences may be considered as "experimental errors". This means that: (i) the structure of evidence has been associated with a sufficiently small set of common and non-essential characteristics (that is, a sufficiently "small" *explanandum* has been identified), and (ii) the relationship between these characteristics and the set of essential and common characteristics forming the *explanans* has become a repetitive one (on the assumption of given sample variety)[10]. Bacon's technique for the careful recording of experimental observations is essentially a way in which the scientist manages to keep track of critical evidence by using the criterion of similarity and "close difference" (*absentiae in proximo*)[11]. What is relevant in our context is that empirical uniformity is not sufficient to the introduction of a causal argument. Uniformity should be decomposed, and a hierarchical logical structure should be introduced. In a Baconian experiment, similarity features are unambiguously classified into: (i) essential and common characteristics, and (ii) non-essential common characteristics. This means that the causal structure is associated with a fixed hierarchy of similarity features. In a non-Baconian (or "open-ended" experiment) the situation is different. Process stabilisation is not a primary aim, so that similarity features that are essential under certain conditions may turn out to be non-essential (accessory) under different conditions. The hierarchy of similarity features may change as the causal focus shifts from one *explanandum* to another. This point of view suggests a different pragmatic status of Baconian *versus* non-Baconian (or "open-ended") experiments. The former are in-depth explorations into the structure of a particular causal relationship: the causal focus is not changed; controllability and repetitiveness "test" the causal connection in a variety of closely monitored environments. Non-Baconian experiments, on the other hand, are explorations into the space of possible (virtual) causal structures. In this case, the causal focus is allowed to change, controllability is only imperfect, and repetitiveness may give way to chance and serendipity.

Either approach presupposes a definite attitude to the relationship between causality and chance. In a "stabilised" Baconian experiment, chance is associated with knowledge of a particular probability distribution, and surprise is induced by low-probability events, or by the realisation that a known distribution no longer applies. A causal relationship can be of the probabilistic type, but chance is often associated with a situation in which both states of the world and probabilities are known (weak uncertainty). A non-Baconian, open-ended, experiment suggests a very different causal setting. Here future states of the world and probabilities may be unknown, and the principal role of chance would be that of generating new causal hypotheses[12]. Surprise (such as an unexpected similarity between distant phenomena) may induce a different hierarchy of similarity features. For example, apparently accessory features may turn out to be essential common characteristics (or vice versa), thus inducing a change of the *explanandum* and a new causal relationship. Chance is no longer a "perturbation" in a stationary causal process, but an essential heuristic tool for the analysis of causality[13].

The above perspective suggests that a Baconian experiment tends to be "universal": experimental trust is associated with the diffusion of common standards, and controllability generates trust by means of shared tools and practices. On the other hand, an "open-ended" experiment is inherently "local", in the sense that its primary aim is not repetitiveness but discovery. For example, the hierarchy of similarity features may change as one moves from one experiment to another (see above). This means that, in this case, experiments are primarily means by which a diverse natural (or social) universe may be explored. In this case too controllability may be essential, but its role is not to reduce experimental "noise". Its primary aim would be to detect unexpected anomalies and to suggest new explanatory hypotheses and theories. In other words, controllability would be a means by which: (i) a more focussed *explanandum* may be identified, and (ii) a new *explanans* may be discovered. The former would generally derive from a change in the hierarchy of similarity features. The latter is often associated with the switch to a different set of initial and boundary conditions.

The above framework calls attention to the inherent duality of experimental practice. Controllability and repetitiveness may be essential to the creative exploration of the natural or social world. But the discovery of interesting anomalies (and unexpected similarity) also presupposes the ability to switch from one focal point to another as the exploration unfolds[14]. Experiments in social science (including economics) presuppose a subtle investigation of latent characteristics and attitudes. In a nutshell, social heuristics is often associated with social maieutics. This suggests that, in the case of experimental economics, the ability to look for unexpected patterns of congruence is at least as important as parameter controllability (and manipulability) in order to explore the diversity of rationality and co-ordination[15].

Department of Economics and Institute of Advanced Studies,
University of Bologna, Bologna, Italy and
Caius College and Clare Hall, University of Cambridge,
Cambridge, United Kingdom

NOTES

[1] Patrick Suppes has discussed a related distinction between "meanings of rationality" respectively associated with the (Bayesian) maximisation of expected utility and the Aristotelian theory of "context-dependent" justification (see Suppes 1984, pp. 184–221).
[2] The *inferential basis* relevant to any given situation may be defined as the set of cognitive premises (background information, tacit assumptions, computational tools and techniques) that give shape to any given problem space and make it conducive to the implementation of certain cognitive strategies more than others. It may be argued that any given inferential basis is associated with some degree of "inductive intuition" (Carnap 1968) and reflects the individual (or collective) awareness that, under any given set of circumstances, a particular set of premises is preferred over its alternatives.
[3] In a similar vein, Ariel Rubinstein has recently argued that a dialogue can be seen "as a mechanism designed to extract information from debaters" (Rubinstein 2000, p. 46).
[4] Amartya Sen has noted that "[t]he real options we have about our identity are always limited by our looks, our circumstances, and our background and history" (Sen 1999, pp. 17–18).
[5] The relevance of ephemeral knowledge (knowledge in flux) as a source of creative co-ordination has been suggested by Barbara Ravelhofer (2002).

[6] I am indebted to Professor Selten for pointing out in workshop discussion the special character of boundary conditions in economics, and its implications for the epistemological structure of economic experiments .

[7] A related discussion of the implications of the non-equilibrium conception of agents in biology may be found in Kauffman (2001).

[8] Focal points may be defined as *foci* of convergent expectations at the level of social interaction (see Schelling 1960, Chapter IV).

[9] These may be defined as agents' beliefs about the most likely social reaction(s) to one's own actions within an interactive set-up (see also Greif 1997).

[10] The above argument is based upon John Maynard Keynes' reconstruction of the structure of evidence in terms of essential and common characteristics (A characteristics), common and non-essential characteristics (B characteristics) and dissimilar characteristics (C characteristics) (See Keynes 1921, pp. 219–220; see also Scazzieri 2001b).

[11] As noted by John Maynard Keynes, Bacon's *Novum Organon* emphasises that evidence should be recorded by distinguishing between (i) a table of all cases in which a given characteristic is found (the table *essentiae et praesentiae*), and (ii) a table of all cases identical to those of table (i) but excluding the above characteristic (the table *declinationis sive absentiae in proximo*). (see Keynes 1921, Chapter xxiii; see also Bacon 2000, 1st edn 1620).

[12] This setting suggests the epistemic situation of "structural ignorance" and "case-based decisions" recently considered by I. Gilboa and D. Schmeidler (2001).

[13] Chance and surprise are important features of cognitive progress associated with experimental practice. In particular, the measurement of surprise calls attention to the relationship between controlled experience and the formulation of new hypotheses: "a surprising sample is one which not only has low probability on the best-supported hypothesis; it also has high probability on an alternative hypothesis" (Edwards 1992, p. 217). Controllability of experience often enhances the likelihood that unexpected anomalies be discovered. In this way, surprise is not alien to a structured experimental practice, and the identification of previously neglected patterns of evidence is made easier by the "sharpened eye" associated with controllability (manipulability) and repetitiveness (see also Fisher 1959).

[14] I am grateful to Alessandro Birolini and Fiorenzo Stirpe, both at the Institute of Advanced Study of the University of Bologna, for seminar discussions in which they have called attention to the inherent duality of experimental practice in technological and life sciences. Carlo Poni has suggested to me in private conversation that experimental practice generally requires means to achieve a reduction of complexity, and that complexity reduction in a particular domain of experience is often associated with complexity increase in other domains of experience. It seems to me that this particular dynamics of complexity highlights one important reason for the relationship between controllability and chance considered above.

[15] John Hicks wrote that, in the case of economics, "the data, with which we are presented, have already been processed, by those directly concerned with them; they have already been classified, for particular purposes, which may not be at all the purposes for which we economists want to use them" (Hicks 1986, p. 99). This point of view is complementary to the foregoing analysis, and suggests that a critical function of experiments could be that of generating new and unexplored items of evidence.

REFERENCES

Bacon, F. 2000. *Novum Organon* (English title: *The New Organon;* 1st edn 1620), edited by Lisa Jardine and Michael Silverthorne. Cambridge: Cambridge University Press.

Carnap, R. 1968. "Inductive Logic and Inductive Intuition". In I. Lakatos (ed.), *The Problem of Inductive Logic*. Amsterdam: North Holland, vol. II, pp. 258–67.

Edwards, A.W.F. 1992. *Likelihood. Expanded Edition*. Baltimore and London: The Johns Hopkins University Press (1st edn 1972).

Fisher, R. A. 1959. *Statistical Methods and Scientific Inference*. Edinburgh: Oliver and Boyd.

Gilboa, I. and Schmeidler D. 2001. *A Theory of Case-Based Decisions*. Cambridge: Cambridge University Press.

Greif, A. 1997. "Cultural Beliefs as a Common Resource in an Integrated World". In P. Dasgupta, K.-G. Maler and A. Vercelli (eds.), *The Economics of Transnational Commons*. Oxford: Clarendon Press.

Hendry, D.F. 2000. "On Detectable and Non-detectable Structural Change". *Structural Change and Economic Dynamics* 11: 45–65.

Hicks, J. 1986. "Is Economics a Science?". In M. Baranzini and R. Scazzieri (eds.), *Foundations of Economics. Structures of Inquiry and Economic Theory*. Oxford and New York: Basil Blackwell, pp. 91–101.

Hirschman, A. 1982. *Shifting Involvements*. Princeton: Princeton University Press.

Kauffman, S. 2001. "Prolegomenon to a General Biology". In A.R. Damasio, A. Harrington, J. Kagan, B.S. Mcewen, H. Moss, R. Shaikh (eds.), *Unity of Knowledge. The Convergence of Natural and Human Science*. In *Annals of the New York Academy of Sciences*, vol. 935, pp. 18–36.

Keynes, J.M. 1921. *A Treatise on Probability*. London: Macmillan.

Ravelhofer, B. 2002. "Kinetic Knowledge and Technical Invention: a Perspective from Performing Arts". Institute of Advanced Study, University of Bologna. Paper presented in the *Creativity and Invention Lecture Series*, 15 April 2002.

Rubinstein, A. and Zhou, L. 1999. "Choice Problems with a 'Reference' Point". *Mathematical Social Sciences* 37: 205–9.

Scazzieri, R. 2000. "Knowledge Economies and Knowledge Systems: A Conceptual Framework". In F. Facchini (ed.), *Scienza e Conoscenza: Verso un Nuovo Umanesimo*, Bologna: Editrice Compositori, pp. 211–222.

Scazzieri, R. 2001a. "Patterns of Rationality and the Varieties of Inference", *Journal of Economic Methodology* 8: 105–110.

Scazzieri, R. 2001b. "Analogy, Causal Patterns and Economic Choice". In M.C. Galavotti, P. Suppes and D. Costantini (eds.), *Stochastic Causality*. Stanford, California: CSLI Publications, pp. 123–139.

Scazzieri, R. 2002. "Practical Rationality and Probabilistic Co-ordination". Paper presented at the meeting *Rationality and Pragmatism: Workshop in Honour of Patrick Suppes* (University of Bologna, 22-23 March 2002).

Schelling, T.C. 1960. *The Strategy of Conflict*. Cambridge, Massachusetts: Harvard University Press.

Selten, R. 2003. "Emergence and Future of Experimental Economics". This volume, pp. 63–70.

Sen, A. 1999. *Reason Before Identity*. The Romanes Lectures for 1998 delivered before the University of Oxford on 17 November 1998. Oxford: Oxford University Press.

Simon, H. 1983. *Reason in Human Affairs*. Oxford: Basil Blackwell.

Suppes, P. 1984. *Probabilistic Metaphysics*. Oxford: Basil Blackwell.

Suppes, P. 1999. "Lezione dottorale" [Honorary Degree Lecture]. In *Laurea Honoris Causa a Patrick Suppes*. University of Bologna, Facoltà di Lettere e Filosofia, 2 December 1999, pp. 19–38.

Valeriani, L. 1804. *Lezione Inaugurale di Pubblica Economia*. Bologna: a S. Tommaso D'Aquino.

Walpole, H. 1754. "Letter to Mann, 28 January 1754". In H.Walpole, *Letters of Horace Walpole, Earl of Oxford, to Sir Horace Mann, British Envoy at the Court of Tuscany. Now First Published from the Originals in the Possession of the Earl of Waldegrave*. Edited by Lord Dover. London: R. Bentley, 1833.

GERD GIGERENZER

WHERE DO NEW IDEAS COME FROM?
A HEURISTICS OF DISCOVERY
IN THE COGNITIVE SCIENCES*

Scientific inquiry can be viewed as "an ocean, continuous everywhere and without a break or division", in Leibniz's words (1690/1951, p. 73). Hans Reichenbach nonetheless divided this ocean into two great seas, the context of discovery and the context of justification. Philosophers, logicians, and mathematicians claimed justification as a part of their territory and dismissed the context of discovery as none of their business, or even as "irrelevant to the logical analysis of scientific knowledge" (Popper 1935/1959, p. 31). Their sun shines over one part of the ocean and has enlightened us on matters of justification, but the other part of the ocean still remains in a mystical darkness where imagination and intuition reigns, or so it is claimed. Popper, Braithwaite, and others ceded the dark part of the ocean to psychology and, perhaps, sociology; but few psychologists have fished in these waters. Most did not dare or care.

In this article, I will argue that discovery can be understood by heuristics (not a logic) of discovery. I will propose a heuristic of discovery that makes use of methods of justification, thereby attempting to bridge the artificial distinction between the two. Furthermore, I will attempt to demonstrate that this discovery heuristic may not only be of interest for an a posteriori understanding of theory development, but also be useful for understanding limitations of present-day theories and research programs and for the further development of alternatives and new possibilities. The discovery heuristic that I call the "tools-to-theories" heuristic (see Gigerenzer 1991, 2000) postulates a close connection between the shining and the dark part of Leibniz's ocean: scientists' tools for justification provide the metaphors and concepts for their theories.

The power of tools to shape, or even to become, theoretical concepts is an issue largely ignored in both the history and philosophy of science. Inductivist accounts of discovery, from Bacon to Reichenbach and the Vienna School, focus on the role of data, but do not consider how the data are generated or processed. Nor do the numerous anecdotes about discoveries, such as Newton watching an apple fall in his mother's orchard while pondering the mystery of gravitation, Galton taking shelter from a rainstorm during a country outing when discovering correlation and regression toward mediocrity, and the stories about Fechner, Kekulé, Poincaré, and others, which link discovery to beds, bicycles, and bathrooms. What unites these anecdotes is the focus on the vivid but prosaic circumstances; they report the setting in which a discovery occurs, rather than analyzing the process of discovery.

The question "Is there a logic of discovery?" and Popper's (1935/1959) conjecture that there is none have misled many into assuming that the issue is whether there exists a logic of discovery or only idiosyncratic personal and accidental reasons that explain the "flash of insight" of a particular scientist (Nickles 1980). I do not think that formal logic and individual personality are the only alternatives, nor do I believe that either of these is a central issue for understanding discovery.

The process of discovery can be shown, according to my argument, to possess more structure than thunderbolt guesses but less definite structure than a monolithic logic of discovery, of the sort Hanson (1958) searched for, or a general inductive hypothesis-generation logic (e.g., Reichenbach 1938). The present approach lies between these two extremes; it looks for structure beyond the insight of a genius, but does not claim that the tools-to-theories heuristic is (or should be) the only account of scientific discovery. The tools-to-theories heuristic applies neither to all theories in science nor to all cognitive theories; it applies to a specific group of cognitive theories developed after the so-called cognitive revolution, in the last three decades.

Nevertheless, similar heuristics have promoted discovery in physics, physiology, and other areas. For instance, it has been argued that once the mechanical clock became the indispensable tool for astronomical research, the universe itself became understood as a kind of mechanical clock, and God as a divine watchmaker. Lenoir (1986) shows how Faraday's instruments for recording electric currents shaped the understanding of electrophysiological processes by promoting concepts such as "muscle current" and "nerve current".

Thus, this discovery heuristic boasts some generality both within cognitive psychology and within science, but this generality is not unrestricted. Since there has been little research in how tools of justification influence theory development, the tools-to-theories heuristic may be more broadly applicable than I am able to show in this article. If my view of heuristics of discovery as a heterogeneous bundle of search strategies is correct, however, this implies that generalizability is in principle bounded.

What follows has been inspired by Herbert Simon's notion of heuristics of discovery, but goes beyond his attempt to model discovery with programs such as BACON that attempt to induce scientific laws from data (e.g., Langley, Simon, Bradshaw, and Zytkow 1987). My focus is on the role of the tools that process and produce data, not the data themselves, in the discovery and acceptance of theories.

HOW METHODS OF JUSTIFICATION SHAPE THEORETICAL CONCEPTS

The tools-to-theories heuristic is twofold:

(1) *Discovery*. Scientific tools, once entrenched in a scientist's daily practice, suggest new theoretical metaphors and theoretical concepts.

(2) *Acceptance*. Once proposed by an individual scientist (or a group), the new theoretical metaphors and concepts are more likely to be accepted by the scientific community if the members of the community are also users of the new tools.

By *tools* I mean both analytical and physical methods that are used to evaluate given theories. Analytical tools can be either empirical or non-empirical. Examples of analytical methods of the empirical kind are tools for data processing such as statistics; examples of the non-empirical kind are normative criteria for the evaluation of hypotheses such as logical consistency. Examples of physical tools of justification are measurement instruments such as clocks. In this article, I will focus on analytical rather than physical tools of justification, and among these, on techniques of statistical inference and hypothesis testing. My topic here will be theories of mind, and how social scientists discovered them after the emergence of new tools for data analysis, rather than of new data.

In this context, the tools-to-theories heuristic consists in the discovery of new theories by changing the conception of the mind through the analogy of the tool. The result can vary in depth from opening new general perspectives, albeit mainly metaphorical, to sharp discontinuity in specific cognitive theories caused by the direct transfer of scientist's tools into theories of mind.

This article will deal with two tools that have turned into cognitive theories: inferential statistics and the digital computer.

COGNITION AS INTUITIVE STATISTICS

I begin with a brief history. In American psychology, the study of cognitive processes was suppressed in the early 20th century by the allied forces of operationalism and behaviorism. The operationalism and the inductivism of the Vienna School, inter alia, paved the way for the institutionalization of inferential statistics in American experimental psychology between 1940 and 1955 (Gigerenzer 1987a; Toulmin and Leary 1985). In experimental psychology, inferential statistics became almost synonymous with scientific method. Inferential statistics, in turn, provided a large part of the new concepts of mental processes that have fueled the so-called cognitive revolution since the 1960s. Theories of cognition were cleansed of terms such as restructuring and insight, and the new mind came to be portrayed as drawing random samples from nervous fibers, computing probabilities, calculating analyses of variance, setting decision criteria, and performing utility analyses.

After the institutionalization of inferential statistics, a broad range of cognitive processes – conscious and unconscious, elementary and complex – were reinterpreted as involving "intuitive statistics". For instance, W. P. Tanner and his coworkers assumed in their theory of signal detectability that the mind "decides" whether there is a stimulus or only noise, just as a statistician of the Neyman-Pearson school decides between two hypotheses (Tanner and Swets 1954). In his causal attribution theory, Harold H. Kelley (1967) postulated that the mind attributes a cause to an effect in the same way as behavioral scientists have come to do, namely, by performing an analysis of variance and testing null hypotheses. These two influential theories show the breadth of the new conception of the "mind as an intuitive statistician" (Gigerenzer 2000; Gigerenzer and Murray 1987). They also exemplify cognitive theories that were suggested not by new data, but by new tools of data analysis.

In what the following, I shall give evidence for three points. First, the discovery of theories based on the conception of the mind as an intuitive statistician caused discontinuity in theory rather than being merely a new, fashionable language: it radically changed the kind of phenomena reported, the kinds of explanations looked for, and even the kinds of data that were generated. This first point illustrates the profound power of the tools-to-theories heuristic to generate quite innovative theories. Second, I will provide evidence for the "blindness" or inability of researchers to discover and accept the conception of the mind as an intuitive statistician before they became familiar with inferential statistics as part of their daily routine. The discontinuity in cognitive theory is closely linked to the preceding discontinuity in method, that is, to the institutionalization of inferential statistics in psychology. Third, I will show how the tools-to-theories heuristic can help us to see the limits and possibilities of current cognitive theories that investigate the mind as an "intuitive statistician".

Discontinuity in cognitive theory development

What has been called the "cognitive revolution" (Gardner 1985) is more than the overthrow of behaviorism by mentalist concepts. The latter have been a continuous part of scientific psychology since its emergence in the late 19th century, even coexisting with American behaviorism during its heyday (Lovie 1983). The cognitive revolution did more than revive the mental; it has changed our concepts of what the mental means, often dramatically. One source of this change is the tools-to-theories heuristic, with its new analogy of the mind as an intuitive statistician. To show the discontinuity within cognitive theories, I will briefly discuss two areas in which an entire statistical technique, not only a few statistical concepts, became a model of mental processes: stimulus detection and discrimination, and causal attribution.

What intensity must a 440-Hz tone have to be perceived? How much heavier than a standard stimulus of 100 gms must a comparison stimulus be, in order for a perceiver to notice a difference? How can the elementary cognitive processes involved in those tasks, known today as "stimulus detection" and "stimulus discrimination" be understood? Since Herbart (1816), such processes have been explained by using a threshold metaphor: detection occurs only if the effect an object has on our nervous system exceeds an absolute threshold, and discrimination between two objects occurs if the excitation from one exceeds that from the other by an amount greater than a differential threshold. E. H. Weber and G. T. Fechner's laws refer to the concept of fixed thresholds; Titchener (1896) saw in differential thresholds the long sought-after elements of mind (he counted approximately 44,000); and classic textbooks such as Brown and Thomson's (1921) and Guilford's (1954) document methods and research.

Around 1955, the psychophysics of absolute and differential thresholds was revolutionized by the new analogy between the mind and the statistician. W. P. Tanner and others proposed a "theory of signal detectability" (TSD), which assumes that the Neyman-Pearson technique of hypothesis testing describes the processes involved in detection and discrimination. Recall that in Neyman-Pearson statistics, two sampling distributions (hypotheses H0 and H1) and a decision criterion (which is a likelihood ratio) are defined, and then the data observed are transformed into a likelihood ratio and compared with the decision criterion. Depending on which side of the criterion the data fall, the decision to "reject H0 and accept H1" or "accept H0 and reject H1" is made. In straight analogy, TSD assumes that the mind calculates two sampling distributions for "noise" and "signal plus noise" (in the detection situation) and sets a decision criterion after weighing the cost of the two possible decision errors (Type I and Type II errors in Neyman-Pearson theory, now called "false alarms" and "misses"). The sensory input is transduced into a form that allows the brain to calculate its likelihood ratio, and depending on whether this ratio is smaller or larger than the criterion, the subject says: "no, there is no signal" or "yes, there is a signal." Tanner (1965) explicitly referred to his new model of the mind as a "Neyman-Pearson detector", and in unpublished work, his flow charts included a drawing of a homunculus statistician performing the unconscious statistics in the brain (Gigerenzer and Murray 1987, pp. 43–53).

The new analogy between mind and statistician replaced the century-old concept of a fixed threshold by the twin notions of observer's attitudes and observer's sensitivity. Just as Neyman–Pearson technique distinguishes between a subjective part (e.g., selection of a criterion dependent on cost-benefit considerations) and a mathematical part, detection and discrimination became understood as involving both subjective processes, such as attitudes and cost-benefit considerations, and sensory processes. Swets, Tanner, and Birdsall (1964, p. 52) considered this link between attitudes and sensory processes to be the "main thrust" of their theory. The analogy between technique and mind made new research questions thinkable, such as "How can the mind's decision criterion be manipulated?" A new kind of data even emerged: two types of errors were generated in the experiments, false alarms and misses, just as the statistical theory distinguishes two types of error.

As far as I can tell, the idea of generating these two kinds of data (errors) was not common before. The discovery of TSD was not motivated by new data; rather, the new theory motivated a new kind of data. In fact, in their seminal paper, Tanner and Swets (1954, p. 401) explicitly admit that their theory "appears to be inconsistent with the large quantity of existing data on this subject", and proceed to criticize the "form of these data."

The Neyman-Pearsonian technique of hypothesis testing was subsequently transformed into a theory of a broad range of cognitive processes, ranging from recognition in memory (e.g., Murdock 1982;

Wickelgreen and Norman 1966) to eyewitness testimony (e.g., Birnbaum 1983) to discrimination between random and nonrandom patterns (e.g., Lopes 1982).

My second example concerns theories of causal reasoning. In Europe, Albert Michotte (1946/1963), Jean Piaget (1930), the Gestalt psychologists, and others had investigated how certain temporal spatial relationships between two or more visual objects, such as moving dots, produced phenomenal causality. For instance, the subjects were made to "perceive" that one dot launches, pushes, or chases another. After the institutionalization of inferential statistics, Harold H. Kelley (1967) proposed in his "attribution theory" that the long-sought laws of causal reasoning are, in fact, the tools of the behavioral scientist: R. A. Fisher's analysis of variance (ANOVA). Just as the experimenter has come to infer a causal relationship between two variables by calculating an analysis of variance and performing an F-test, the man-in-the-street infers the cause of an effect by unconsciously making the same calculations. By the time Kelley discovered the new metaphor for causal inference, about 70% of all experimental articles already used ANOVA (Edgington 1974).

The theory was quickly accepted in social psychology; Kelley and Michaela (1980) reported more than 900 references in one decade. The vision of the Fisherian mind radically changed the understanding of causal reasoning, the problems posed to the subjects, and the explanations looked for. I list a few discontinuities that reveal the fingerprints of the tool.

(1) ANOVA needs repetitions or numbers as data in order to estimate variances and covariances. Consequently, the information presented to the subjects in studies of causal attribution consists of information about the frequency of events (e.g., McArthur 1972), which played no role in either Michotte's or Piaget's work.

(2) Whereas Michotte's work still reflects the broad Aristotelian conception of four causes (see Gavin 1972), and Piaget (1930) distinguished 17 kinds of causality in children's minds, the Fisherian mind concentrates on the one kind for which ANOVA is used as a tool (similar to Aristotle's "material cause").

(3) In Michotte's view, causal perception is direct and spontaneous and needs no inference, as a consequence of largely innate laws that determine the organization of the perceptual field. ANOVA, in contrast, is used in psychology as a technique for inductive inferences from data to hypotheses, and the focus in Kelley's attribution theory is consequently on the data-driven, inductive side of causal perception.

The latter point illustrates that the specific use of a tool, that is, its practical context rather than its mathematical structure, can also shape theoretical conceptions of mind. To elaborate on this point, let us assume that Harold Kelley had lived one and a half centuries earlier. In the early 19th century, significance tests (similar to those in ANOVA) were already being used by astronomers (Swijtink 1987). However, they used their tests to reject data, so-called outliers, and not to reject hypotheses. At least provisionally, the astronomers assumed that the theory was correct and mistrusted the data, whereas the ANOVA mind, following the current statistical textbooks, assumes the data to be correct and mistrusts the theories. So, to our 19th-century Kelley, the mind's causal attribution would have seemed expectation-driven rather than data-driven: the statistician homunculus in the mind would have tested the data and not the hypothesis.

These two areas – detection and discrimination, and causal reasoning – may be sufficient to illustrate some of the fundamental innovations in the explanatory framework, in the research questions posed, and in the kind of data generated. The spectrum of theories that model cognition after statistical inference ranges from auditory and visual perception to recognition in memory, and from speech perception to thinking and reasoning (Gigerenzer and Murray 1987; Gigerenzer 1991, 1994).

To summarize: the tools-to-theories heuristic can account for the discovery and acceptance of a group of cognitive theories in apparently unrelated subfields of psychology, all sharing the view that cognitive processes can be modeled by statistical hypothesis testing. Among these are several highly innovative and influential theories that have radically changed our understanding of what "cognitive" means.

Before the institutionalization of inferential statistics

There is an important test case for the present hypothesis (a) that familiarity with the statistical tool is crucial to the discovery of corresponding theories of mind, and (b) that the institutionalization of the tool within a scientific community is crucial to the broad acceptance of those theories. That test case is the era prior to the institutionalization of inferential statistics. Theories that conceive of the mind as an intuitive statistician should have a very small likelihood of being discovered, and even less likelihood of being accepted. The two strongest tests are cases where (a) someone proposed a similar conceptual analogy, and (b) someone proposed a similar probabilistic (formal) model. The chances of theories of the first kind being accepted should be small, and the chances of a probabilistic model being interpreted as "intuitive statistics" should be similarly small. I know of only one case each, which I will analyze after first defining what I mean by the term "institutionalization of inferential statistics".

Statistical inference has been known for a long time, but not used as theories of mind. In 1710, John Arbuthnot proved the existence of God using a kind of significance test; as mentioned above, astronomers used significance tests in the 19th century; G. T. Fechner's statistical text *Kollektivmasslehre* (1897) included tests of hypotheses; W. S. Gosset (using the pseudonym "Student") published the t-test in 1908; and Fisher's significance-testing techniques, such as ANOVA, as well as Neyman-Pearsonian hypothesis-testing methods have been available since the 1920s (Gigerenzer et al. 1989). Bayes' theorem has been known since 1763. Nonetheless, there was little interest in these techniques in experimental psychology before 1940 (Rucci and Tweney 1980).

The statisticians' conquest of new territory in psychology began in the 1940s. By 1942, Maurice Kendall could comment on the statisticians' expansion: "They have already overrun every branch of science with a rapidity of conquest rivaled only by Attila, Mohammed, and the Colorado beetle" (p. 69). By the early 1950s, half of the psychology departments in leading American universities offered courses on Fisherian methods and had made inferential statistics a graduate program requirement. By 1955, more than 80% of the experimental articles in leading journals used inferential statistics to justify conclusions from the data (Sterling 1959), and editors of major journals made significance testing a requirement for the acceptance of articles submitted (e.g., Melton 1962).

I shall therefore use 1955 as a rough date for the institutionalization of the tool in curricula, textbooks, and editorials. What became institutionalized as the logic of statistical inference was a mixture of ideas from two opposing camps, those of R. A. Fisher, on the one hand, and Jerzy Neyman and Egon S. Pearson (the son of Karl Pearson) on the other.

Discovery and rejection of the analogy

The analogy between the mind and the statistician was first proposed before the institutionalization of inferential statistics, in the early 1940s, by Egon Brunswik at Berkeley (e.g., Brunswik 1943). As Leary (1987) has shown, Brunswik's probabilistic functionalism was based on a very unusual blending of scientific traditions, including the probabilistic world view of Hans Reichenbach and members of the Vienna School, and Karl Pearson's correlational statistics.

The important point here is that in the late 1930s Brunswik changed his techniques for measuring perceptual constancies, from calculating (non-statistical) "Brunswik ratios" to calculating Pearson correlations, such as "functional" and "ecological" validities. In the 1940s, he also began to think of the organism as "an intuitive statistician", but it took him several years to spell out the analogy in a clear and consistent way (Gigerenzer 1987b, 2001).

The analogy is this: the perceptual system infers its environment from uncertain cues by (unconsciously) calculating correlation and regression statistics, just as the Brunswikian researcher does when (consciously) calculating the degree of adaptation of a perceptual system to a given environment. Brunswik's "intuitive statistician" was a statistician of the Karl Pearson School, like the Brunswikian researcher. Brunswik's "intuitive statistician" was not well adapted to the psychological science of the time, however, and the analogy was poorly understood and generally rejected (Leary 1987).

Brunswik's analogy came too early to be understood and accepted by his colleagues of the experimental discipline; it came before the institutionalization of statistics as the indispensable method of scientific inference, and it came with the "wrong" statistical model, correlational statistics. Correlation was an indispensable method not in experimental psychology, but rather in its rival discipline, known as the Galton-Pearson program, or, as Lee Cronbach (1957) put it, the "Holy Roman Empire" of "correlational psychology".

The schism between the two disciplines had been repeatedly taken up in presidential addresses before the APA (Dashiell 1939; Cronbach 1957) and had deeply affected the values and the mutual esteem of psychologists (Thorndike 1954). Brunswik could not succeed in persuading his colleagues from the experimental discipline to consider the statistical tool of the competing discipline as a model of how the mind works. Ernest Hilgard (1955), in his rejection of Brunswik's perspective, did not mince words: "Correlation is an instrument of the devil" (p. 228).

Brunswik, who coined the metaphor of "man as intuitive statistician", did not survive to see the success of his analogy. It was accepted only after statistical inference became institutionalized in experimental psychology, and with the new institutionalized tools rather than (Karl) Pearsonian statistics serving as models of mind. Only in the mid-1960s, however, did interest in Brunswikian models of mind emerge (e.g. Brehmer and Joyce 1988; Hammond and Stewart 2001).

Probabilistic models without the "intuitive statistician"

My preceding point was that the statistical tool was accepted as a plausible analogy of cognitive processes only after its institutionalization in experimental psychology. My second point is that although some probabilistic models of cognitive processes were advanced before the institutionalization of inferential statistics, they were not interpreted using the metaphor of the "mind as intuitive statistician". The distinction I draw here is between probabilistic models that use the metaphor and ones that do not. The latter kind is illustrated by models that use probability distributions for perceptual judgment, assuming that variability is caused by lack of experimental control, measurement error, or other factors that can be summarized as experimenter's ignorance. Ideally, if the experimenter had complete control and knowledge (such as Laplace's superintelligence), all probabilistic terms could be eliminated from the theory. This does not hold for a probabilistic model that is based on the metaphor. Here, the probabilistic terms model the ignorance of the mind rather than that of the experimenter. That is, they model how the "homunculus statistician" in the brain comes to terms with a fundamental uncertain world. Even if the experimenter had complete knowledge, the theories would remain probabilistic, since it is the mind that is ignorant and needs statistics.

The key example is L. L. Thurstone, who in 1927 formulated a model for perceptual judgment that was formally equivalent to the present-day theory of signal detectability (TSD). But neither Thurstone nor his followers recognized the possibility of interpreting the formal structure of their model in terms of the "intuitive statistician". Like TSD, Thurstone's model had two overlapping normal distributions, which represented the internal values of two stimuli and which specified the corresponding likelihood ratios, but it never occurred to Thurstone to include the conscious activities of a statistician, such as the weighing of the costs of the two errors and the setting of a decision criterion, in his model. Thus neither Thurstone nor his followers took the – with hindsight – small step to develop the "law of comparative judgment" into TSD. When Duncan Luce (1977) reviewed Thurstone's model 50 years later, he found it hard to believe that nothing in Thurstone's writings showed the least awareness of this small, but crucial step. Thurstone's perceptual model remained a mechanical, albeit probabilistic, stimulus response theory without a homunculus statistician in the brain. The small conceptual step was never taken, and TSD entered psychology by an independent route.

In summary: there are several kinds of evidence for a close link between the institutionalization of inferential statistics in the 1950s and the subsequent broad acceptance of the metaphor of the mind as an intuitive statistician: (1) the general failure to accept, and even, to understand Brunswik's "intuitive statistician" before the institutionalization of the tool, and (2) the case of Thurstone, who proposed a probabilistic model that was formally equivalent to one important present-day theory of "intuitive statistics", but was never interpreted in this way; the analogy was not yet seen. Brunswik's case illustrates that tools may act at two levels: first, new tools may suggest new cognitive theories to a scientist. Second, the degree to which these tools are institutionalized within the scientific community to which the scientist belongs can prepare (or hinder) the acceptance of the new theory. This close link between tools for justification, on the one hand, and discovery and acceptance, on the other, reveals the artificiality of the discovery/justification distinction. Discovery does not come first, and justification afterwards. Discovery is inspired by justification.

Heuristics of discovery may help in understanding limitations and possibilities of current research programs

The preceding analysis of discovery is of interest not only for a psychology of scientific discovery and creativity (e.g., Gardner 1988; Gruber 1981; Tweney, Dotherty, and Mynatt 1981), but also for the evaluation and further development of current cognitive theories. The general point is that institutionalized tools such as statistics are not theoretically inert. Rather, they come with a set of assumptions and interpretations that may be smuggled Trojan-horse fashion into the new cognitive theories and research programs. One example was mentioned above: the formal tools of significance testing are interpreted in psychology as tools for rejecting hypotheses, assuming that the data are correct, whereas in other fields and at other times the same tools were interpreted as tools for rejecting data (outliers), assuming that the hypotheses were correct. The latter use of statistics is practically extinct in experimental psychology (although the problem of outliers routinely emerges), and therefore also absent in theories that liken cognitive processes to significance testing. In cases like these, analysis of discovery may help to reveal "blind spots" associated with the tool and, as a consequence, new possibilities for cognitive theorizing.

There are several assumptions that became associated with the statistical tool in the course of its institutionalization in psychology, none of them being part of the mathematics or statistical theory proper. The first assumption can be called "There is only one statistics". Textbooks on statistics for psychologists (usually written by non-mathematicians) generally teach statistical inference as if only one logic of inference existed. Since the 1950s and 1960s, almost all texts teach a mishmash of R. A. Fisher's ideas tangled with those of Jerzy Neyman and Egon S. Pearson, but without acknowledgment. The fact that Fisherians and Neyman-Pearsonians could never agree on a logic of statistical inference is not mentioned in the textbooks, nor are the controversial issues that divide them. Even alternative statistical logics for scientific inference are rarely discussed (Gigerenzer 1993, 2001). For instance, Fisher (1955) argued that concepts such as Type II error, power, the setting of a level of significance before the experiment, and its interpretation as a long-run frequency of errors in repeated experiments are concepts inappropriate for scientific inference – at best they could be applied to technology (his pejorative example was Stalin's). Neyman, for his part, declared that some of Fisher's significance tests are "worse than useless" (since their power is less than their size; see Hacking 1965, p. 99).

I know of no textbook written by psychologists for psychologists that mentions and explains this and other controversies about the logic of inference. Instead, readers are presented with an intellectually incoherent mix of Fisherian and Neyman-Pearsonian ideas, but a mix presented as a seamless, uncontroversial whole: the logic of scientific inference (for more details see Gigerenzer et al. 1989, chs. 3 and 6).

This assumption that "statistics is statistics is statistics" – characteristic of the practical context in which the statistical tool has been used, not of the mathematical theories – reemerges at the theoretical level in current cognitive psychology, just as the tools-to-theories heuristic would lead us to expect (Gigerenzer 1991). For instance, research on so-called "cognitive illusions" assumes that there is one, and only one, correct answer to statistical reasoning problems. As a consequence, other answers are considered to reflect reasoning fallacies, attributed to shabby mental software. Some of the most prominent reasoning problems, however, such as the cab problem (Tversky and Kahneman 1980, p. 62), do not have only one answer; the answer depends on the theory of statistical inference and the assumptions applied. Birnbaum (1983), for example, shows that the "only correct answer" to the cab problem claimed by Tversky and Kahneman, based on Bayes' rule, is in fact only one of several reasonable answers – different ones are obtained, for instance, if one applies the Neyman-Pearson theory (Gigerenzer and Murray 1987, ch. 5).

A second assumption that became associated with the tool during its institutionalization is "there is only one meaning of probability". For instance, Fisher and Neyman-Pearson had different interpretations of what a level of significance means. Fisher's was an epistemic interpretation, that is, that the level of significance informs us about the confidence we can have in the particular hypothesis under test, whereas Neyman's was a strictly frequentist and behavioristic interpretation, which claimed that a level of significance tells us nothing about a particular hypothesis, but about the long-run relative frequency of wrongly rejecting the null hypothesis if it is true. Although the textbooks teach both Fisherian and Neyman-Pearsonian ideas, these alternative views of what a probability (such as level of significance) could mean are generally neglected – not to speak of the other meanings, subjective and objective, that have been proposed for the formal concept of probability (Hacking 1965).

Many of the so-called cognitive illusions were demonstrated using a subjective interpretation of probability, specifically, asking people about the probability they assign to a single event. When researchers instead began to ask people for judgments of frequencies, these apparently stable reasoning errors – the conjunction fallacy, the overconfidence bias, for example – largely or completely disappeared (Gigerenzer 1994, 2000a). Untutored intuition seems to be capable of making conceptual distinctions of the sort statisticians and philosophers make, such as between judgments of subjective probability and those of frequency (e.g., Cohen 1986; Lopes 1981; Teigen 1983). And these results suggest that the important research questions to be investigated are "How are different meanings of 'probability' cued in every-day language?" and "How does this affect judgment?" rather than "How can we explain the alleged bias of 'overconfidence' by some general deficits in memory, cognition, or personality?"

To summarize: assumptions entrenched in the practical use of statistical tools – which are not part of the mathematics – can reemerge in research programs on cognition, resulting in severe limitations in these programs. This could be avoided by pointing out these assumptions, which, in turn, may even lead to new research questions (Gigerenzer 2000).

I now extend my analysis from techniques of statistical inference to another tool, the computer (Gigerenzer and Goldstein 1996). In the first part, I argue that a conceptual divorce between intelligence and calculation around 1800, motivated by economical transformations, made mechanical computation (and ultimately the computer) conceivable. The tools-to-theories heuristic comes into play in the second part, where we show how the computer, after becoming a standard laboratory tool in this century, was proposed, and with some delay accepted, as a model of mind. Thus, we travel in a full circle from mind to computer and back.

MIND AS COMPUTER

Act I: From mind to computer

The president of the Astronomical Society of London, Henry Colebrooke (1825) summed up the significance of Charles Babbage's (1791–1871) work: "Mr. Babbage's invention puts an engine in place of the computer." This seems a strange statement about the man who is now praised for having invented the computer. But at Babbage's time, the computer was a human being, in this case someone who was hired for exhaustive calculations of astronomical and navigational tables.

How did Babbage ever arrive at the idea of putting a mechanical computer in place of a human one? A divorce between intelligence and calculation, as Daston (1994) has argued, made it possible for Babbage to conceive this idea.

In the Enlightenment, calculation was not considered a rote, mechanical thought process. In contrast, philosophers of the time held that intelligence and even moral sentiment were, in their essence, forms of calculation (Daston 1988, 1994). Calculation was the opposite of the habitual and the mechanical, remote from the realm of menial labor. For Condillac, d'Alembert, Condorcet, and other Enlightenment philosophers, the healthy mind worked by constantly taking apart ideas and sensations into their minimal elements, then comparing and rearranging these elements into novel combinations and permutations. Thought was a combinatorial calculus, and great thinkers were proficient calculators. In the eulogies of great mathematicians, for instance, prodigious mental reckoning was a favorite topic – Gauss' brilliant arithmetic was perhaps the last of these stock legends. Calculation was the essence of moral sentiment, as well. Even self-interest and greed, as opposed to dangerous passions, by their nature of being calculations, were at least predictable and thereby thought to reinforce the orderliness of society (Daston 1994).

The computer as a factory of workers

By the turn of the 19th century, calculation was shifting from the company of *hommes éclairés* and savants to that of the unskilled work force. Extraordinary mental arithmetic became associated with the idiot savant and the sideshow attraction. Calculation grew to be seen as dull, repetitive work, best performed by patient minds that lacked imagination. Women ultimately staffed the "bureaux de calculs" in major astronomical and statistical projects (despite being earlier accused of vivid imaginations and mental restlessness, see Daston 1992). Talent and genius ceased to be virtuoso combinatorics and permutations, and turned into romantic, unanalyzable creations. Thereby, the stage became set for the neo-romanticism in 20th century philosophy of science that declared creativity as mystical, and the context of discovery as "irrelevant to the logical analysis of scientific knowledge" (Popper 1935/1959, p. 31).

Daston (1994) and Schaffer (1994) argue that one force in this transformation was the introduction of large-scale division of labor in manufacturing, as evidenced in the automatic system of the English machine-tool industry and in the French government's large-scale manufacturing of logarithmic and trigonometric tables for the new decimal system in the 1790s. During the French revolution, the engineer Gaspard Riche de Prony organized the French government's titanic project for the calculation of 10,000 sine values to the unprecedented precision of 25 decimal places and some 200,000 logarithms to 14 or 15 decimal places. Inspired by Adam Smith's praise of the division of labor, Prony organized the project in a hierarchy of tasks. At the top were a handful of excellent mathematicians, including Adrien Legendre and Lazare Carnot, who devised the formulae; in the middle were seven or eight persons trained in analysis; and at the bottom were seventy or eighty unskilled persons knowing only the rudiments of arithmetic, who performed millions of additions and subtractions. These "manufacturing" methods, as Prony called them, pushed calculation away from intelligence and towards work. The terms "work" and "mechanical" were linked both in England and in France until the middle of the 19th century (Daston 1994). Work concerned the body but not the mind; in large-scale manufacturing, each worker did only one task throughout his entire life.

Once it was shown that elaborate calculation could be carried out by an assemblage of unskilled workers, each knowing very little about the large computation, it became possible for Babbage to conceive of replacing these workers with machinery. Babbage's view of the computer bore a great resemblance to a factory of unskilled human workers. When Babbage talked about the parts of his Analytical Engine, the arithmetic computation and the storage of numbers, he called these the "mill" and the "store", respectively (Babbage 1812/1994, p.23). The metaphor came from the textile industry. In the textile industry, yarns were brought from the store to the mill where they were woven into fabric, which was then sent back to the store. In the Analytical Engine, numbers were brought from the store to the arithmetic mill for processing, and the results were returned to the store. Commenting on this resemblance, Lady Lovelace stated, "we may say most aptly that the Analytical Engine weaves algebraic patterns just as the Jaquard loom weaves flowers and leaves" (Babbage 1812/1994, p.27).[1] In his chapter on the "division of mental labor", Babbage explicitly refers to the French government's program for the computation of new decimal tables as the inspiration and foundation of a general science of machine intelligence.

To summarize the argument: during the Enlightenment, calculation was the distinctive activity of the scientist and the genius, and the very essence of the mental life. New ideas and insights were assumed to be the product of the novel combinations and permutations of ideas and sensations. In the first decades of the 19th century, numerical calculation was separated from the rest of intelligence and demoted to one of the lowest operations of the human mind. Once calculation became the repetitive task of an army of unskilled workers, Babbage could envision mechanical computers replacing human computers. Pools of human computers and Babbage's mechanical computer manufactured numbers in the same way as the factories of the day manufactured their goods.[2]

The computer as a brain

Babbage once reported a dream: that all tables of logarithms could be calculated by a machine. However, this dream did not turn into a reality during his lifetime. He never could complete any of the three machines he had started to build. Modern computers, such as the ENIAC and the EDVAC at the University of Pennsylvania, came about during and after the Second World War. Did the fathers of computer science see the mind as a computer? We (Gigerenzer and Goldstein 1996) argued that the contemporary analogy stating that the mind is a computer was not yet established before the "cognitive revolution" of the 1960s. As far as we can see, there were two groups willing to draw a parallel between the human and the computer, but neither used the computer as a theory of mind. One group, which tentatively compared the nervous system to the computer, is represented by the Hungarian mathematician John von Neumann (1903–1957). The other group, which investigated the idea that machines might be capable of thought, is represented by the English mathematician and logician Alan Turing (1912–1954).

Von Neumann, known as the father of the modern computer, wrote about the possibility of an analogy between the computer and the human nervous system. It seems that his reading of a paper by Warren McCulloch and Walter Pitts called "A logical calculus of the ideas immanent in nervous activity" triggered his interest in information processing in the human brain soon after its publication in 1943 (Asaray 1990). The paper begins with the statement that because of the all-or-none character of the nervous system, neural events can be represented by means of propositional logic. The McCulloch-Pitts model did not deal with the structure of neurons, which were treated as "black boxes". The model was largely concerned with the mathematical rules governing the input and output of signals. In the 1945 EDVAC (the "Electronic Discrete Variable Computer" at the University of Pennsylvania) report, von Neumann described the computer as being built from McCulloch and Pitt's idealized neurons rather than from vacuum tubes, electromechanical relays, or mechanical switches. To understand the computer in terms of the human nervous system appeared strange to many, including the chief engineers of the ENIAC project, Eckert and Mauchly (Aspray 1990, p. 173). However, von Neumann hoped that his theory of natural and artificial automata would both improve understanding of the design of computers and the human nervous system. His last work, for the Silliman Lectures, which owing to illness he could neither finish nor deliver, was largely concerned with pointing out similarities between the nervous system and computer, the neuron and the vacuum tube – but adding cautionary notes on their differences (von Neumann 1958).

What was the reception of von Neumann's tentative analogy between the nervous system and the computer? His intellectual biographer, William Aspray (1990, p. 181) concludes that psychologists and physiologists were less than enthusiastic about the McCulloch-Pitts model; Seymor Papert spoke of "a hostile or indifferent world" (McCulloch 1965, p. xvii) and McCulloch himself admitted the initial lack of interest in their work (p. 9).

The computer as a mind

Von Neumann and others searched for a parallel between the machine and the human on the level of hardware. Alan Turing (1950), in contrast, thought the observation that both the modern digital computer and the human nervous system are electrical was based on a "very superficial similarity" (p. 439). He pointed out that the first digital computer, Babbage's Analytical Engine, was purely mechanical (as opposed to electrical), and that the important similarities to the mind are in function rather than in hardware. Turing discussed the question of whether machines can think, rather than the question of whether the mind is like a computer. Thus, he was looking in the direction opposite to that of psychologists after the cognitive revolution and, consequently, he did not propose any theories of mind. For example, the famous Turing Test is about whether a machine can imitate a human mind, but not vice versa.

Turing argued that it would be impossible for a human to imitate a computer, as evidenced by human's inability to perform complex numerical calculations quickly. Turing also discussed the question of whether a computer could be said to have a free will, a property of humans. Many years later, cognitive psychologists, under the assumptions that the mind is a computer and that computers lack free will, instead pondered the question of whether humans could be said to have one. A similar story to this is that Turing (1969), in a paper written in 1947 but published only years after his death, contemplated teaching machines to be intelligent using the same principles used to teach children. The analogy of the computer as a mind was reversed again after the cognitive revolution, as McCorduck (1979) points out, when from Massachusetts Institute of Technology (MIT) psychologists tried to teach children with the very methods that had worked for computers.

Turing (1969) anticipated much of the new conceptual language and even the very problems Allen Newell and Herbert Simon were to attempt, as we will see in the second part of this paper. With amazing prophecy, Turing suggested that nearly all intellectual issues can be translated into the form "find a number n such that ...", that is, that "search" is the key concept for problem solving, and that Whitehead and Russell's (1935) *Principia Mathematica* might be a good start for demonstrating the power of the machine (McCorduck 1979, p. 57).

Not only did Turing's life end early and under tragic circumstances, but his work had practically no influence on artificial intelligence in Britain until the mid-1960s (McCorduck 1979, p.68). Neither von Neumann nor his friends were persuaded to look beyond similarities between cells and diodes to functional similarities between humans and computers.

To summarize: two groups compared humans and computers before the cognitive revolution. One of these groups, represented by von Neumann, spoke tentatively about the computer as a brain, but warned about taking the analogy too far. The other group, represented by Turing, asked whether the computer has features of the human mind but not vice versa, that is, did not attempt to design theories of mind through the analogy of the tool.

Before the second half of the century, the mind was not yet a computer. However, a new incarnation of the Enlightenment view of intelligence as a combinatorial calculus was on the horizon.

Act II: From computer to mind

In this section, we see how a tools-to-theories explanation accounts for the new conception of the mind as a computer, focusing on the discovery and acceptance of Herbert Simon and Allen Newell's brand of information-processing psychology. We will try to reconstruct the discovery of Newell and Simon's (1972) information-processing model of mind and its (delayed) acceptance by the psychological community in terms of the tools-to-theories heuristic.

Discovery

Babbage's mechanical computer was preceded by human computers. Similarly, Newell and Simon's first computer program, the Logic Theorist, was also preceded by a human computer. Before the Logic Theorist was up and running, Newell and Simon reconstructed their computer program out of human components (namely, Simon's wife, children and several graduate students), to see if it would work. Newell wrote up the subroutines of the Logic Theorist (LT) program on index cards:

> To each member of the group, we gave one of the cards, so that each person became, in effect, a component of the LT computer program – a subroutine – that performed some special function, or a component of its memory. It was the task of each participant to execute his or her subroutine, or to provide the contents of his or her memory, whenever called by the routine at the next level above that was then in control.
>
> So we were able to simulate the behavior of the LT with a computer consisting of human components ... The actors were no more responsible than the slave boy in Plato's Meno, but they were successful in proving the theorems given them. (Simon 1991, p. 207)

The parallels to Prony's *bureaux de calculs* and the large-scale manufacturing of the new factories of the early 19th century are striking. At essence is a division of labor, where the work is carried out by a hierarchy of humans – each requiring little skill, and repeating the same routine again and again. Complex processes are achieved by an army of workers who never see but a little piece of the larger picture.[3]

However, between Prony's human computer and Simon's human computer, there is an important difference. Prony's human computer and Babbage's mechanical computer (which was modeled upon it) performed numerical calculations. Simon's human computer did not. Simon's humans matched symbols, applied rules to symbols, and searched through lists of symbols. In short, they performed what is now generally known as symbol manipulation.

The reader will recall from the first part of this paper that the divorce between intelligence and numerical calculation made it possible for Babbage to replace the human computer by a mechanical one. In the 21st century, intelligence and calculation are still divorced. Given this divorce and the early conception of the computer as a fancy number cruncher, it is not surprising that the computer never suggested itself as a theory of mind. We argue that an important precondition for the view of mind as a computer is the realization that computers are symbol manipulation devices, in addition to being numerical calculators. Newell and Simon were among the first to realize this. In the interviews with Pamela McCorduck (1979, p. 129), Allen Newell recalls, "I've never used a computer to do any numerical processing in my life." Newell's first use of the computer at RAND corporation – a prehistoric card-programmed calculator hooked up to a line printer – was calculating and printing out symbols representing airplanes for each sweep of a radar antenna.

The symbol-manipulating nature of the computer was important to Simon because it corresponded to some of his earlier views on the nature of intelligence:

The metaphor I'd been using, of a mind as something that took some premises and ground them up and processed them into conclusions, began to transform itself into a notion that a mind was something which took some program inputs and data and had some processes which operated on the data and produced output (McCorduck 1979, p. 127).

It is interesting to note that twenty years after seeing the computer as a symbol manipulating device, Newell and Simon came forth with the explicit hypothesis that a physical symbol system is necessary and sufficient for intelligence.

The Logic Theorist generated proofs for theorems in symbolic logic, specifically, the first twenty-five or so theorems in Whitehead and Russell's (1935) *Principia Mathematica*. It even managed to find a proof more elegant than the corresponding one in the *Principia*.

In the summer of 1958, psychology was given a double-dose of the new school of information-processing psychology. One was the publication of the *Psychological Review* article "Elements of a Theory of Human Problem Solving" (Newell, Shaw and Simon 1958). The other was the Research Training Institute on the Simulation of Cognitive Processes at the RAND institute, which we will discuss later.

The *Psychological Review* paper is an interesting document of the transition between the view that the Logic Theorist is a tool for proving theorems in logic (the artificial intelligence view), and an emerging view that the Logic Theorist is a model of human reasoning (the information processing view). In fact, the authors go back and forth between both views, expounding on the one hand that "the program of LT [Logic Theorist] was not fashioned directly as a theory of human behavior; it was constructed in order to get a program that would prove theorems in logic" (p. 154), but on the other hand that the Logic Theorist "provides an explanation for the processes used by humans to solve problems in symbolic logic" (p. 163). The evidence provided for projecting the machine into the mind is mainly rhetorical. For instance, the authors spend several pages arguing for the resemblance between the methods of the Logic Theorist and concepts such as "set", "insight", and "hierarchy" described in the earlier psychological literature on human problem solving.

In all fairness, despite the authors' claim, the resemblance to these earlier concepts as they were used in the work of Karl Duncker, Wolfgang Köhler and others is slight. New discoveries, by definition, clash with what has come before, but it is often a useful strategy to hide the amount of novelty and claim historical continuity. When Tanner and Swets, four years earlier (also in the *Psychological Review*) proposed that another scientific tool, Neyman-Pearsonian techniques of hypothesis testing, would model the cognitive processes of stimulus detection and discrimination, their signal detection model also clashed with earlier notions, such as the notion of a sensory threshold. Tanner and Swets (1954, p. 401), however, chose not to conceal this schism between the old and the new theories, explicitly stating that their new theory "appears to be inconsistent with the large quantity of existing data on this subject." As evidenced in this paper, there is a different historical continuity in which Simon and Newell's ideas stand: the earlier Enlightenment view of intelligence as a combinatorial calculus.

Conceptual change

Newell et al. (1958) tried to emphasize the historical continuity of what was to become their new information-processing model of problem solving, as did Miller, Galanter and Pribram in their *Plans and the Structure of Behavior* (1960) when they linked their version of Newell and Simon's theory to many great names such as William James, Frederic Bartlett, and Edward Tolman. We believe that these early claims for historical continuity served as protection: George Miller, who was accused by Newell and Simon as having stolen their ideas and gotten them all wrong, said "I had to put the scholarship into the book, so they would no longer claim that those were their ideas. As far as I was concerned they were old familiar ideas" (Baars 1986, p. 213). In contrast to this rhetoric, we will here emphasize the discontinuity introduced by the transformation of the new tool into a theory of mind.

The new mind

What was later called the "new mental chemistry" pictured the mind as a computer program:

> The atoms of this mental chemistry are symbols, which are combinable into larger and more complex associational structures called lists and list structures. The fundamental "reactions" of the mental chemistry employ elementary information processes that operate upon symbols and symbol structures: copying symbols, storing symbols, retrieving symbols, inputting and outputting symbols, and comparing symbols. (Simon 1979, p. 63).

This atomic view is certainly a major conceptual change in the views on problem solving compared to the theories of Köhler, Wertheimer, and Duncker. However, it bears much resemblance to the combinatorial view of intelligence of the Enlightenment philosophers.[4]

The different physical levels of a computer lead to Newell's cognitive hierarchy, which separates the knowledge-level, symbol-level, and register-transfer levels of cognition. As Arbib (1993) points out, the seriality of 1971-style computers is actually embedded in Newell's cognitive theory.

One of the major concepts in computer programming that made its way into the new models of the mind is the decomposition of complexity into simpler units, such as the decomposition of a program into a hierarchy of simpler subroutines, or into a set of production rules. On this analogy, the most complex processes in psychology, such as scientific discovery, can be explained through simple subprocesses. Thus, the possibility of the logic of scientific discovery, the existence of which Karl Popper so vehemently disclaimed, has returned in the analogy between computer and mind (Langley et al. 1987).

The first general statement of Newell and Simon's new vision of mind appeared in their 1972 book *Human Problem Solving*. In this book, the authors argue for the idea that higher-level cognition proceeds much like the behavior of a production system, a formalism from computer science (and before that, symbolic logic) which had never been used in psychological modeling before. They speak of the influence of programming concepts on their models:

Throughout the book we have made use of a wide range of organizational techniques known to the programming world: explicit flow control, subroutines, recursion, iteration statements, local naming, production systems, interpreters, and so on. ...We confess to a strong premonition that the actual organization of human programs closely resembles the production system organization (Newell and Simon 1972, p. 803).

We will not attempt to probe the depths of how Newell and Simon's ideas of information processing changed theories of mind; the commonplace usage of computer terminology in the cognitive psychological literature since 1972 is a reflection of this. How natural it seems for present-day psychologists to speak of cognition in terms of encoding, storage, retrieval, executive processes, algorithms, and computational cost.

New experiments, new data

The tools-to-theories heuristic implies that new theories need not be a consequence of new experiments and new data. Furthermore, new tools can transform the kinds of experiments performed and data collected. I have described this consequence of the tools-to-theories heuristic when statistical tools turned into theories of mind.

A similar story is to be told with the conceptual change brought about by Newell and Simon – it mandated a new type of experiment, which in turn involved new kinds of subjects, data, and justification. In academic psychology of the day, the standard experimental design, modeled after the statistical methods of Ronald A. Fisher, involved many subjects and randomized treatment groups. The 1958 *Psychological Review* paper uses the same terminology of "design of the experiment" and "subject", but radically changes their meanings. There are no longer groups of human or animal subjects. There is only one subject: an inanimate being named the Logic Theorist. There is no longer an experiment in which data are generated by either observation or measurement. Experiment takes on the meaning of simulation.

In this new kind of experiment, the data are of an unforeseen type: computer printouts of the program's intermediate results. These new data, in turn, require new methods of hypothesis testing. How did Newell and Simon determine if their program was doing what minds do? There were two methods: for Newell and Simon, simulation was a form of justification itself, that is, a theory that is coded as a working computer program shows that the processes it describes are, at the very least, sufficient to perform the task, or, in the more succinct words of Simon, "A running program is the moment of truth" (1992, p. 155). Furthermore, a stronger test of the model was made by comparing the computer's output to the think-aloud protocols of human subjects.

Although this was all a methodological revolution in the experimental practice of the time, some important parallels exist between the new information-processing approach and the turn-of-the-century German approach to studying mental processes. These parallels concern the analysis of individual subjects (rather than group means), the use of think-aloud procedures, and the status of the subject. In early German psychology, as well in American psychology of the time (until around the 1930s), the unit of analysis was the individual person, and not the average of a group (Danziger 1990). The two most prominent kinds of data in early German psychology were reaction times and introspective reports. Introspective reports have been frowned upon ever since the inception of American behaviorism, but think-aloud protocols, their grandchildren, are back (as are reaction times).

Finally, in the tradition of the Leipzig (Wundt) and Würzburg (Külpe) schools, the subject was more prestigious and important than the experimenter. Under the assumption that the thought process is introspectively penetrable, the subject, not the experimenter, was assumed to provide the theoretical description of the thought process. In fact, the main experimental contribution of Külpe, the founder of the Würzburg school, was to serve as a subject; and it was often the subject who published the paper. In the true spirit of these schools, Newell and Simon named their subject, the Logic Theorist, as a co-author of a paper submitted to the *Journal of Symbolic Logic*. Regrettably, the paper was rejected (as it contained no new results from modern logic's point of view), and the Logic Theorist never tried to publish again.

Acceptance

The second dose of information processing (after the *Psychological Review* paper) administered to psychology was the Research Training Institute on the Simulation of Cognitive Processes at the RAND institute, organized by Newell and Simon. The institute held lectures and seminars, taught IPL-IV programming, and demonstrated the Logic Theorist, the General Problem Solver, and the EPAM model of memory on the RAND computer. In attendance were some individuals who would eventually develop computer simulation methods of their own, including George Miller, Robert Abelson, Bert Green, and Roger Shepard.

An early, but deceptive harbinger of acceptance for the new information processing-theory was the publication, directly after the summer institute, of *Plans and the Structure of Behavior* (Miller, Galanter, and Pribram 1960), written mostly by George Miller. Despite the 1959 dispute with Newell and Simon (mentioned earlier) over the ownership and validity of the ideas within it, this book drew a good deal of attention from all areas of psychology.

It would seem the table was set for the new information processing psychology; however, it did not take hold. Simon complained of the psychological community who took only a "cautious interest" in their ideas. The "acceptance" part of the tools-to-theories thesis can explain this: computers were not yet entrenched in the daily routine of psychologists, as we will show.

No familiar tools, no acceptance

We take two institutions as case studies to demonstrate the part of the tools-to-theories hypothesis that concerns acceptance: the Center for Cognitive Studies at Harvard, and Carnegie Mellon University. The former never came to fully embrace the new information-processing psychology. The latter did, but after a considerable delay. Tools-to-theories might explain both phenomena.

George Miller, the co-founder of the Center at Harvard, was certainly a proponent of the new information-processing psychology. As we mentioned, his book *Plans and the Structure of Behavior* was so near to Newell and Simon's ideas that it was at first considered a form of theft, although the version of the book that did see the presses is filled with citations recognizing Newell, Shaw, and Simon. Given Miller's enthusiasm, one might expect the Center, partially under Miller's leadership, to blossom into information-processing research. It never did. Looking at the Harvard University Center for Cognitive Studies *Annual Reports* from 1963–1969, we found only a few symposia or papers dealing with computer simulation.

Although the Center had a PDP-4C computer, and the reports anticipated the possibility of using it for cognitive simulation, as late as 1969 it never happened. The reports mention that the computer served to run experiments, to demonstrate the feasibility of computer research, and to draw visitors to the laboratory. However, difficulties involved with using the tool were considerable. The PDP saw 83 hours of use on an average week in 1965–66, but 56 of these were spent on debugging and maintenance. In the annual reports are several remarks of the type "It is difficult to program computers ... Getting a program to work may take months." They even produced a 1966 technical report called

"Programmanship, or how to be one-up on a computer without actually ripping out its wires."

What might have kept the Harvard computer from becoming a metaphor of the mind was that the researchers could not integrate this tool into their everyday laboratory routine. The tool instead turned out to be a steady source of frustration. As tools-to-theories suggests, this lack of entrenchment into everyday practice accounted for the lack of acceptance of the new information-processing psychology. Simon (1979) has taken notice of this:

> Perhaps the most important factors that impeded the diffusion of the new ideas, however, were the unfamiliarity of psychologists with computers and the unavailability on most campuses of machines and associated software (list processing programming languages) that were well adapted to cognitive simulation. The 1958 RAND Summer Workshop, mentioned earlier, and similar workshops held in 1962 and 1963, did a good deal to solve the first problem for the 50 or 60 psychologists who participated in them; but workshop members often returned to their home campuses to find their local computing facilities ill-adapted to their needs (Simon 1979, p. 356).

At Carnegie Mellon University, Newell, Simon, a new enthusiastic department head, and a very large National Institute of Mental Health (NIMH) grant were pushing "the new IP [information processing] religion" (Simon 1994). Even this concerted effort failed to proselytize the majority of researchers within their own department. This again indicates that entrenchment of the new tool into everyday practice was an important precondition for the spread of the metaphor of the mind as a computer.

Acceptance of theory follows familiarity with tool

In the late 1950s, at Carnegie Mellon, the first doctoral theses involving computer simulation of cognitive processes were being written (Simon, personal communication). However, this was not representative of the national state of affairs. In the mid-1960s, a small number of psychological laboratories, including at Carnegie Mellon, Harvard, Michigan, Indiana, MIT, and Stanford were built around computers (Aaronson, Grupsmith and Aaronson 1976, p. 130). As indicated by the funding history of NIMH grants for cognitive research, the amount of computer-based research tripled over the next decade: in 1967, only 15% of the grants being funded had budget items related to computers (e.g., programmer salaries, hardware, supplies). By 1975, this figure had increased to 46%. The late 1960s saw a turn towards mainframe computers, which lasted until the late 1970s when the microcomputer began its invasion of the laboratory. In the

1978 *Behavioral Research Methods and Instrumentation* conference, microcomputers were the issue of the day (Castellan 1981, p. 93). By 1984, the journal *Behavioral Research Methods and Instrumentation* appended the word "Computers" to its title to reflect the broad interest in the new tool. By 1980, the cost of computers had dropped an order of magnitude from their cost in 1970 (Castellan 1981, 1991). During the last two decades, computers have become the indispensable research tool of the psychologist.

Once the tool became entrenched into everyday laboratory routine, a broad acceptance of the view of the mind as a computer followed. In the early 1970s, information-processing psychology finally caught on at Carnegie Mellon University. In the 1973 edition of the Carnegie Symposium on Cognition, every CMU-authored article mentions some sort of computer simulation. For the rest of the psychological community, who were not as familiar with the tool, the date of broad acceptance was years later. In 1979, Simon estimated that from about 1973 to 1979, the number of active research scientists working in the information-processing vein had "probably doubled or tripled" (Simon 1979).

This does not mean that the associated methodology became accepted as well. It clashed too strongly with the methodological ritual that was institutionalized during the 1940s and 1950s in experimental psychology. We use the term "ritual" here for the mechanical practice of a curious mishmash between Fisher's and Neyman-Pearson's statistical techniques that was taught to psychologists as the sine qua non of scientific method (Gigerenzer 1993, 2000). Most psychologists assumed, as the textbooks had told them, that there was only one way to do good science. But their own heroes – Fechner, Wundt, Pavlov, Köhler, Bartlett, Piaget, Skinner, Luce, to name a few –had never used this "ritual"; some had used experimental practices that resembled the newly proposed methods used to study the mind as computer.

Pragmatics

Some of my experimental colleagues have objected to my analysis of how statistical tools turned into theories of mind. They argue that tools are irrelevant in discovery, and that my tool-to-theories examples are merely illustrations of psychologists being quick to realize that the mathematical structure of a tool (such as analysis of variance, or the digital computer) is precisely that of the mind. It is not easy to convince someone who believes (in good Neoplatonic fashion) that today's theory of mind exactly fits the nature of the mind, that such a splendid theory might mirror something other than pure and simple reality. If it were true that tools have no role in discovery, and that the new theories just happen to mirror the mathematical structure of the tool, then the pragmatics of a tool's use (which is independent of the mathematical structure) would find no place in the new theories. However, the case of statistical tools in the first half of this article provides evidence that not only the new tool, but also its pragmatic uses are projected into the mind. The tools-to-theories heuristic cannot be used to defend a spurious Neoplatonism.

The same process of projecting the pragmatic aspects of a tool's use onto a theory can be shown for the view of the mind as a computer. One example is Levelt's model of speaking (Levelt 1989). The basic unit in Levelt's model, which he calls the "processing component", corresponds to the computer programmer's concept of a subroutine. Gigerenzer and Goldstein (1996) argued that Levelt's model not only borrowed the subroutine as a tool, but also borrowed the pragmatics of how subroutines are constructed.

A subroutine (or "subprocess") is a group of computer instructions, usually serving a specific function, that are separated from the main routine of a computer program. It is common for subroutines to perform often-needed functions, such as extracting a cube root or rounding a number. There is a major pragmatic issue involved in writing subroutines that centers around what is called the principle of isolation (Simon 1986). The issue is whether subroutines should be black boxes or not. According to the principle of isolation, the internal workings of the subroutine should remain a mystery to the main program, and the external program should remain a mystery to the subroutine. Subroutines built without respect to the principle of isolation are "clear boxes" that can be penetrated from the outside and escaped from the inside. To the computer, of course, it makes no difference whether the subroutines are isolated or not. Subroutines that are not isolated work just as well as those that are. The only difference is a psychological one. Subroutines that violate the principle of isolation are harder to read, write and debug, from a person's point of view. For this reason, introductory texts on computer programming stress the principle of isolation as the essence of good programming style.

The principle of isolation – a pragmatic feature of using subroutines – has a central place in Levelt's model, where the processing components are "black boxes" and constitute what Levelt considers to be a definition of Fodor's notion of "informational encapsulation" (Levelt 1989, p. 15). In this way, Levelt's psychological model embodies a maxim of good computer-programming methodology: the principle of isolation. That this pragmatic feature of the tool shaped a theory of speaking is not an evaluation of the quality of the theory. Our point concerns origins, not validity. However, in this case, this pragmatic feature of the subroutine has not always served the model well: Kita (1993) and Levinson (1992) have attacked Levelt's model at its Achilles' heel – its insistence on isolation.

To summarize: I have drawn a full circle from theories of mind to computers and back to theories of mind. The argument was that economic changes – the large-scale division of labor in manufacturing and in the *"bureaux de calculs"* – corresponded with the breakdown of the Enlightenment conception of the mind, in which calculation was the distinctive essence of intelligence. Once calculation was separated from the rest of intelligence and relegated to the status of a dull and repetitive task, Babbage could envision replacing human computers by mechanical ones. Both human and mechanical computers manufactured numbers as the factories of the day manufactured goods. In the 20th century, the technology became available to make Babbage's dream a reality. Computers became indispensable scientific tools for everything from number crunching to simulation. Our focus was on the work by Herbert Simon and Allen Newell and their colleagues, who proposed the tool as a theory of mind. Their proposal reunited mere calculation with what was now called "symbol processing", returning to the Enlightenment conception of mind. After computers found a place in nearly every psychological laboratory, broad acceptance of the metaphor of the mind as computer followed.[5] Now that the metaphor is in place, many find it hard to see how the mind could be anything else: to quote Philip Johnson-Laird (1983, p. 10), "The computer is the last metaphor; it need never be supplanted".

Discovery reconsidered

New technologies have been a steady source of metaphors of mind: "In my childhood we were always assured that the brain was a telephone switchboard. ('What else could it be?')", recalls John Searle (1984, p. 44). The tools-to-theories heuristic is more specific than general technology metaphors. Scientists' tools for justification, not just any tools, are used to understand the mind. Holograms are not social scientists' tools, but computers are, and part of their differential acceptance as metaphors of mind by the psychological community may be a result of psychologists' differential familiarity with these devices in research practice.

The examples of discovery I gave in this paper are modest instances, compared with the classical literature in the history of science treating the contribution of a Copernicus or a Darwin. But in the narrower context of recent cognitive psychology, however, the theories discussed above count as among the most influential. In this more prosaic context of discovery, the tools-to-theories heuristic can account for a group of significant theoretical innovations.

Also, as I have argued, this discovery heuristic can both open and foreclose new avenues of research, depending on the interpretations attached to the statistical tool. My focus was on analytical tools of justification, and I have not dealt with physical tools of experimentation and data processing. Physical tools, once familiar and considered indispensable, also may become the stuff of theories. This holds not only for the hardware (like the software) of the computer, but also for theory innovation beyond recent cognitive psychology. Smith (1986) argued that Edward C. Tolman's use of the maze as an experimental apparatus transformed Tolman's conception of purpose and cognition into spatial characteristics, such as cognitive maps. Similarly, he argued that Clark L. Hull's fascination with conditioning machines has shaped Hull's thinking of behavior as if it were machine design.

The tools-to-theories heuristic connects the contexts of discovery and justification, and shows that the commonly assumed, fixed temporal order between discovery and justification – discovery first, justification second – is not a necessary one. I have discussed cases of discovery where tools for justification came first, and discovery followed. Let me conclude with some reflections on how the present view stands in relation to major themes in scientific discovery.

Data-to-theories reconsidered

Should we continue telling our students that new theories originate from new data, if only because "little is known about how theories come to be created", as J. R. Anderson introduces the reader to his Cognitive Psychology (1980, p. 17)? Holton (1988) noted the tendency among physicists to reconstruct discovery with hindsight as originating from new data, even if this is not the case. His most prominent example is Einstein's special theory of relativity, which was and still is celebrated as an empirical generalization from Michelson's experimental data – by such eminent figures as R. A. Millikan and H. Reichenbach, as well as by textbook writers. As Holton demonstrated with first-hand documents, the role of Michelson's data in the discovery of Einstein's theory was slight, a conclusion shared by Einstein himself.

The strongest claim for an inductive view of discovery came from the Vienna Circle's emphasis on sensory data (reduced to the concept of "pointer readings"). Carnap (1928/1969), Reichenbach (1938), and others focused on what they called the "rational reconstruction" of actual discovery rather than on actual discovery itself, in order to screen out the "merely" irrational and psychological. For instance, Reichenbach reconstructed Einstein's special theory of relativity as being "suggested by closest adherence to experimental facts", a claim that Einstein rejected, as mentioned above (see Holton 1988, p. 296). It seems fair to say that all attempts to logically reconstruct discovery in science have failed in practice (Blackwell 1983, p. 111). The strongest theoretical disclaimer concerning the possibility of a logic of discovery came from Popper, Hempel, and other proponents of the hypothetico-deductive account, resulting in the judgment that discovery, not being logical, occurs irrationally. Theories are simply "guesses guided by the unscientific" (Popper 1935/1959, p. 278). But rational induction and irrational guesses are not exhaustive of scientific discovery, and the tools-to-theories heuristic explores the field beyond.

Scientist' practice reconsidered.

The tools-to-theories heuristic is about scientists' practice, that is, the analytical and physical tools used in the conduct of empirical research. This practice has a long tradition of neglect. The very philosophers who called themselves logical empiricists had, ironically, no interest in the empirical practice of scientists. Against their reduction of observation to pointer reading, Kuhn (1970) has emphasized the theory-ladenness of observation. Referring to perceptual experiments and Gestalt switches, he says, "scientists see new and different things when looking with familiar instruments in places they have looked before ..." (p. 111). Both the logical empiricists and Kuhn were highly influential on psychology (see Toulmin and Leary 1985), but neither view has emphasized the role of tools and experimental conduct. Their role in the development of science has been grossly underestimated until recently (Danziger 1985, 1987, 1990; Lenoir 1988).

Through the lens of theory, we are told, we can understand the growth of knowledge. But there is a recent move away from a theory-dominated account of science that pays attention to what really happens in the laboratories. Hacking (1983) argued that experimentation has a life of its own, and that not all observation is theory-laden. Galison (1987) analyzed modern experimental practice, such as in high-energy physics, focusing on the role of the fine-grained web of instruments, beliefs, and practice that determine when a "fact" is considered to be established and when experiments end. Both Hacking and Galison emphasized the role of the familiarity experimenters have with their tools, and the importance and relative autonomy of experimental practice in the quest for knowledge. This is the broader context in which the present tools-to-theories heuristic stands: the conjecture that theory is inseparable from instrumental practices.

In conclusion, my argument is that discovery in recent cognitive psychology can be understood beyond mere inductive generalizations or lucky guesses. More than that, I argue that for a considerable group of cognitive theories, neither induction from data nor lucky guesses played an important role. Rather, these innovations in theory can be accounted for by the tools-to-theories heuristic, as can conceptual problems and possibilities in current theories. Scientists' tools are not neutral. In the present case, the mind has been recreated in their image.

Max Planck Institute for Human Development,
Center for Adaptive Behavior and Cognition
Berlin, Germany.

NOTES

[1] The Jaquard loom was a general-purpose device which was loaded with a set of punched cards and could weave infinite varieties of patterns. Factories in England were equipped with hundreds of these machines, and Babbage was one of the "factory tourists" of the 1830s and 1840s.

[2] Calculation became dissociated from and opposed to not only the human intellect, but also moral impulse. Madame de Staël, for instance, used the term "*calcul*" only in connection with the "egoism and vanity" of those opportunists who exploited the French Revolution for their own advantage and selfishness (Daston 1994).

[3] The Manhattan Project at Los Alamos, where the atomic bomb was constructed, housed another human computer. Although the project could draw on the best technology available, in the early 1940s mechanical calculators (such as the typewriter-sized Marchant calculator) could do nothing but addition, subtraction, multiplication, and, with some difficulty, division. Richard Feynman and Nicholas Metropolis arranged a pool of people (mostly scientists' wives who were getting paid three-eighths salary), each of whom repetitively performed a small calculation (such as cubing a number) and passed the result on to another person, who incorporated it into yet another computation (Gleick 1992).

[4] In fact, the new view was directly inspired by the 19th-century mathematician George Boole, who, in the very spirit of the Enlightenment mathematicians such as the Bernoullis and Laplace, set out to derive the laws of logic, algebra, and probability from what he believed to be the laws of human thought (Boole 1854/1958). Boole's algebra culminated in Whitehead and Russell's (1935) *Principia Mathematica*, describing the relationship between mathematics and logic, and in Claude E. Shannon's seminal work (his master's thesis at MIT in 1937), which used Boolean algebra to describe the behavior of relay and switching circuits (McCorduck 1979, p.41).

[5] Our reconstruction of the path "from mind to computer and back" also provides an explanation for one widespread type of resistance against the computer metaphor of mind. The post-Enlightenment divorce between intelligence and calculation still holds to this day, and for those who still associate the computer with mere calculation (as opposed to symbol processing) the mind as a computer is a contradiction in itself.

REFERENCES

Aaronson, D., Grupsmith, E. and Aaronson, M. 1976. "The Impact of Computers on Cognitive Psychology". *Behavioral Research Methods & Instrumentation* 8: 129–138.

Anderson, J. R. 1980. *Cognitive Psychology and Its Implications*. San Francisco, CA: Freeman.

Anderson, J. R., and Milson, R. 1989. "Human Memory: An Adaptive Perspective". *Psychological Review* 96: 703–719.

Arbib, M. A. 1993. "Allen Newell, Unified Theories of Cognition". *Artificial Intelligence* 59: 265–283.

Aspray, W. 1990. *John von Neumann and the Origins of Modern Computing*. Cambridge, MA: MIT Press.

Baars, B. J. 1986. *The Cognitive Revolution in Psychology*. New York: Guilford Press.

Babbage, C. 1994. *Passages from the Life of a Philosopher* (M. Campbell-Kelley, ed.). Piscataway, NJ: IEEE Press. (Original work written c. 1812).

Birnbaum, M. H. 1983. "Base Rates in Bayesian Inference: Signal Detection Analysis of the Cab Problem". *American Journal of Psychology* 96: 85–94.

Blackwell, R. J. 1983. "Scientific Discovery: The Search for New Categories". *New Ideas in Psychology* 1: 111–115.

Boole, G. 1958. *An Investigation of the Laws of Thought*. New York: Dover. (Original work published in 1854).

Brehmer, B. and Joyce, C. R. B. (eds.) 1988. *Human Judgment: The SJT View*. Amsterdam: North-Holland.

Brown, W. and Thomson, G. H. 1921. *The Essentials of Mental Measurement*. Cambridge, UK: Cambridge University Press.

Brunswik, E. 1943. "Organismic Achievement and Environmental Probability". *Psychological Review* 50: 255–272.

Carnap, R. 1969. *The Logical Structure of the World*. (R. A. George, Trans.). Berkeley: University of California Press (Original work published in 1928).

Castellan, N. J. 1981. "On-line Computers in Psychology: The Last 10 Years, the Next 10 years – The Challenge and the Promise". *Behavioral Research Methods & Instrumentation* 13: 91–96.

Castellan, N. J. 1991. "Computers and Computing in Psychology: Twenty Years of Progress and Still a Bright Future". *Behavior Research Methods, Instruments, & Computers* 23: 106–108.

Cohen, L. J. 1986. *The Dialogue of Reason*. Oxford: Clarendon Press.

Colebrooke, H. 1825. "Address on Presenting the Gold Medal of the Astronomical Society to Charles Babbage". *Memoirs of the Astronomical Society* 1: 509–512.

Cronbach, L. J. 1957. "The Two Disciplines of Scientific Psychology". *American Psychologist* 12: 671–684.

Danziger, K. 1985. "The Methodological Imperative in Psychology". *Philosophy of the Social Sciences* 16: 1–13.

Danziger, K. 1987. "Statistical Method and the Historical Development of Research Practice in American Psychology". In L. Krüger, G. Gigerenzer, and M. S. Morgan (eds.), *The Probabilistic Revolution. Vol. II: Ideas in the Sciences*. Cambridge, MA: MIT Press, pp. 35–47.

Danziger, K. 1990. *Constructing the Subject: Historical Origins of Psychological Research*. Cambridge, UK: Cambridge University Press.

Dashiell, J. F. 1939. "Some Rapprochements in Contemporary Psychology". *Psychological Bulletin* 36: 1–24.

Daston, L. 1988. *Classical Probability in the Enlightenment*. Princeton, NJ: Princeton University Press.

Daston, L. 1992. "The Naturalized Female Intellect". *Science in Context* 5: 209–235.

Daston, L. 1994. "Enlightenment Calculations". *Critical Inquiry* 21: 182–202.

de Finetti, B. 1989. "Probabilism". *Erkenntnis* 31: 169–223. (Original work published in 1931).

Edgington, E. E. 1974. "A New Tabulation of Statistical Procedures Used in APA Journals". *American Psychologist* 29: 25–26.

Fechner, G.T. 1897. *Kollektivmasslehre* [The Measurement of Collectivities]. Leipzig: W. Engelmann.

Fisher, R. A. 1955. "Statistical Methods and Scientific Induction". *Journal of the Royal*

Statistical Society (B) 17: 69–78.

Galison, P. 1987. *How Experiments End*. Chicago: Chicago University Press.

Gardner, H. 1985. *The Mind's New Science*. New York: Basic Books.

Gardner, H. 1988. "Creative Lives and Creative Works: A Synthetic Scientific Approach". In R. J. Sternberg (ed.), *The Nature of Creativity*. Cambridge, UK: Cambridge University Press, pp. 298–321.

Gavin, E. A. 1972. "The Causal Issue in Empirical Psychology from Hume to the Present with Emphasis upon the Work of Michotte". *Journal of the History of the Behavioral Sciences* 8: 302–320.

Gigerenzer, G. 1987a. "Probabilistic Thinking and the Fight against Subjectivity". In L. Krüger, G. Gigerenzer, and M. S. Morgan (eds.), *The Probabilistic Revolution. Vol. II: Ideas in the Sciences*. Cambridge, MA: MIT Press, pp. 11–33.

Gigerenzer, G. 1987b. "Survival of the Fittest Probabilist: Brunswik, Thurstone, and the Two Disciplines of Psychology". In L. Krüger, G. Gigerenzer, and M. S. Morgan (eds.), *The Probabilistic Revolution: Vol. II: Ideas in the Sciences*. Cambridge, MA: MIT Press, pp. 49–72.

Gigerenzer, G. 1991. "From Tools to Theories: A Heuristic of Discovery in Cognitive Psychology". *Psychological Review* 98: 254–267.

Gigerenzer, G. 1992. "Discovery in Cognitive Psychology: New Tools Inspire New Theories". *Science in Context* 5: 329–350.

Gigerenzer, G. 1993. "The Superego, the Ego, and the Id in Statistical Reasoning". In G. Keren and G. Lewis (eds.), *A Handbook for Data Analysis in the Behavioral Sciences: Methodological Issues*. Hillsdale, NJ: Lawrence Erlbaum Associates, pp. 313–339.

Gigerenzer, G. 1994. "Where Do New Ideas Come from?". In M. A. Boden (ed.), *Dimensions of Creativity*. Cambridge, MA: MIT Press, pp. 53–74.

Gigerenzer, G. 2000. *Adaptive Thinking: Rationality in the Real World*. New York: Oxford University Press.

Gigerenzer, G. 2001. "Ideas in Exile: The Struggles of an Upright Man". In K.R. Hammond and T.R. Stewart (eds.), *The Essential Brunswik: Beginnings, Explications, Applications*. New York: Oxford University Press, pp. 445–452.

Gigerenzer, G., and Goldstein, D. G. 1996. "Mind as Computer: The Birth of a Metaphor". *Creativity Research Journal* 9: 131–144.

Gigerenzer, G. and Murray, D. J. 1987. *Cognition as Intuitive Statistics*. Hillsdale, NJ: Lawrence Erlbaum Associates.

Gigerenzer, G., Swijtink, Z., Porter, T., Daston, L., Beatty, J., and Krüger, L. 1989. *The Empire of Chance: How Probability Changed Science and Everyday Life*. Cambridge, UK: Cambridge University Press.

Gleick, J. 1992. *Genius. The Life and Science of Richard Feynman*. New York: Pantheos.

Good, I. J. 1971. "46656 Varieties of Bayesians". *The American Statistician* 25: 62–63.

Gooding, D., Pinch, T. and Schaffer, S. (eds.) 1989. *The Uses of Experiment. Studies in the Natural Sciences*. Cambridge: Cambridge University Press.

Gruber, H. 1981. *Darwin on Man, a Psychological Study of Scientific Creativity*. 2nd ed. Chicago: University of Chicago Press.

Guilford, J. P. 1954. *Psychometric Methods*. 2nd ed. New York: McGraw-Hill.

Hacking, I. 1965. *Logic of Statistical Inference*. Cambridge, UK: Cambridge University Press.

Hacking, I. 1975. *The Emergence of Probability*. Cambridge, UK: Cambridge University Press.

Hacking, I. 1983. *Representing and Intervening*. Cambridge, UK: Cambridge University Press.

Hammond, K.R. and Stewart, T.R. (eds.) 2001. *The Essential Brunswik: Beginnings, Explications, Applications*. New York: Oxford University Press.

Hanson, N. R. 1958. *Patterns of Discovery*. Cambridge, UK: Cambridge University Press.

Harvard University Center for Cognitive Studies. 1963. *Third annual report*.

Harvard University Center for Cognitive Studies. 1964. *Fourth annual report*.

Harvard University Center for Cognitive Studies. 1966. *Sixth annual report*.

Harvard University Center for Cognitive Studies. 1968. *Eighth annual report*.

Harvard University Center for Cognitive Studies. 1969. *Ninth annual report*.

Heims, S. 1975. "Encounter of Behavioral Sciences with New Machine-Organism Analogies in the 1940's". *Journal of the History of the Behavioral Sciences* 11: 368–373.

Herbart, J. F. 1816. *Lehrbuch zur Psychologie*. Hamburg and Leipzig: G. Hartenstein.

Hilgard, E. R. 1955. "Discussion of Probabilistic Functionalism". *Psychological Review* 62: 226–228.

Holton, G. 1988. *Thematic Origins of Scientific Thought*. 2nd ed. Cambridge, MA: Harvard University Press.

Johnson-Laird, P. N. 1983. *Mental Models*. Cambridge, UK: Cambridge University Press.

Kahneman, D., Slovic, P. and Tversky, A. (eds.) 1982. *Judgment under Uncertainty: Heuristics and Biases*. Cambridge, UK: Cambridge University Press.

Kelley, H. H. 1967. "Attribution Theory in Social Psychology". In D. Levine (ed.), *Nebraska Symposium on Motivation*. Vol. 15. Lincoln: University of Nebraska Press, pp. 192–238

Kelley, H. H. and Michaela, I. L. 1980. "Attribution Theory and Research". *Annual Review of Psychology* 31: 457–501.

Kendall, M. G. 1942. "On the Future of Statistics". *Journal of the Royal Statistical Society*

105: 69–80.

Kita, S. 1993. *Language and Thought Interface: A Study of Spontaneous Gestures and Japanese Mimetics*. Unpublished doctoral dissertation, The University of Chicago.

Kuhn, T. 1970. *The Structure of Scientific Revolutions*. 2nd ed. Chicago: University of Chicago Press.

Langley, P., Simon, H. A., Bradshaw, G. L. and Zytkow, J. M. 1987. *Scientific Discovery*. Cambridge, MA: MIT Press.

Leary, D. E. 1987. "From Act Psychology to Probabilistic Functionalism: The Place of Egon Brunswik in the History of Psychology". In M. S. Ash and W. R. Woodward (eds.), *Psychology in Twentieth-century Thought and Society*. Cambridge, UK: Cambridge University Press, pp. 115–142.

Leibniz, G. W. von 1951. "The Horizon of Human Doctrine". In P. P. Wiener (ed.), *Selections*. New York: Scribner's, pp.73–77 (Original work published 1690).

Lenoir, T. 1986. "Models and Instruments in the Development of Electrophysiology, 1845–1912". *Historical Studies in the Physical Sciences* 17: 1–54.

Lenoir, T. 1988. "Practice, Reason, Context: The Dialogue Between Theory and Experiment". *Science in Context* 2: 3–22.

Levelt, W. J. M. 1989. *Speaking: From Intention to Articulation*. Cambridge, MA: MIT Press.

Levinson, S. 1992. "How to Think in order to Speak Tzeltal". Unpublished manuscript. Cognitive Anthropology Group, Max Planck Institute for Psycholinguistics, Nijmegen, The Netherlands.

Lopes, L. L. 1981. "Decision Making in the Short Run". *Journal of Experimental Psychology: Human Learning and Memory* 7: 377–385.

Lopes, L. L. 1982. "Doing the Impossible: A Note on Induction and the Experience of Randomness". *Journal of Experimental Psychology: Learning, Memory, and Cognition* 8: 626–636.

Lopes, L. L. 1991. "The Rhetoric of Irrationality". *Theory and Psychology* 1: 65–82.

Lovie, A. D. 1983. "Attention and Behaviorism – Fact and Fiction". *British Journal of Psychology* 74: 301–310.

Luce, R. D. 1977. "Thurstone's Discriminal Processes Fifty Years Later". *Psychometrika* 42: 461–489.

McArthur, L. A. 1972. "The How and What of Why: Some Determinants and Consequents of Causal Attribution". *Journal of Personality and Social Psychology* 22: 171–193.

McCorduck, P. 1979. *Machines Who Think*. San Francisco: W. H. Freeman and Company.

McCulloch, W. S. 1965. *Embodiments of Mind*. Cambridge, MA: MIT Press.

Melton, A. W. 1962. "Editorial". *Journal of Experimental Psychology* 64: 553–557.

Michotte, A. 1963. *The Perception of Causality*. London: Methuen (Original work published 1946).

Miller, G. A., Galanter, E. and Pribram, K. H. 1960. *Plans and the Structure of Behavior.* New York: Holt, Reinhart and Winston.

Mises, R. von 1957. *Probability, Statistics, and Truth.* London: Allen and Unwin.

Murdock, B. B., Jr. 1982. "A Theory for the Storage and Retrieval of Item and Associative Information". *Psychological Review* 89: 609–626.

Neumann, J. von 1958. *The Computer and the Brain.* New Haven, CT: Yale University Press.

Newell, A., Shaw, J. C. and Simon, H. A. 1958. "Elements of a Theory of Human Problem Solving". *Psychological Review* 65: 151–166.

Newell, A. and Simon. H. A. 1972. *Human Problem Solving.* Englewood Cliffs, NJ: Prentice-Hall.

Neyman, J. 1937. "Outline of a Theory of Statistical Estimation Based on the Classical Theory of Probability". *Philosophical Transactions of the Royal Society (Series A)* 236: 333–380.

Neyman, J. and Pearson, E. S. 1928. "On the Use and Interpretation of Certain Test Criteria for Purposes of Statistical Inference. Part I". *Biometrika* 20A: 175–240.

Nickles, T. 1980. "Introductory Essay: Scientific Discovery and the Future of Philosophy of Science". In T. Nickles (ed.), *Scientific Discovery, Logic, and Rationality.* Dordrecht: Reidel, pp. 1–59.

Piaget, J. 1930. *The Child's Conception of Causality.* London: Kegan Paul.

Popper, K. 1959. *The Logic of Scientific Discovery.* New York: Basic Books. (Original work published in 1935).

Reichenbach, H. 1938. *Experience and Prediction.* Chicago: University of Chicago Press.

Rucci, A. J. and Tweney, R. D. 1980. "Analysis of Variance and the 'Second Discipline' of Scientific Psychology: A Historical Account". *Psychological Bulletin* 87: 166–184.

Schaffer, S. 1992. "Disappearing Acts: On Gigerenzer's 'Where Do New Ideas Come from?'" Unpublished Manuscript.

Schaffer, S. 1994. "Babbage's Intelligence: Calculating Engines and the Factory System". *Critical Inquiry* 21: 203–227.

Searle, J. 1984. *Minds, Brains and Science.* Cambridge, MA: Harvard University Press.

Simon, H. A. 1969. *The Sciences of the Artificial.* Cambridge, MA: MIT Press.

Simon, H. A. 1979. "Information Processing Models of Cognition". *Annual Review of Psychology* 30: 363–96.

Simon, H. A. 1991. *Models of My Life.* New York: Basic Books.

Simon, H. A. 1992. "What is an 'Explanation' of Behavior?". *Psychological Science* 3: 150–161.

Simon, H. A. 1994. personal communication.

Simon, H. A. and Newell, A. 1986. "Information Processing Language V on the IBM 650". *Annals of the History of Computing* 8: 47–49.

Smith, L. D. 1986. *Behaviorism and Logical Positivism*. Stanford, CA: Stanford University Press.

Sterling, T. D. 1959. "Publication Decisions and Their Possible Effects on Inferences Drawn from Tests of Significance or vice versa". *Journal of the American Statistical Association* 54: 30–34.

Swets, J. A., Tanner, W. D. and Birdsall, T. G. 1964. "Decision Processes in Perception". In J. A. Swets (ed.), *Signal Detection and Recognition in Human Observers*. New York: Wiley, pp. 3–57.

Swijtink, Z. G. 1987. "The Objectification of Observation: Measurement and Statistical Methods in the Nineteenth Century". In L. Krüger, L. J. Daston and M. Heidelberger (eds.), *The Probabilistic Revolution. Vol. 1: Ideas in History*. Cambridge, MA: MIT Press, pp. 261–285

Tanner, W. P., Jr. 1965. *Statistical Decision Processes in Detection and Recognition* (Technical Rep.). Ann Arbor: University of Michigan, Sensory Intelligence Laboratory, Department of Psychology.

Tanner, W. P., Jr. and Swets, J. A. 1954. "A Decision-making Theory of Visual Detection". *Psychological Review* 61: 401–409.

Teigen, K. H. 1983. "Studies in Subjective Probability IV: Probabilities, Confidence, and Luck". *Scandinavian Journal of Psychology* 24: 175–191.

Thorndike, R. L. 1954. "The Psychological Value Systems of Psychologists". *American Psychologist* 9: 787–789.

Thurstone, L. L. 1927. "A Law of Comparative Judgement". *Psychological Review* 34: 273–286.

Titchener, E. B. 1896. *An Outline of Psychology*. New York: Macmillan.

Toulmin, S. and Leary, D. E. 1985. "The Cult of Empiricism in Psychology, and Beyond". In S. Koch (ed.), *A Century of Psychology as Science*. New York: McGraw-Hill, pp. 594–617.

Turing, A. M. 1950. "Computing Machinery and Intelligence". *Mind* 59: 433–460.

Turing, A. M. 1969. "Intelligent Machinery". In B. Meltzer and D. Michie (eds.), *Machine Intelligence, 5*. Edinburgh: Edinburgh University Press.

Tversky, A. and Kahneman, D. 1974. "Judgment under Uncertainty: Heuristics and Biases". *Science* 185: 1124–1131.

Tversky, A. and Kahneman, D. 1980. "Causal Schemata in Judgments under Uncertainty". In M. Fishbein (ed.), *Progress in Social Psychology. Vol. 1*. Hillsdale, NJ: Lawrence Erlbaum Associates, pp. 49–72.

Tversky, A. and Kahneman, D. 1982. "Judgments of and by Representativeness". In D. Kahneman, P. Slovic, and A. Tversky (eds.), *Judgment under Uncertainty: Heuristics and Biases*. Cambridge, UK: Cambridge University Press, pp. 84–98.

Tversky, A. and Kahneman, D. 1983. "Extensional versus Intuitive Reasoning: The Conjunction Fallacy in Probability Judgment". *Psychological Review* 90: 293–315.

Tweney, R. D., Dotherty, M. E., and Mynatt, C. R. (eds.) 1981. *On Scientific Thinking*. New York: Columbia University Press.

Whitehead, A.N., and Russell, B. 1935. *Principia Mathematica* (2nd ed., Vol. 1). Cambridge, UK: Cambridge University Press.

Wickelgreen, W. A. and Norman, D. A. 1966. "Strength Models and Serial Position in Short-term Recognition Memory". *Journal of Mathematical Psychology* 3: 316–347.

* This article is based on Gigerenzer, G. 1991. "From Tools to Theories. A Heuristics of Discovery in Cognitive Psychology". *Psychological Review* 98: 254–267; Gigerenzer, G., and Goldstein, D. G. 1996. "Mind as Computer: The Birth of a Metaphor". *Creativity Research Journal* 9: 131–144; and on chapters 1 and 2 in my book *Adaptive Thinking: Rationality in the Real World*. Oxford University Press, 2000. For permission to reprint parts of these texts, I am grateful to the American Psychological Association, to Lawrence Erlbaum Associates, and to Oxford University Press.

DAVID PAPINEAU

COMMENTS ON GERD GIGERENZER

I am very grateful to have the opportunity to comment on Gerd Gigerenzer's rich and informative paper. I shall restrict myself to two areas. First, I would like to say something about the general contrast between the context of discovery and the context of justification. Second, I would like to consider the normative implications of Gigerenzer's examples, focusing in particular on his suggestion that the Neyman-Pearson theory of statistical testing was the model for theories of cognitive "signal detection".

Discovery and Justification

What exactly is the contrast between the context of discovery and the context of justification? Although this contrast is philosophically familiar, it is by no means straightforward. Standardly, "discovery" is taken to be something to do with the initial formulation of hypotheses, whereas "justification" is taken to relate to the checking of those hypotheses against the empirical evidence. But this is only the start of a proper explanation, and there are hidden complexities behind this initial gloss.

As a preliminary point, let me query the term "discovery". In normal English, "discovery" is a success verb. You can't "discover" something unless it is there to be discovered. In the context of intellectual "discoveries", you can only discover something if it is a truth. So, for example, in standard English, Stahl didn't *discover* that combustion is the emission of phlogiston, since combustion is no such thing.

However, if "discovery" refers to the initial formulation of hypotheses, then Stahl did indeed "discover" the phlogiston theory of combustion, since he originally formulated this theory. Given this contrast between standard English and usage in philosophy of science, I prefer to stick with standard English. Of course, specialists can define their terms as they wish. But in this case there is a danger that the philosophers' usage will surreptitiously foster sceptical views of science, by suggesting the "scientific discovery" requires only the formulation of a plausible theory, and doesn't require getting the facts right. (For further discussion of the strategic distortion of English by philosophers of science, particularly in the Popperian school, see David Stove 1982.)

So in what follows I shall use the term "invention" instead of "discovery", to refer to the initial formulation of a theory. In this sense, Stahl undoubtedly invented the theory that combustion involves the emission of phlogiston, though this didn't amount to a discovery, since the facts didn't bear out his theory. (For this terminology I am indebted to my old colleague at Macquarie University, Robert McLaughlin. *Cf.* McLaughlin 1982.)

So let us agree that the "context of *invention*" is something to do with the initial formulation of hypotheses, whereas "the context of *justification*" relates to the checking of those hypotheses against the evidence. Now, this contrast is widely assumed to line up with another one, namely the contrast between *descriptive* and *normative* analyses of science. Where justificatory practices are taken to raise normative issues (what attitude *ought* we to take to this theory in the light of this evidence?), strategies for invention are taken to raise only descriptive issues (since there is no right and wrong about which ideas are put forward, prior to their evaluation against the evidence).

However, I see no good reason to cut things up in this way. On the contrary, I would say that the contrast between invention and justification is orthogonal to that between normative and descriptive issues. In particular, I see no reason not to analyse invention from a normative point of view. After all, if the point of science is to identify true theories, then it is a good idea to propose for further testing just those theories with a reasonable probability of being true, and to shun those with no such chance. Given this, we should normatively approve strategies for invention which generate theories with some non-negligible initial probability of truth, and condemn those which generate theories which can be seen to be hopeless from the start.

In making this last point, I have spoken as a realist about science, that is, as someone who takes the aim of science to be the identification of true theories. But the point generalizes to any "product-orientated" philosophy of science, that is, any philosophy which takes the aim of science to be the production of theories with a certain characteristic. Realists take the relevant characteristic to be truth, but others may think of science as aiming at empirically adequate theories, or at technologically useful theories, or at theories which are consonant with religion, or whatever. Any such product-orientated view of science will have reason to take a normative attitude to the context of invention as well as to the context of justification. For any such view of science will want invention strategies that generate theories which have a reasonable chance of delivering, if not truth, then empirical adequacy, or technological usefulness, or consonance with religion, or whatever.

It seems to me that only a purely "process-orientated" philosophies of science could regard invention as an entirely non-normative matter. A "process-orientated" philosophy of science wouldn't take the aim of science to be the production of theories with a certain characteristic, but simply to be good scientific practice itself. On such a view, science proceeds properly as long as it processes theories correctly, whatever other characteristics those theories display. Since such a view of science doesn't care about theories as such, but only about their processing, it will not regard the initial generation of theories in the context of invention as something which can be done well or badly.

Are there any purely "process-orientated" philosophies of science? At first sight it may seem as if classical Popperianism fills the bill. After all, Popper doesn't demand that good science should end up with true, or even empirically adequate theories, provided it rigorously practices the method of conjecture and refutation. (Indeed Popper expects that the best theories will characteristically prove neither true nor empirically adequate.) But even Popper demands something more of theories than that they are processed properly after their initial formulation. He also wants theories that are contentful and novel. To this extent, he is interested in products, and correspondingly has every reason to view the context of invention normatively: that is, he has every reason to condemn invention strategies that produce theories without substantial content, or which simply recapitulate previous theories.

The only example of a purely process-orientated theory of science I can think of is a Bayesianism which puts no normative constraints on which theories get which prior probabilities, but insists only that those priors are updated by conditionalization (*cf.* Howson and Urbach 1989, ch. 11.j). On such a view, there will indeed be no scope for normative considerations in the context of invention, since it will be an arbitrary matter which theories you attach which prior probabilities to.

However, I shall not consider such purely process-orientated theories of science any further in these remarks. To my mind, it is not far short of absurd to claim that science has no aim beyond the encouragement of proper scientific practice. If it is worth practising science properly, this must surely be because science is good for something beyond itself. So from now on I shall assume a product-orientated account of science. (Moreover, I shall simplify the discussion by adopting the additional realist assumption that the relevant product is theoretical truth. As it happens, I am sympathetic to this realist view, but the most of the points I make will generalize to other product-orientated views of science. So non-realist readers with a product-orientated view of science can substitute their favoured scientific products, like empirical adequacy, for my references to truth in what follows.)

I have argued that normative issues arise within the context of invention, as well as within the context of justification, on the grounds that science needs strategies of invention that generate theories with a non-negligible chance of being true. Some readers will no doubt be curious about how such strategies are possible. How can the inventors of theories judge the likelihood of those theories being true, prior to comparing them with the empirical data?

One possible answer to this question would be that such judgements can be made on purely a priori grounds. After all, the underdetermination of theories by empirical data implies that *any* effective system of theory choice will need judgements of prior probability, if it is to reach definite eventual conclusions about which theories are justified (*cf.* Howson and Urbach 1989, ch. 4.k). In line with this, we might posit some purely innate faculty that will allow us to tell, prior to any engagement with empirical data, which theories have any initial probability of truth.

However, there is no need to take such a strongly aprioristic line on prior judgments of probability. It will serve well enough if initial judgements of prior probability are informed by *background knowledge* as well as pure reason, where background knowledge is understood to include the distilled upshot of old empirical data. From this perspective, then, the context of invention will be informed by past empirical data, and can be expected to rule out, as unworthy of further consideration, theories which contradict established facts, or theories which posit mechanisms which our experience of the world shows are never realized. (Of course, this appeal to background knowledge raises a threat of regress: how did the claims in background knowledge get to be justified, except with the help of prior probabilities informed by pre-existing background knowledge? But it is by no means obvious that this threat is vicious. For discussion of some related issues, see Papineau 1993, ch. 5.)

In defence of taking the context of invention to depend in this way on background knowledge, and hence on old empirical data, note that the purely aprioristic alternative view of invention is in danger of ruling out any inductivist account of invention as incoherent. Thus consider the kind of inductivist view discussed in the penultimate section of Gigerenzer's paper, where he considers the view "that new theories originate from new data". Now, we may in the end wish to reject such inductive accounts of invention, along with Gigerenzer, on the grounds that they portray the process of invention as more mechanical than it is. But it would be unfortunate, I take it, if such inductive theories of invention were ruled out from the start as self-contradictory, on the grounds that "invention" is restricted to evaluation of theories which does not depend in any way on empirical data. So we will do better to allow that invention can depend, in various ways, on the empirical data that are already available at the time of invention, and understand the "context of justification" as referring specifically to the checking of hypotheses against *further* empirical evidence, that is, against evidence not yet available at the time of invention.

The normative implications of the tools-to-theories hypothesis

Gigerenzer argues that an important source of theories in twentieth-century cognitive science was provided by a "tools-to-theories" heuristic. "Tools" here refers to methods that are used to evaluate theories. The examples Gigerenzer focuses on are statistical techniques and digital computers. The idea is thus that cognitive scientists have tended to develop theories which are made in the image of these evaluative tools.

I am happy to agree that Gigerenzer makes a convincing and illuminating case for his tools-to-theories hypothesis (with caveats on matters of detail to be entered below). What I want now to consider is whether this tools-to-theories hypothesis shows that the practice of twentieth-century cognitive science should be viewed as normatively reprehensible from a realist point of view (or from the point of view of other product-orientated account of science). Gigerenzer himself seems to think that his hypothesis does have such negative normative implications. ("It is not easy to convince someone who believes (in good Neoplatonic fashion) that today's theory of mind exactly fits the nature of mind, that such a theory might mirror something other than reality pure and simple... The tools-to-theories heuristic cannot be used to defend such a spurious Neoplatonism". This volume, p. 127)

Let me start by considering the context of invention. Given my remarks in the last section, I can't of course take the standard line that invention is a matter of psychology rather than norms, so any stimulus to the invention of theories is as good as any other. For I have argued that a strategy for invention is good only insofar as it generates theories with some initial chance of truth, rather than hopeless theories. So there is an issue of whether the tools-to-theories hypothesis is consistent with this requirement.

Well, perhaps the tools-to-theories hypothesis would not be so consistent, if the *only* influence on which theories get onto the agenda of cognitive science were that those theories should made in the image of some evaluative tool. After all, there seems no obvious reason why theories made in such an image should have a significant chance of yielding a true account of some cognitive faculty. Why should the workings of the human cognitive system mirror the evaluative tools that happen to be used by twentieth-century cognitive scientists?

However, Gigerenzer offers no reason to suppose that analogies with the structure of evaluative tools were the *only* influence on the invention of the theories in question. In particular, he says nothing to rule out the possibility that these theories also attracted the interest of cognitive scientists because they struck them, given their background knowledge, as having the ring of possible truth. Note that there is nothing in this thought to stop the availability of the relevant evaluative tools from being *necessary* conditions for the invention of the theories in question. It is quite possible that these theories would never have occurred to cognitive scientists, were it not for the evaluative tools to provide a stimulus. My point is only that the evaluative tools can be necessary without being sufficient and that their necessity leaves it open that some significant initial plausibility, judged against background knowledge, may also have been necessary for the relevant theories to get onto the cognitive science agenda.

So we can accept Gigerenzer's tools-to-theories hypothesis without drawing negative normative implications for invention strategies in cognitive science. Provided background-knowledge-based judgements of initial plausibility also played a role in invention, Gigerenzer's hypothesis gives us no reason to doubt that invention strategies in cognitive science were well-designed to produce theories with non-negligible initial probabilities of truth.[1]

What now of the context of justification itself? Gigerenzer agues that the tools-to-theories heuristic affects not just "discovery", but also *acceptance:* he points out that some of the theories he considers were in fact initially proposed at earlier dates, yet only won general acceptance after the relevant tools became common institutional currency within cognitive science. At first sight, this may certainly look normatively damning. How can cognitive scientists possibly be tracking the truth, if they can be persuaded to believe given theories by institutional developments which have no apparent connection with the subject matter of those theories?

But in fact an analogy of the point just made about invention applies to justification too. It would indeed be damning if the institutional developments in question were *sufficient* to determine theory acceptance. But their being necessary leaves it open that other factors might also have been necessary, and in particular that proper empirical support might have been necessary too. Gigerenzer does not say very much on how the theories he considers actually fared against the empirical evidence. Still, it is compatible with his observations that theory acceptance required, not just the familiarity of the tools which provided models, but adequate empirical support as well. On this conception, then, the familiarity of the tools would have been required before cognitive scientists were ready to accept the theory (perhaps because the theory would not have been widely understood otherwise), but actual acceptance would not have followed unless the theory also had adequate empirical support. Clearly, this conception gives us no reason to doubt that justification strategies in cognitive science fulfil the normative requirement of being well-designed to produce theories that are true.[2]

For all I have said, some readers may feel there remains something normatively worrying about the tools-to-theories hypothesis. I have argued that there is nothing wrong with modelling your theories on evaluative tools, provided (a) you don't take those theories seriously unless they are also consonant with background knowledge, and provided (b) you don't actually believe them unless they also have adequate empirical support. However, it is possible to query whether theories which mirror evaluative tools are ever likely to satisfy these further stipulations. Why *should* a theory drawn in the image of some arbitrary evaluative tool turn out to be consonant with background knowledge, or with further experimental data? Surely that would be a complete freak. And this might make one suspect that the enthusiasm of cognitive scientists for the relevant theories in fact derived entirely from institutional fashions, untempered by consonance with background knowledge or positive empirical support.

I have some sympathy for this line of thought. However, it seems to me that Gigerenzer's own discussion suggests an interesting reply: perhaps the proponents of the theories in question didn't blindly mimic the analogy with evaluative tools, but also developed their theories with some sensitivity to considerations of empirical plausibility. My thought here is that the relevant evaluative tools may have provided rough frameworks, but that the theories at issue were then further articulated in ways empirically appropriate to their real subject matter. This would then make it less surprising that we should then end up with theories that fit the empirical facts.

Thus consider Gigerenzer's discussion of "signal detection theory". As Gigerenzer tells the story, the empirical theory of signal detection was modelled on the principles of "Neyman-Pearson statistics". The relevant principles, explains Gigerenzer, involve a decision criterion for deciding between two alternative hypotheses H_0 and H_1 ("noise" vs "signal plus noise"), where the criterion will draw a line across possible sample data (sensory input), distinguishing those data points which require "reject H_0, accept H_1" from those which require "accept H_0, reject H_1". This decision criterion will be constructed by weighing the cost of the two possible errors, the Type I error of "rejecting H_0, accepting H_1" when H_0 is true, and the Type II error of "accepting H_0, rejecting H_1" when H_1 is true.

Now, this seems an eminently plausible approach to signal detection, and Gigerenzer seems happy to allow that it led to much empirically successful research. It is a further question, however, how closely it is really modelled on the Neyman-Pearson account of statistical inference. It turns out, on closer examination, that cognitive scientists like W.P. Tanner were not drawing their inspiration from Neyman-Pearson methodology itself, but rather a bastardised version of that methodology specifically adapted to suit the empirical needs of signal detection theory.

Thus consider the problem facing a visual system that needs to decide whether its sensory input signifies a tiger (H_1) or merely background noise (H_0). We can indeed expect the design of this system to be influenced by the relative "cost of the two possible decision errors (Type I and Type II errors in Neyman-Pearson theory, now called 'false alarms' and 'misses')", as Gigerenzer puts it (this volume, p. 104). But note that the design of the system will *also* need to take into account the *prior probability of a tiger* (that is, the prior probability of H_1). After all, organisms who are frequently threatened by tigers will do well to be relatively quick to decide for H_1, while those who live in areas where tigers are scarce will be well advised to opt for H_1 only on stronger data, thereby saving themselves the cost of too much unnecessary fleeing. However, this appeal to prior probabilities of hypotheses is decidedly antithetical to the whole spirit of Neyman-Pearson theory. It would not be too far from the truth to say that the whole motivation for the elaborate Neyman-Pearson approach to statistical testing was the misguided hope of avoiding any appeal to prior probabilities of statistical hypotheses.

Perhaps it could be countered that Neyman-Pearson theory locates any appeal to the prior probability of hypotheses within the "subjective processes, such as attitudes and cost-benefit considerations" (Gigerenzer, this volume, p. 104) which lie behind the selection of an objective decision criterion. I myself think that this is reading more into Newman-Pearson methodology than the history supports (*cf.* Neyman and Pearson 1967). But, even if we concede the point, there remain a further feature of the Neyman-Pearson system which would be inappropriate in any signal detection mechanism, and which the cognitive scientists discussed by Gigerenzer understandably seem to have ignored in constructing their signal detection theory.

This is the asymmetry, central to Neyman-Pearson theory, between null hypotheses (usually designated H_0) and alternative hypotheses (H_1). Gigerenzer presents Neyman-Pearson statistics as treating H_0 and H_1 quite symmetrically, with the one "accepted" and the other "rejected", "depending on which side of the criterion the data fall" (this volume, p. 104). However, standard Newman-Pearson theory treats alternative and null hypotheses quite differently. First, a significance level is chosen so as to fix a very low probability (normally 0.01 or 0.05) of a Type I error, that is, of rejecting the *null* hypothesis when it is true. Then, subject to this constraint, the decision criterion is drawn so as to minimize a Type II error, that is, so as to minimize the danger of letting the null hypothesis stand when the alternative is true. In most cases, the chance of a Type II error will be much higher than that of a Type I error, since the aim of minimizing both pull against each other, and once the chance of a Type I error is set at 1% or 5%, the minimum for a Type II error may well be of the order of 50%. So it makes a big difference which hypothesis is treated as null and which alternative: your chance of wrongly "rejecting" a given hypothesis will be very low (0.01 or 0.05) if it is designated as the null hypothesis, but standardly much higher if it is viewed as the alternative.

Because of this, Neyman-Pearsonians need to treat the terminology of "reject" and "accept" with some care. It makes reasonable sense (modulo normal priors) to "reject" a null hypothesis if you get a sample which would only occur 1% or 5% of the time were the hypothesis true. But it's not nearly so clear that it makes sense to "accept" the null hypothesis whenever it isn't so rejected, given that you may well be doing this 50% of the time when the alternative hypothesis H_1 obtains. Standard textbooks on statistical inference are sensitive to this point. For example, my old *Statistics* text in "Schaum's Outline Series" (Spiegel 1961), carefully specifies, after saying you should reject H_0 if the sample statistic falls in the rejection region, that you should "accept the hypothesis (*or if desired make no decision at all*) otherwise" (p. 169, my italics).

Of course, Newman-Pearson methodology is a theoretical mess, and it is doubtful that there is any good way of making sense of it (*cf.* Howson and Urbach 1989, ch. 7.) And, given this, it is scarcely surprising that psychology textbooks explaining the logic of statistical inference present "an intellectually incoherent mix" of different ideas, as Gigerenzer puts it (this volume, p. 111).

Still, this does not alter my underlying point that theorists of signal detection like W.P. Tanner managed to develop a plausible approach precisely by diverging from Neyman-Pearson thinking when empirical considerations demanded. So, for instance, they allowed that the prior probabilities of H_0 and H_1 were significant in determining where the signal detection system should locate its decision criterion. And they treated these two hypotheses symmetrically, constructing the decision criterion in such a way that the system could make definite decisions between H_0 and H_1, rather than being forced to suspend judgement when the sensory input fell outside the rejection region of the null hypothesis H_0.

This is the signal detection theory described by Gigerenzer, and, as I said, he allows that it led to fruitful empirical research. But it is by no means a theory constructed by slavishly mimicking the evaluative tool that cognitive science acquired from Neyman and Pearson. At first pass, that evaluative tool was ill-suited to the needs of signal detection theory, and it had to be significantly transformed before it could give rise to a plausible empirical theory.

These last remarks were prompted by the worry that it would be surprising if cognitive theories modelled on arbitrary evaluative tools should turn out to be consonant with background knowledge, or with further experimental data. However, if the example of signal detection theory is typical, we can see why this is not so surprising after all. For that theory wasn't simply a blind copy of the relevant tool, but a reworking that owed as much to the empirical demands of its subject matter as to the tool on which it was nominally modelled.

Department of Philosophy, King's College London
London, United Kingdom

NOTES

[1] Nor is there anything worrying in the circumstance, highlighted by Gigerenzer, that tools for *justification* are playing a significant role in the context of *invention*. Given that background-knowledge-based judgements of initial plausibility also play a role, there is no need to place limits on where invention draws its inspiration from, and the structure of tools seems as good a source of ideas as any.

[2] I have argued that Gigerenzer's tools-to-theory hypothesis does not in itself demand a negative evaluation of epistemological practice in cognitive science. However, this doesn't mean that there aren't other reasons for such a negative evaluation. Thus we may well be doubtful about the enthusiasm with which cognitive science has taken up successive technological models of the mind (the mind is a . . . telephone exchange, cybernetic system, digital computer, neural net). The limited shelf-time of these models is clear testimony to their lack of serious empirical grounding. But note that the reason for distrusting these models is not that they are drawn from extraneous sources – my discussion of the tools-to-theories hypothesis shows that this is not in itself damning – but rather that a standard meta-induction shows that these fashions in cognitive science outstrip their epistemological warrant.

REFERENCES

Howson, C. and Urbach, P. 1989. *Scientific Reasoning*. La Salle, Illinois: Open Court.

McLaughlin, R. 1982. "Invention and Induction: Laudan, Simon and the Logic of Discovery". *Philosophy of Science* 49: 198–210.

Newman, J. and Pearson, E. 1967. *Joint Statistical Papers*. Cambridge: Cambridge University Press.

Papineau, D. 1993. *Philosophical Naturalism*. Oxford: Blackwell.

Spiegel, M. 1961. *Statistics*. New York: Schaum Publishing Co.

Stove, D. 1982. *Popper and After*. Oxford: Pergammon Press.

JEANNE PEIJNENBURG

ON THE CONCEPT OF DISCOVERY
COMMENTS ON GERD GIGERENZER

Professor Gigerenzer has given us a stimulating account of the genesis of two recent cognitive theories, viz. the theory that the mind works like a statistician and the theory that it works like a computer. With lucid examples and convincing arguments he showed that in the creation of both theories a so-called "tools-to-theories heuristic" played a role. Inferential statistics constitutes the tool which led to the theory that the mind is an intuitive statistician, computer programming is the tool which prompted the theory of the mind as a computer. Moreover, Gigerenzer stressed that the two cognitive theories are more likely to be accepted if the tools that led to them are generally used, a claim that is very plausible indeed. It is certainly very likely that, the more commonly inferential statistics are used, and the more common computers become, the more effective the computer metaphor and the statistician metaphor will be.

In short, I find the tools-to-theories hypothesis plausible and valuable, and I think that Gigerenzer's arguments for it are solid and to the point. However, I did not come all the way to Bertinoro just to praise the previous speaker. I also wish to make two critical remarks.

The first remark applies to an additional function that Gigerenzer ascribes to the tools-to-theories heuristic. According to Gigerenzer, this heuristic acts as a sort of bridge between the context of discovery and the context of justification. For the two tools that we are talking about, namely inferential statistics and the digital computer, simultaneously play a role in both contexts. On the one hand, they have taken root in the context of justification, since they provide us with means to evaluate scientific theories. On the other hand, they are entrenched in the context of discovery, for they inspire us on our way to the point where new theories are created:

This close link between tools for justification, on the one hand, and discovery and acceptance, on the other, reveals the artificiality of the discovery/justification distinction. Discovery does not come first, and justification afterwards. Discovery is inspired by justification. (This volume, p. 110).

But this is rather odd. As Gigerenzer sees it, the fact that a tool taken from the context of justification is used in the context of discovery, constitutes a bridge between the context of discovery and the context of justification. But if that were true, then the fact that a falling apple led Newton to his ideas about gravitation would constitute a bridge between the context of discovery and the context of orchard maintenance. This is of course strange, and more is needed to bridge the contexts of discovery and justification than only a haphazard common feature.

That was my first point. The rest of my commentary will be devoted to the second point, which is that there exists a shift of meaning in the use of the term "context of discovery". In particular, Gigerenzer's use of the term is quite different from that by the person who actually coined it, viz. Hans Reichenbach.

Reichenbach introduced the term "context of discovery" in *Experience and Prediction* –the original German word is "Entdeckungszusammenhang". In *Experience and Prediction* Reichenbach does not elaborate on the context of discovery; he devotes not much more than one paragraph to it. The reason for that we know: as a not yet naturalised epistemologist Reichenbach believes that the context of discovery should be studied by an empirical psychologist rather than by a philosopher. However, there is no doubt about what he thinks psychological research into the context of discovery should reveal. It should reveal which heuristics do in fact lead to discoveries. That is to say, it should lay bare the thought processes, not of people who *tried* to make discoveries, but rather of people who actually *made* them. For Reichenbach, finding the secret of the context of discovery means finding the route to scientific achievements, not retreading the numerous paths to scientific failures. He is thus only referring to the success part of the context of discovery, not to the part that led to scientific error, falsehood, and misunderstanding.

Moreover, Reichenbach seems to have had a particular *kind* of scientific success in mind. The examples he mentions are Newton's discovery of the law of gravitation, Einstein's explanation of the precession of the perihelion of Mercury, the deflection of a light trajectory through a gravitational field, and the equivalence of mass and energy. All these discoveries are, to be sure, taken from physics.

In addition, when Reichenbach talks about the context of discovery, he is referring to the strictly subjective processes that take place in the creative mind of a single individual. He talks about "traces of subjective motives" ("spuren der subjektiven Motive") and about the actual thought processes that occurred in individuals such as Newton or Einstein shortly before they had their ideas about gravitation.

However, when Gigerenzer is talking about the context of discovery, he seems to be contemplating something rather different from what Reichenbach had in mind. At least three differences demand attention.

First, Gigerenzer does not seem to be particularly interested in the personal thoughts of one particular individual. His context of discovery is more of the sociological kind. Time and again he points to historical and sociological circumstances in order to explain how certain ideas about the mind could be embraced by a scientific community, rather than could occur in a single individual.

Second, while Reichenbach talks about discoveries in physics, Gigerenzer talks about discoveries in psychology and philosophy of mind. The question is whether we can put the discoveries in physics and those in psychology on a par. Bluntly assuming that we can, as Gigerenzer seems to do, runs the risk of reanimating the old ideal of the unity of science, albeit now within the context of discovery. This takes us to the third difference.

There seems to be a great difference in character between the idea that the mind is a computer and, for instance, the discovery of the law of gravitation. Apart from the fact that the former comes from psychology whereas latter stems from physics, there are more telling differences. For example, the computer metaphor is a new variation on a familiar theme; it is a link in an entire chain of pictures of the mind. Gigerenzer himself mentions the pictures of the mind as a hologram or as a telephone switchboard, but of course there are many more analogies. The Dutch psychologist Douwe Draaisma recently listed a few famous ones: Plato pictured the mind as a wax tablet, medieval philosophers described it as a codex or book, the sixteenth century English hermeticist Robert Fludd saw the mind as an Elizabethan theatre, still others described it as a treasure-chest, an aviary, a storehouse, a filing cabinet, a pendulum clock, a water organ, a photo camera or a film director (Draaisma 2000). As Draaisma notes, each of these analogies has been defended with great gusto, sometimes because the defenders in question were impressed by the new technology, sometimes because they themselves were personally involved in its development (like von Neumann, or Turing, or Newell and Simon). However, all this is quite different from the discovery of the law of gravitation. Firstly, the law of gravitation is not a metaphor or an analogy. Secondly, it is not so easily describable as a link in an entire chain of similar discoveries. Thirdly, it does not saddle us with difficulties in distinguishing successful from unsuccessful applications. A weak side of the computer metaphor, as of any metaphor, is that it can have correct and incorrect applications alike. When does the computer metaphor lead to success and when does it lead us astray? That is the key question, for which there is no counterpart in the gravitation case.

In addition, there seems to be a fourth difference between the computer metaphor and the law of gravitation. The law of gravitation has been a source for numerous testable hypotheses, such as Kepler's three laws and the hypothesis that falling apples and moving planets are quantitatively on a par. But as Draaisma notes, the computer metaphor was less fruitful in producing testable hypotheses about the mind. In so far as such hypotheses were indeed formulated, they benefited more from technological and methodological developments than from the computer metaphor as such, i.e., from the general idea that the mind is a computer (Draaisma 2000, p. 159).

Given all these differences between Reichenbach's and Gigerenzer's uses of the term "context of discovery" (individual vs sociological, physics vs psychology, and several differences in the character of the discovery at hand), it is perhaps fair to say that Gigerenzer is talking about the "context of invention" rather than the "context of discovery". The difference between discoveries and inventions is, of course, a difference in degree. They might all be put on one scale. At the one extreme of the scale are the pure discoveries, mostly of empirically detectable entities like planets, molecules, and black holes. At the other we find the pure inventions: fables, fairy tales and other fiction or science fiction.

The reason for calling the computer metaphor and the statistician metaphor inventions rather than discoveries is not that they are unscientific – on the contrary. For example, non-Platonists will put mathematical findings on the invention-side, but they will not deny that mathematics is very important from a scientific point of view. The reason is simply that the statistician metaphor and the computer metaphor stand closer to one extreme of the scale than to the other. Whatever exactly the difference between discoveries and inventions might be (Nickles 1980 contains various suggestions), it tends to get blurred in the title of Gigerenzer's contribution, "Where Do New Ideas Come From? A Heuristics of Discovery in the Cognitive Sciences". For there the word "discovery" suggests that by "new ideas" are meant "ideas-as-discoveries". But the new ideas that Gigerenzer talks about, viz. the computer metaphor and the statistician metaphor, are more "ideas-as-inventions".

I have been arguing that Gigerenzer uses the term "context of discovery" differently than does Reichenbach (or Popper or many others for that matter).

Now Gigerenzer's reply could simply be: "Well, so what? I am neither Hans Reichenbach nor Karl Popper, and we are not living in the 1930's anymore. Why should I use my terms in the way Reichenbach and Popper used theirs?"

But if this were Gigerenzer's reply, my answer in turn would be twofold. First, I would say that the shifts of meaning I have been tracing are not entirely innocent. For they make it look as though, with the tools-to-theories heuristic, we finally have opened up the old context of discovery. But I doubt whether we did so, for Gigerenzer has not been talking about the context of discovery in the old sense.

Second, I would point out that Gigerenzer himself warns us for the very sort of meaning-shifts that I have been ferreting out. One of the finest parts of his paper deals with the observation that institutionalized tools are not theoretically inert (*cf.* the section "Heuristics of discovery may help in understanding limitations and possibilities of current research programs", pp. 110–113). Such tools have hanging from them a whole set of assumptions and interpretations, many of which are implicit and thus not crystal clear to the scientists who are using the tools. As an example Gigerenzer mentions statistics in psychology. Once the old statistical tool is brought to bear upon a new cognitive theory, it brings with it a number of assumptions that were not clear at the outset. In the same vein, I think, Gigerenzer's use of the term "context of discovery", in his own theory of where new ideas come from, brings with it a number of assumptions that were not clear at the beginning. What I tried to do here was to make these assumptions more explicit.

Moreover, I tried to do this in the manner which Gigerenzer himself recommends on p. 112 and which he actually follows in the second part of his paper (where he travels from mind to computer and back), namely by the method of meaning-investigation. Just as Gigerenzer recommends an investigation of the meaning of "probability", and actually investigates the meaning-shifts in the concept "intelligence", I have investigated the meaning of the concept "discovery", and in doing that I discovered (or invented) some meaning-shifts.

Faculty of Philosophy, University of Groningen
Groningen, The Netherlands

REFERENCES

Draaisma, D. 2000. *Metaphors of Memory. A History of Ideas About the Mind.* Cambridge: Cambridge University Press. Originally published in Dutch as *De Metaforenmachine – een geschiedenis van het geheugen* by Historische Uitgeverij, 1995. Translation from the Dutch by Paul Vincent.

Gigerenzer, G. 2003. "Where Do New Ideas Come From? A Heuristics of Discovery in Cognitive Sciences". This volume, pp. 99–139.

Nickles, Th. (ed.). 1980. *Scientific Discovery, Logic, and Rationality.* Dordrecht: D. Reidel Publishing Company.

Reichenbach, H. 1953. *Experience and Prediction. An Analysis of the Foundations and Structure of Knowledge.* Chicago, Ill.: University of Chicago Press.

URSULA KLEIN

STYLES OF EXPERIMENTATION

Ian Hacking's dictum that "experimentation has many lives of its own" (Hacking 1983, 165) has been a landmark in the ongoing discourse about experimentation. What has often been overlooked in this discourse is the little word "many." But how diverse is experimentation? Or is diversity an obsession of historians of science only, which is of little interest for philosophers? Another question that has come up again and again in recent discussions about experiments concerns conceptual and theoretical issues. Until around twenty years ago, philosophers and historians of science shared the view that experiments are a method – the "experimental method"– employed for testing and justifying scientific theories or for clarifying and solving problems left open by an extant theory. Today, many, if not most, historians of science and science study scholars reject this view. Detailed studies in the history of science have shown that experiments performed with the explicit methodological goal of testing, further developing and justifying an extant theory are comparatively rare. But there has also been a constant confusion about experimental goals and the role of concepts and theories in scientific experimentation. Though hardly anybody wishes to claim plainly that scientific experimentation is identical to handicraft, factory labor, or action by trial and error, and that experimenters' goals do not also differ from those of artisans, instrument makers, amateur constructors and so on, only few historians of science and science study scholars have seriously attempted to analyze and reflect about these differences. What are the intellectual preconditions for experimentation? What are its goals if not exploring and justifying an extant theory?

In this paper, I present two historical cases that shed some light on these questions.[1] Both of my case studies come from the history of the oldest experimental science, chemistry. The first exemplifies a style of experimentation largely ignored by philosophers and historians of science, which I term the pluricentred style of "experimental analysis." The second case, organic or carbon chemistry, which emerged in Europe from the late 1820s onward, is an example of an "experimental culture." "Experimental cultures" are the dominant style of experimenting in present sciences and have been at the center of interest in recent discussions about experimentation.[2] My most important philosophical criterion for the distinction between these two styles of experimentation is the way of production, individuation and definition of the objects of inquiry.[3] This criterion is related to another one, namely the internal dynamics of experimental inquiry. The pluricentred style of eighteenth-century experimental analysis is further distinguished from experimental philosophy of the time. With respect to this second distinction, which is not discussed in detail in the paper, the line is defined mainly by the role played by scientists' concern about issues of justification and along the axis uniformity/plurality of nature.

1. EXPERIMENTAL ANALYSIS

Throughout the eighteenth century and at the beginning of the nineteenth century, chemists subjected thousands of plant and animal tissues to chemical operations that they called "analysis."[4] In the late seventeenth and early eighteenth centuries, the procedure of plant and animal analysis was mainly distillation. For example, pressed seeds or ground leaves of a particular plant specimen were gently heated in a retort and thereby separated into different distillation products, which were received in distillation flasks and identified as water, acid "spirits," oils, and volatile salts (the residue in the retort being further separated into "earths" and "fixed salts"). An enormous effort went into these experiments. In France, for example, the Paris Academy of Sciences shortly after its foundation, in 1670 launched a project about the experimental analysis of plant species by distillation. When the Academy's secretary Fontenelle reported about these experiments in 1719, he enumerated more than 1400 plant specimens that had been subjected to these experiments. The project was then seen as a failure, and chemists began to turn to extractions by solvents (such as water, alcohol, ether, oils, etc.). In the decades to follow, chemists analyzed countless plant specimens by this new extraction method. They invested a lot of time and effort in improving the techniques of extraction by trying out new solvents, or by studying the effect of combinations of various solvents, the role of temperature and so on. Even after 1830, when chemists' style of experimentation in organic chemistry began to transform into an experimental culture, they continued these experiments, but more at the margins of the new experimental culture.

What were the goals of these experimental analyses? An apparently simple answer is the following: chemists wanted to acquire knowledge about particular plants and their composition from simple chemical components. These goals were coupled with at least four different layers of concepts and scientific objects, which had different historical, practical, and intellectual roots: first, the natural-historical concept and object of plant species, linked to plant anatomy and botanical classification; second, chemical concepts and objects of inquiry, such as "chemical composition", "chemical component", and chemical "analysis"; third, the philosophical (or metaphysical) concept of principles or elements; fourth medical and pharmaceutical goals. A historical reconstruction of chemists' experimental goals thus requires an analysis of the meaning of these conceptual layers and of the formation of scientific objects coupled with them. In the following historical parts of my analysis I will omit the medical and pharmaceutical objectives and concentrate on the first three layers of concepts.

The first layer of concepts: plant species and their anatomy

Unlike the type of organic chemistry that developed from the late 1820s onward, plant chemistry prior to 1830 overlapped considerably with natural history. It was concerned with all natural historical aspects of plants, including their anatomy and physiology. The different "plant species", which had been identified and classified by botanists, and examples of which had been collected by travelers and naturalists, grown and bred by gardeners, and stored and ordered into boxes and closets by owners of natural historical cabinets, were its immediate objects of inquiry. Experimental analysis was performed to study the components of these entities. Viewed from the natural historical perspective, it was a continuation of botanical anatomy with chemical means; hence "anatomy" was a term that was often used synonymously with "analysis" in this context. Other terms used by the historical actors were "experimental history", "analytical history" and "chemical history" (Boerhaave 1741, 2: 78, 86, 89).[5]

Chemists did not only take over botanical taxonomy for identifying plant specimens subjected to chemical analysis, they also identified and classified the extracted substance components largely in a natural-historical mode, that is based on their observable properties (including so-called chemical properties) and their natural origin. Moreover, the natural historical goal of acquiring refined knowledge about plant species engendered questions such as: do the substance components of plants have properties that differ in kind from components of minerals? Are these substance components carriers of "properties" of life? In the late eighteenth century the vast majority of European chemists gave an affirmative answer to these questions. The term "organic" component or substance that became a substitute for "plant" and "animal" substance at the time clearly set apart the components of living and of non-living or "inorganic" beings.

The acquisition of knowledge about particular plant species and about their composition from organic substance components was the first goal of experimental analysis of plants in the eighteenth and early nineteenth centuries. This is important to note with respect to experimental philosophy of the early eighteenth century and physics from the late eighteenth century onward, which have been paradigmatic for our philosophical understanding of the use of experiments. To put it more general and provocative: in experimental analysis the "laboratory style" was coupled with the "taxonomic style" of natural historical reasoning. It was not linked with the "style of hypothetical modeling" as has been proposed by Hacking. Experimental analysis was spurred by the interest in particulars and the plurality of things, rather than in primary causes and reductive theory. Furthermore, particular plant specimens were observable objects rather than unobservable theoretical entities presupposed by Hacking's "laboratory style" (Hacking 1992b, 6, 11).

The second layer of concepts:
"chemical analysis", "chemical components", "purity"

The framework of experimental analysis of plants was additionally shaped by a particular conceptual system, which chemists had built in the chemistry of pure laboratory substances at the end of the seventeenth century and the beginning of the eighteenth century. Hence, terms like "composition", "component", and "analysis" also had meanings that were specific to chemistry only. This specific chemical meaning was determined by the entire conceptual system to which these terms belonged. Apart from composition, component, and analysis the most important concepts of this system (hereafter termed conceptual system of "chemical compound and reaction") were: chemical compound, affinity, immediate constituent, synthesis, reaction, and pure substance. As a consequence for experimentation, the components chemists were seeking to extract had to be "pure" substances, that is, individual chemical compounds rather than impure mixtures of compounds. This is an important difference from natural historical enterprises outside plant (and animal) chemistry. Most of chemists' practical effort of experimental analysis through extraction was devoted to this specific *chemical* goal that originated in the chemical laboratory practice: the purity of the extracted substances.

I have written elsewhere about the emergence of the conceptual system of chemical compound and reaction.[6] Two aspects are important in our context. First, it referred to entities produced in the chemical laboratory, and was mainly the result of chemists' reflection on a system of experimental marks produced in specific experiments, which were afterwards conceptualized as "analysis" and "re-synthesis". The most basic assumptions were that the homogeneous chemical compounds consisted of preformed building blocks, which were simpler substances. These substance components (not atoms or corpuscles!) could be obtained through experimental chemical analysis. Further, chemists believed that all kinds of chemical transformation were merely recombinations of these preformed substance components guided by chemical affinity. I only want to hint at Paracelsian alchemy – at its concepts of transmutation, homogenous mixtures, inner essences, arcana, etc. – to indicate that there was a profound conceptual rupture between these earlier alchemical concepts and the conceptual network of chemical compound and reaction, built in the decades before and after 1700.

Second, this conceptual system was built not in plant or animal chemistry, but in the domain of chemistry which investigated purified natural or artificially synthesized salts, acids, alkalis and metals. Under largely the same physical conditions such materials displayed a stable reproducible pattern of behavior in repeated series of chemical operations. Salts, such as vitriol, common salt, saltpeter, and so on were heated in distillation vessels (or mixed with another ingredient and heated) and thus transformed into powerful solvents: the mineral acids. Mixtures of a mineral acid with a metal, or a so-called metal calx (later metal oxide) or an alkali yielded salts, reproduced natural or new artificial ones. These salts could be easily decomposed again in subsequent chemical operations into the two original ingredients, the acid and the metal or alkali. Such reversible chemical transformations, first, of two initial substances into one homogeneous new substance, and, second, the recovering of the two initial substances from that new substance – composition and decomposition, as it were – were also performed with metals and metal alloys. By contrast, it was not possible to re-synthesize a plant tissue from the components extracted from it. Nonetheless, eighteenth-century chemists extended aspects of the conceptual system of chemical compound and reaction established and entrenched in the chemistry of pure laboratory substances to plant chemistry. Hence, their search for pure substance components and their view that these materials were pre-existing building blocks of plants which were merely isolated, but not created, by experimental analysis.

During the eighteenth century, the conceptual system of compound and reaction, or aspects of it, became a constitutive element of any chemical experimentation, including the experimental analysis of plant tissues. Yet it was never tested or explicitly justified by chemists. Rather, it functioned as an indispensable internal prerequisite of experimentation – with respect to its interpretative, productive and intervening features. It was part of a system of material production of marks, as well as of their reading and interpretation. Hence, this conceptual system was so fundamental that it makes sense to characterize it a constitutive element of the coming into being of chemists' experimental objects. Borrowing a term from Bachelard, one might also speak of a "reified" intellectual element of experimentation.[7] It was only in the twentieth century, with the emergence of quantum chemistry, that this intellectual fundament of chemical experimentation was abolished.

The third layer of concepts: chemical "principles"

In the late seventeenth and early eighteenth centuries, chemists often interpreted the materials extracted from plants and other natural bodies as "elements" or "principles". A few decades later, they still used these terms alongside terms like "plant substance" or "plant materials" and "immediate principles". There is hardly any other term more intimately related to fundamental theory than "principles". This raises the question of whether this aspect of experimental analysis was tied to theoretical goals, to the justification or development of theory.

The term "principle", as it was used around 1700, referred to materials obtained from a natural body through dry distillation, such as water, oil and spirits found in the receiver of the distillation apparatus. Yet the meaning of that term was by no means exhausted by its empirical reference. "Principle" or "element" also, and primarily, meant a constituent of natural, corporeal and observable bodies, which per definition was not a corporeal body or observable matter itself. In keeping with the Aristotelian or Paracelsian theoretical tradition of elements or principles, chemists assumed three, four or five of such fundamental ontological entities. The material referents of "principle", which could be observed in the receiver or retort, were considered not as the true pure principles but as sufficiently purified representatives of them that allowed to further explore their properties. Thus, with respect to the last decades of the seventeenth and the first two decades of the eighteenth centuries it is appropriate to claim that chemists' experiments also pursued the philosophical goal of exploring fundamental ontological principles.

This type of philosophical goal of experimentation vanished after ca. 1720. But it vanished not via public debates or any other form of explicit discourse but slowly faded away. Philosophical goals played less and less a role in experimental reports, and instead were banished to the theoretical parts of textbooks. Chemists still believed in Aristotelian or Paracelsian types of elements, but they became convinced that chemical art was unable to extract these most fundamental entities from natural bodies and to separate them from each other. Experimental failure was explained away by arguing that only nature could operate directly on principles, create natural "mixts" from them and decompose these mixts into principles. The general decline of Aristotelian philosophy at European universities during the eighteenth century may have contributed to this development.

In the middle of the eighteenth century, Continental European chemists broadly accepted the assumption that experimental analysis yielded different kinds of components of natural bodies. First, simple substances that could not be further decomposed by any chemical experiment but were not simple in the traditional philosophical sense, since they were constituted by the true, ontologically fundamental principles; these substances were called "simple substances" or "principles" in a new, more operational sense. Second, there were also components that were simpler than the body they were extracted from but could still be decomposed by further chemical operations. The latter were called "immediate principles."[8] Chemists' ontology allowed to reconciliate the metaphysical meaning of principle and the more operational one, but only the latter was relevant for experimentation, as has been analyzed in the section above.

The "more" operational concept of principle, however, had also a metaphysical dimension; that is, it was not operational in the strict sense of modern philosophy. The simple substances and immediate principles extracted from a natural body were conceived parts which carried qualities that engendered the properties of the entire body. The whole was viewed as fully determined by its parts. This key postulate of mechanical philosophy was preserved wholesale in the chemical revolution of the last third of the eighteenth century. It was never challenged by experiments before the introduction of structure theory in the 1860s; and it was dismissed only in the twentieth century with the acceptance of quantum chemistry. The decline of this metaphysics of analysis, which began in the second half of the nineteenth century, cannot be explained by any crucial experiment. Nor can it be explained by any theoretical program elaborated, tested or even justified by experiments. It is even an open question to what extent it can be explained as a result of reflection on a long tradition of experimentation. The creation of new conceptual networks crossing the boundaries between twentieth century chemistry and physics as well as transformations in the broader philosophical and cultural context may have played a role as important as experiments and other forms of experience did.

A pluricentred style of experimentation

While in the late seventeenth and early eighteenth centuries, experimental analysis in plant and animal chemistry also pursued the philosophical goal of justifying and refining an Aristotelian or Paracelsian concept of elements or principles, in the second half of the eighteenth century this kind of goals faded away. The experimental goals in plant and animal chemistry in the second half of the eighteenth century and the first two decades of the nineteenth century were framed by three conceptual layers, which originated in different practices. First, the natural historical concept of plant species and the practice of collecting, observing, and ordering exemplars of them. Second, the chemical conceptual system of compound and reaction and the laboratory practice which investigated chemically pure components of substances. Third, concepts stemming from pharmaceutical and medical practice linked with the goal of isolating and further studying the component of plants that carried its medical virtues. Pure chemical components also had a dual nature in themselves: they were visible, sometimes palpable and smelling things, which stimulated epistemical and commercial interests for their own sake, and simultaneously they were signals of an invisible object: the composition of plants from its chemically pure organic constituents.

On the one hand, experimental analysis studied the properties of visible, palpable or otherwise observable natural objects, which had been previously individuated and circumscribed in practices outside the space of experimentation, in everyday life, artisanal and commercial practice, scientific expeditions, natural historical cabinets and so on. On the other hand its objects of inquiry were also circumscribed by experiments and the signals produced by them. The experimentally isolated and purified components of plant or animal tissues could take on a life of their own beyond their function as signals. They were material artifacts that had an impact on the material inventory of the laboratory and potentially also on technological and natural sites outside the laboratory. Yet, compared to experimental cultures that developed in the nineteenth century and became the dominant style of experimentation in the twentieth century, the productivity of the pluricentered style of experimental analysis was restricted by non-experimental parameters. The chemical components isolated in experimental analysis formed an epistemic unit with the natural plant and animal specimen they were extracted from. They were also defined by a philosophy that was concerned with elements, principles or immediate principles of natural bodies only. The overall question of what kind of bodies existed in nature and how they were composed restrained another type of question that became more dominant in the nineteenth century. This second type of question was concerned with processes and laws that existed not only in nature but also in experimental art and with possibilities to produce novel objects by experimental technologies.

From the seventeenth century until the middle of the nineteenth century the pluriconditioned, multilayered objects of experimental analysis thrived in a culture of savants that was not yet transformed into a specialized expert culture or scientific discipline, but incorporated many different knowledge traditions and practices. Experiments were only one form of experience in that pluricentered culture. In accordance with the general objectives of natural history and of the pharmaceutical and other workshop traditions, the goal of experimental analysis was the investigation of the plurality of things, which was complemented by classification. The laboratory style and the taxonomic style of inquiry were two sides of the same coin. After ca. 1720, experimental analysis was no longer concerned with fundamental principles of nature and final causes, but became rather independent of global philosophy. Moreover, it did also not aim at justification or refinement of other forms of theory. This is an important distinction with respect to the use of experiments in experimental philosophy of the time. In that area many experiments were actually involved strongly in theoretical controversies and strategies of justification.[9] Yet experimental analysis implied conceptual elements as a prerequisite.

Furthermore, viewed from the perspective of the experimental culture of organic chemistry, which followed and partly replaced the pluricentered style of experimental analysis in plant and animal chemistry, another aspect needs to be emphasized. A series of experiments performed for investigating a particular plant specimen and the properties of its immediate principles was completed when all of the types of operations that were usually done to this end had been performed. This is an analogy to experiments performed in experimental philosophy for theoretical goals.

The experiments came to an end when all of the available experimental means for the defined goal had been mobilized. Though in case of experimental analysis in plant and animal chemistry, experimental procedures were nearly endlessly repeated and extended to numerous different plant specimens, there was no substantial development in the type of object of inquiry and the type of questions asked about it. Experimentation had not the kind of autonomous, internal dynamics that became characteristic of modern experimental cultures. It was embedded in and constrained by objectives engendered outside experimentation and hence not open-ended in the same sense as experimental cultures.

How far can this be generalized? Though future studies need to provide more detailed empirical evidence, I suggest that the pluricentered style of experimental analysis, as defined above, also covers some areas of anatomy, microscopy and physiology of the late eighteenth and early nineteenth centuries. Moreover, this style of experimentation still exists in some present scientific disciplines and sub-disciplines. For example, the scientific objects of ecology are defined not only by experiments but also by a broad variety of different forms of experience. Many experiments in ecology analyze special features of such pluridetermined objects. They are embedded in a broad set of empirical strategies, which limit their use and internal dynamics. Or in research on animal behavior, field observation still plays an important role apart from experimentation. Here, as in ecology, experiments are one way only of empirical research, and the objects of inquiry are circumscribed in a space that exceeds experimentation. Meteorology and climate research are another example for present sciences in which experiments, mostly experimental analysis of previously isolated phenomena, play a rather limited role compared to systematic observation and computer generated simulation. In all three cases, the goal of experimentation is analyzing natural phenomena and exploring their features rather than justifying or developing a theory. Furthermore, in all three cases the interest in the plurality of natural objects and processes plays an important or even dominant role.

2. EXPERIMENTAL CULTURES

In the type of organic chemistry that developed after 1830, the study of plant and animal species and of their chemical components moved to the margins of experimental practice.[10] Chemists working in this field now focused their experimental research on chemical processes, that is, on chemical reactions of organic compounds and on the possibilities of synthesis. This type of experimental research also allowed conclusions about the constitution (later structure) of organic compounds. Interest in particulars, the plurality of things and their classification remained a major objective of experimentation, but the type of particulars had changed. The vast majority of "organic" substances no longer were "organic" in the traditional sense – that is, natural substances produced in and extracted from plants or animals – but synthetic materials that did not exist in nature outside the chemical laboratory. Such synthetic materials were first created as "reaction products", that is, as experimental marks of chemists' new objects of inquiry: the invisible chemical reactions and constitutions of organic substances. But quite exceptionally, this kind of experimental marks could be, and had to be, isolated from apparatus for further processing. Marks which first merely signified another object of inquiry – the invisible chemical reaction – were transformed into things – real chemical materials – which invoked chemists' interest for their own sake. In this way, each experimental mark of an investigated chemical reaction spurred further experimentation and contributed to an exponential increase of synthetic materials that served as new experimental objects of inquiry, both with respect to their visible properties and their invisible constitution and reactions.

This leads to two more general epistemological issues concerning experimental cultures. First, unlike the pluricentered style of experimental analysis in plant and animal chemistry, experimental cultures, as a rule, provide a comprehensive space for the production, individuation, definition and classification of their objects of inquiry. For example, chemical reactions as defined in the laboratory practice of carbon chemistry, or, to name only a few additional examples, molecular structures, genes, ribosomes, and atoms are totally unfamiliar from everyday life or from natural historical practices. Such invisible scientific objects can only be grasped and circumscribed in a mediated way, namely via their visible or otherwise observable effects or marks produced in experiments. This is an important difference between experiments and mere observation: phenomena such as experimental marks are not merely observed and interpreted but they are produced in experimental practices. They require specific social and material settings, techniques and skills for coming into existence. As the invisible objects of inquiry would be mere fantasy or even entirely beyond the horizon of imagination without such experimentally produced signifiers, we can say that these objects are produced by experimenters, epistemically and to some extent materially as well.

Experimenters in modern experimental cultures do not only ask what exists in the world and how the world is structured, but also what it is possible to produce and how we humans can change the world. They are also concerned with scientific objects that are materially produced and individuated in a more literal sense. Examples of production and individuation in this strong sense are stable electric currents in the late eighteenth century, artificial chemical compounds, radioactive elements, and other synthetic materials from the early eighteenth century onwards, some types of submicroscopic particles in the twentieth century, genetically manipulated plants and animals at the beginning of the twenty-first century, and so on. Such kinds of objects are originally entirely non-existent outside the laboratory. Of course, neither in technology nor in experimental cultures "production" means creation ex nihilo and at will. Experimental production reassembles and reshuffles extant forms of matter and redraws boundaries of individual entities by means of specific instruments and techniques. In doing so, it depends on the possibilities and constraints described by the laws of nature.

Second, if it is true that experimental cultures typically produce and individuate their objects of inquiry, rather than merely observing or experimenting on extant ones, and if this production requires specific social, material, and cultural resources that are absent in everyday life and other scientific practices, the following consequences can be drawn easily: new objects of inquiry must first be stabilized, identified and demarcated before any theory about them can be thought of. The experimental marks have to be purified, disentangled and drawn together as a coherent signal for a novel, individual object. It must also be possible to reproduce the experimental marks under controllable conditions. As a rule, at the beginning such performance is tentative and subjected to many revisions, as the series of experiments and their interpretations go on. It is also open-ended: the process may end with the individuation and stabilization of a novel object along with a new concept and inscription referring to it, but it may also end with dismissals, shifts of goals, and epistemic reconfigurations.

There is also no reason to believe that any set of instruments and skills available at a given time may close the possibility of producing novel signals, which may contribute to considerable re-configurations of the novel object. Unexpected and surprising marks may be produced even through slightly different applications of a given instrument, a variation of the preparation of the target, a variation of the experimenter's manipulation, and so on; that is, technological innovation of instruments is not always a necessary precondition for the production of experimental novelty. Novel marks require efforts to fit them with the older ones. The result of such work of representation is a transformed circumscription of the previously introduced scientific object, which may go hand in hand with a shift of scientists' goals. All of these steps contribute to the processual character of experimental culture and its intrinsic, open-ended dynamics.[11]

This epistemological focus on the practice of experimentation forms a stark contrast to the still dominating philosophical view that experiments are a method of testing extant theories. Methods are mostly defined as means whose uses are determined by external goals, in our case by a theory. On this view, an experiment, or a series of experiments, always comes to an end if the test of a theory has been completed. It should be noted that this is a systematic consequence of the epistemological understanding of experiments as mere methods, which should be clearly distinguished from the historical investigations of contingent constraints of the experimental dynamics.

A remark about social aspects

This brings me to another aspect of experimental cultures, which defines its distinction from other kinds of explorative experimentation. Continuous series of systematically varied experiments exploring experimental effects, which were ascribed to a vaguely circumscribed scientific object, were performed repeatedly in experimental philosophy and early physics from the seventeenth century until the first decades of the nineteenth century. For example, when Charles Dufay (1698–1739), André-Marie Ampère (1775–1836) and Michael Faraday (1791–1867) set out to study electrical phenomena and interactions between electricity and magnetism, respectively, they had no concepts at hand that allowed them to individuate and identify the entities that engendered these phenomena.[12] However, such potentially open-ended explorations were repeatedly interrupted when the experimenter turned to theoretical issues or to other activities. There was neither an institutional differentiation between experimentalists and theoreticians at the time nor a social form of professionalization that guaranteed continuous experimental research. The possible open-endedness of experimental explorations was limited by a variety of contingent social factors as well as by biological and intellectual limits of the individual. Local groups of individuals, such as a team in a modern laboratory, transcend individuals' limits to some degree. However, it is only in a trans-local, coherent scientific culture where explorative experimentation can thrive and fully develop its open-ended research dynamics – presupposed the social and politic conditions endorse such development. Social and political institutions and relationships – such as acceptance by the political system and the broader society, stability of financial support, means of communication, pedagogical forms of transmission of knowledge, skills and instruments from one generation to the next, forms of exchange of instruments and materials – are crucial constituents of experimental cultures, which cannot be discussed here for reasons of space. It should be mentioned, however, that the institutional boundaries of experimental cultures are not necessarily identical with the kind of institutional framework that entrench scientific disciplines or sub-disciplines. For example, laboratory-based organic chemistry became a sub-discipline of chemistry only in the second half of the nineteenth century. It was established historically after the formation of an experimental culture of organic chemistry.

The role of theory

Experimental marks need interpretation. Moreover, they need to be recognized as signifiers rather than meaningless noise or impurity. In other words: the role of theory in experimental cultures is up for grabs again.

As has been mentioned in part one of this paper, the conceptual system of chemical compound and reaction was a long-term cognitive framework, built in the decades before and after 1700 and changed only in the twentieth century with the emergence and acceptance of new physical theories and quantum chemistry. Though this cognitive framework had an impact on experimental analysis in plant and animal chemistry, it was only with the emergence of the experimental culture of organic chemistry that it came to working life in all its aspects in this domain of chemistry. Only from the late 1820s onward, the characteristic link between analysis and re-synthesis in inorganic experimentation was complemented by its analogue in organic chemistry. In organic chemistry, however, the experimental investigation of a broad variety of chemical reactions, including synthesis, required modeling.

For the construction of interpretative models of chemical reactions of organic substances and of their constitution, chemists applied Berzelian chemical formulas from the late 1820s onward. Berzelian formulas, such as H^2O for water or H^2SO^4 for sulphuric acid, had been introduced by the Swedish chemist Jacob Berzelius (1779–1848) in 1813 in order to represent the composition of chemical compounds in accordance with his "theory of chemical proportions."[13] This theory overlapped with Daltonian atomism, without being identical with it. Like Dalton's atomic theory, it postulated discrete quantitative portions of chemical elements and compounds that have a substance-specific, invariable, relative combining weight. But unlike Dalton's atomic theory, it did not further define these portions as submicroscopically small, mechanical bodies. Rather, it postulated scale-independent chemical portions, that is, invariable bits of chemical substances which were identified by their characteristic relative combining weight.[14] The "theory of chemical portions", as I would like to term it, forged that specific difference between the submicroscopically small parts of mechanical bodies – the atoms in the natural philosophical tradition – and the scale-independent portions of pure chemical substances by inventing a new sign system. One letter of a Berzelian formula denoted one chemical portion without simultaneously invoking the meaning of "atoms" in the philosophical tradition. Furthermore, Berzelian formulas gave the theory of chemical portions a working life on paper since their syntax allowed easy manipulations.

It is important to note that the Berzelian theory of chemical portions and Berzelian formulas first referred only to inorganic compounds. They were experimentally well underpinned by stoichiometry, where inorganic compounds were the paradigmatic referents. Analyses of organic compounds, however, did not confirm the most relevant stoichiometric law in this respect, the "law of multiple proportions", which allowed small integers only.[15] As a consequence, roughly until 1820 Berzelius and other European chemists doubted that the extension of the theory of chemical portions and of chemical formulas to organic substances was justified. Yet, from the late 1820s onward French and German chemists began to use Berzelian formulas in the very domain where they most lacked empirical evidence.[16]

Now, it might be assumed that chemists' goal of applying Berzelian formulas in the context of organic chemical experiments was the clarification of theoretical problems and the subsequent justification of the theory of chemical portions. This view, however, is not confirmed by historical analysis. What chemists actually did was the following: they transformed the data from experimental quantitative analysis of a given organic compound into a Berzelian formula, regardless the fact that in most cases this transformation did not yield both integer and small numbers, as prescribed by the stoichiometric law of multiple proportions. That is, they rounded down or up and allowed larger numbers. They then used this Berzelian formula as an unquestioned prerequisite for constructing an interpretative model of the reaction under experimental investigation and of the binary constitution of the organic substance. In doing so, chemists manipulated Berzelian formulas on paper and constructed two-part formulas (so-called "rational formulas") to represent the presumed binary constitution of organic substances. The partition of the formula into two parts yielded a visual image of the binarity of the organic compound. Furthermore, in their interpretations of organic chemical reactions chemists constructed formula models in form of formula equations. Formula equations structured the complexity of experimental marks produced in an experimental study of an organic reaction by distinguishing between a main reaction, parallel reactions and successive reactions. Both the models of reaction and of the binary constitution of organic compounds had to fit the general conceptual framework of chemical compound and reaction and the particular experimental marks. Chemists then used these formula models to solve one of their most urgent problems, to create a better, that is "natural", classification of organic substances.

I have analyzed elsewhere in detail examples of this pragmatic application of formulas as tools on paper or "paper tools" for constructing models of reaction and constitution (Klein 2002). These examples demonstrate clearly that chemists used Berzelian formulas as if they were well-confirmed preconditions for modeling. When it turned out that a new Berzelian formula did not work, that is, if it did not fit other formulas and did not allow the construction of interpretative models of chemical reactions of the substance denoted by the new formula, chemists re-focused their attention on the tool, the Berzelian raw formula, and tried to accommodate it by repeating quantitative analysis and/or by rounding the numbers of portions differently.

The goal of chemists' application of Berzelian formulas was not the solution of puzzles arising from the theory of chemical portions and the justification of that theory. Rather, their goal was to construct models that fitted the more traditional conceptual framework of chemical compound and reaction. As chemical formulas proved to be excellent tools for achieving this goal, their application proliferated in the chemical community after 1830. Three decades after their introduction, Berzelian formulas had become theoretical tools that were stabilized to such a degree that they were viewed as emblematic of the laboratory science of chemistry as beakers, distillation columns, and test tubes. What general conclusion can be drawn from this? I propose that experimentalists working in an experimental culture are willing to apply a general theory which was elaborated outside the intellectual space of their culture only if that theory has a pragmatic value for them. As a rule, modern experimentalists are not interested in testing and justifying a theory that did not originate in their own research and whose reference exceeds the range of their scientific objects.[17] I also propose that experimentalists' application of an extant theory as a paper tool may lead in an unintentional way to the refinement and justification of that theory.[18] If the refined theoretical tool works better, that is, if it contributes to the solution of local problems and/or allows the extension of experimental research to new objects, experimentalists believe in its truth or approximate truth. This view is corroborated by the history of the theory of chemical portions after 1830. The formation and acceptance of the concept of valence and of the theory of atomic structure was coupled with chemists' pragmatic application of chemical formulas as paper tools in experimental practice, the adjustment of these tools to locally emerging goals and their further refinement and transformation into type formulas and structural formulas.

Experimental marks and representation

The fact that the objects of inquiry in experimental cultures are, as a rule, unobservable entities that are accessible only in a mediated way, via the marks or signals they produce in a technically shaped space, leads to an epistemological problem, which I want to discuss briefly. A mark – such as a reaction product, a trace of a particle in a cloud chamber, a pattern of x-ray spots on a photographic plate, a pattern of bands in an electrophoretic gel, a number of radioactive decays and so on – is a visible, or otherwise observable, material trace that is recognized and read by experimenters as a signal of the absent scientific object. Experimental marks refer to physical events in the past, and simultaneously they have to be meaningful to the experimenter, and require intellectual skill of counting, reading, and pattern recognition.[19]

The subsequent work of representing a scientific object by means of culturally available sign systems, such as charts, tables, diagrams, formulas, verbal language and so on, has been studied in particular by Bruno Latour and more recently by Peter Galison.[20] The representation of scientific objects, Latour has proposed, proceeds in steps, creating chains of representations or inscriptions. In each step the former inscription is transformed and some new intellectual content added, such as the theoretical concept of chemical portion, while the original referential structure is preserved (Latour 1990). In his studies about the experimental culture of twentieth-century molecular biology, Hans-Jörg Rheinberger has taken over Latour's concept of chains of inscriptions, however, with a decisive distinction. "It is unnecessary", he states referring to Latour, "to distinguish between machines that 'transform matter between one state and another' and apparatuses or 'inscription devices' that 'transform pieces of matter into written documents" (Rheinberger 1997, p. 111). In other words, Rheinberger rejects the assumption that there is any relevant difference between experimentally produced "marks" and the inscriptions or "representations" produced from these marks. In accordance with this, his broad notion of "representation" demolishes the distinction between "presentation" and "representation", and hence covers both experimental marks and the subsequent forms of representation.

This Derridean conceptualization of scientific representation is justified to the extent that representation of a scientific object by means of culturally available sign systems has a referential structure that is analogous to that between experimental marks and the scientific objects they refer to. Signs that are both material, visible traces and have an intellectual content, distinguished as "signifier" and the "signified" in semiotics, refer to other signs constructed earlier in the chain of representations; and the first representation or sign refers to experimental marks, which in turn refer to a previous physical event. Nelson Goodman has analyzed such referential structure in a discussion of the problem of how "explaining the root relation between language and the non-linguistic experience it describes" (Goodman 1997, p. 124). He has compared that relation to what Peirce has called in his triad of types of signs "indices" (i.e. signs physically connected to their objects): "A clue to a better starting point than resemblance", he writes, "lies in the fact that a toot may warn of an oncoming train or that a ray of a dawn foretells the approach of daylight. Here are non-verbal events standing as signals for others." Analogously, he continues, a sensory experience, such as the feeling of warmth, may signal the appearance of a fiery red patch in the visual field. And he concludes: "If experiences comprised of such presentations as shaded patches can signal, there is no mystery about how an irregular black patch or a brief stretch of sound may function the same way. And a statement-event, or other string of word-events, is simply some such patch or stretch" (*ibid.*, 125 f.).

From the broader perspective of Goodman's epistemology or of a Derridean critique of Western metaphysics, it might be appropriate to highlight the likeness of referential structures in physical and semiotic worlds. Yet, if applied consequently in the more modest endeavour of a historical epistemology of experimentation, such approach would relinquish conceptual tools that allow us to reconstruct and explain the specificity of this enterprise.[21] Although it is not possible to disentangle the given and constructed elements of experimental marks, in our historical-epistemological reconstructions of scientists' representational activities it is necessary to differentiate between "experimental marks" that also comprise elements that are given to humans and sign systems that are intellectually and culturally constructed in their entirety. If this differentiation is given up in the history and philosophy of science, the notions of scientific empirical "experience" and of "experiments" no longer have any distinctive meaning at all, but become identical to writing, modeling and theory formation tied to scientists' work with sign systems. Moreover, the Derridean approach relinquishes conceptual tools that allow us to understand the technoscientific consequences of modern experimental cultures. Whether we like it or not, most of the experimentally produced marks in modern experimental cultures are more than intellectually and culturally meaningful signifiers. For example, reaction products are the most important signifiers of the unobservable movements of chemical substances in chemical reactions, but their role is not exhausted by that semiotic function. They are also materials that have an impact on our social and natural world, and they can be powerful physical agencies that demonstrate in their way that our world is not only a world of signs.

Why "culture"?

The concept of "experimental culture" is an attempt to come to grips with the complexity of relationships between experimental manipulations and instruments, experimentally produced marks, conceptual systems, theories embodied by paper tools, chains of representation and interpretative models in the new style of experimentation which began to develop in the nineteenth century. It also takes into account that classification may be closely coupled with experimentation.[22] "Scientific cultures" are networks of collectively shared scientific practices and tools for these practices (material, symbolic, and conceptual), beliefs, habits, styles of communication and various other forms of more or less institutionalized social relationships on a trans-local communal level of a science. The connotations of the term "culture" are particularly useful to underline the historical contingency and local diversity of the kind of scientific networks that is at stake here.[23] There is neither one universal form of a scientific culture nor are the relations between its practices and elements absolutely fixed and static. Tensions and frictions may result, for example, from the different historical origins of the various practices and elements, and from their individual relations to bordering scientific cultures and the broader social and cultural context. As a consequence, the stability of a scientific culture and its coherence are never complete, and shifts of practices and elements may occur all the time. But what conditions the relative coherence, stability and longevity of a scientific culture?

Again, there is no universal answer to this question. In case of the experimental culture of organic chemistry, which began to develop in the late 1820s and thrived in the 1850s and 1860s, collective practices which evolved around two new types of experiments (quantitative analysis and the experimental study of chemical reactions of organic substances and of their constitution), a new sign system and paper tool (Berzelian formulas), and the traditional conceptual system of chemical compound and reaction were constitutive of such coherence and duration. Experiments, along with the conceptual system of chemical compound and reaction described above and the new paper tools, became the material and epistemic scaffolding, as it were, for the experimental formation and representation of all scientific objects and for their classification. "Scaffolding" is an intentionally vague metaphor for those practices and elements of a scientific culture that engender its relative coherence, stability and longevity. Analytic approaches which attempt to disentangle the material, performative, observational, semiotic, and conceptual items in order to study each element separately create historical and philosophical artifacts, often coupled with reductionism. In the experimental culture of carbon chemistry all of these items together constituted experimentation; and experimentation constituted the horizon of possible objects of inquiry. In contrast, plant and animal chemistry was concerned with many scientific objects which were not studied experimentally at all, such as physiology and the medical virtues of extracted plant and animal materials. Moreover, even its objects of experimental inquiry were framed in a pluricentered space that linked laboratory experimentation with artisanal practice, natural historical inquiry, and metaphysical reasoning. Experimentation was only one constituent of scientific objects in plant and animal chemistry apart from many others.

CONCLUSION

We tend to think about modern experimentation as a continuation of seventeenth and eighteenth centuries experimental philosophy, which embedded experiments in an overall philosophical inquiry. In this view, experiments are a method in the service of something else: the refinement of an extant theory and its justification. As experiments are constrained by theoretical goals engendered outside experimentation, this kind of reasoning goes on, it is hard to conceive why they should not be constrained by other external factors as well, such as ethical norms; hence the belief that ethical discourse may have an impact on decisions of modern experimenters. Yet, our most recent experience with gene technology tells a different story. Gene technology seems to be a rolling train that is hard to stop or even to direct. In order to understand this philosophically we need to study the actual practice of experimentation and we need to go back to its historical roots and its historical diversity.

Experimental philosophy was not the only historical route to modern experimentation. The second, and perhaps more influential one, was the pluricentered style of experimental analysis as exemplified in this paper by eighteenth-century plant chemistry. Experimental analysis in eighteenth-century plant chemistry was coupled with natural history and artisanal practices and, in some loose way, with metaphysical reasoning. Its objects were conditioned by these heterogeneous elements. The extension of knowledge about particulars within a collectively accepted conceptual framework and set of techniques was perhaps its main goal, but others, such as the improvement of specific techniques of extraction and commercial interests played a role as well. Because of its peculiar amalgamation of natural historical, artisanal, experimental and philosophical traditions eighteenth-century plant chemistry embodied the Baconian dream of science.

In the decades before and after 1830, elements of this pluricentered style of experimentation were partly excluded from what was to become organic or carbon chemistry, and partly re-organized. In the newly emerging experimental culture of organic chemistry an experimentalization of all kinds of scientific objects and goals took place. Reference was now constituted mainly in the space of the laboratory.[24] Synthetic laboratory materials became the bulk of organic substances; classification was grounded in experimentation; a general theory embodied by chemical formulas was implemented as a paper tool for constructing interpretative models. Pragmatism became the overall orientation, which for a few decades even supplanted chemists' former utilitarianism. New types of questions moved in the center of chemists' interests: What can be produced in the laboratory? What kind of substitutions in organic substances are possible under which conditions? What are the general rules for such substitutions? In the decades after 1830, thousands of new organic substances, which contained alien elements, such as chlorine, bromine or a metal, were produced without any theoretical or utilitarian goal.

As a rule, modern experimental cultures produce and individuate their own objects of inquiry. These scientific objects may coalesce into technological things, become part of the material inventory of the laboratory and powerful agencies in nature and society. Furthermore, experimental cultures lead to an experimentalization of many kinds of activities, which previously had loose connections only with experiments, and they develop their own intrinsic dynamics of production and innovation. Thus, a more complete philosophical understanding of modern experimentation requires detailed analyses of at least the following three items: the technoscientific productivity of experimental cultures, their self-reference and their internal dynamics.

My distinction between the pluricentered style of experimental analysis and experimental cultures emphasizes differences between the material and epistemic productivity of experimentation. Such differences, however, are differences of degree rather than absolute dichotomies. Furthermore, they are not instantiated by each individual member of a scientific culture but are to be understood as differences between averages or patterns of individual experimentation, entrenched by different forms of its social organization. Hence, it is not excluded that experimenters working within a pluricentered culture of experimental analysis may sporadically produce a scientific object which exists exclusively in the laboratory. The so-called ethers in eighteenth-century plant chemistry are an example of such exceptional experimental production that did not alter the overall type of objects explored in plant chemistry (see Klein 2002).

Inversely, from the second half of the nineteenth century onward new interfaces between local experimental systems, technology and industry were established, along with a transfer of material objects from science to industry and vice versa. In the twentieth century, accelerating after World War II, new technoscientific units emerged, often funded or directly organized by the military. Yet this does not imply that commercial and military interests are able to direct experimentation at will. The experimental production and individuation of scientific objects depends strongly on material "imperatives" of the experimental system.[25] The way how intentionality is matched with the materiality of the system has to be studied empirically, exactly because there is no guarantee of such match. The success of experimental production is a result of the intricate interaction of both types of agencies, which does not follow universal rules.

Although experimental cultures are expert systems that have a dynamics of their own, they are not independent of the broader society, policy and culture. Experimental objects, produced and individuated in a space of experimentation, can not only be defined and conceptualized in a variety of different ways, depending on the intellectual history of a given experimental culture, they can also be subjected to many different social and political goals and interests. Furthermore, they can leave the exclusive space of the research laboratory both as material agencies and as cultural emblems. There are many forms of interaction between scientific expert systems and society. Again, such interactions have to be studied locally and empirically. They do not contradict the fact that experimental cultures also have many lives of their own.

Max Planck Institute for the History of Science
Berlin, Germany

NOTES

[1] Experiments may contribute in various ways to conceptual and theoretical development, apart from explicit experimental strategies for clarifying and solving problems defined by an extant theory. This problem will not be discussed in this paper. Elsewhere I have analyzed a case of conceptual development involved in experimentation which was achieved in an unintended way in the course of scientists' attempts to construct particular models by means of a newly introduced sign system (Klein 2001a). The fact that in the long run experiments may contribute in one way or the other to the development of scientific theories does not imply that experimenters perform experiments with such goal.
[2] See, for example, Galison 1987, 1997; Hacking 1983; Heidelberger and Steinle 1998; Latour 1987; Latour and Woolgar 1986; Pickering 1995; Rheinberger 1997.
[3] There are other criteria for this distinction, most notably sociological ones, and the form of interrelation between experimental art, technology and industry.
[4] On this issue, see also Holmes 1989; and Stroup 1990.
[5] Bacon spoke of a "chemical history of vegetables" that he included in his natural history. See Bacon 1879, 4: 254 f, 299.
[6] See Klein 1994a, 1994b, 1996.
[7] Bachelard used the term "reification" with respect to scientists' knowledge that is presupposed to construct a scientific instrument and hence is represented and "embodied" by the constructed instrument. In our case, the reification lies more in the manipulations than in the specificity of instruments, though, of course, the manipulations were bound to specific instruments.

[8] It should be noted that Lavoisier was a heir of this tradition. Contrary to common lore, Lavoisier did not invent (or "discover") the more operational concept of chemical element. What he did was reversing the extant classification of simple substances and compounds.

[9] For an historical account and analysis of the goals of experimental philosophy in a slightly earlier time see Shapin and Schaffer 1985.

[10] For a more comprehensive account of this development see Klein 2002.

[11] For further discussion of open-endedness of experimental cultures see also Rheinberger 1997.

[12] See Steinle 1997, 1998, 2001.

[13] Berzelian formulas first used superscripts in order to emphasize their analogy to algebra. On this issue see Klein 2002.

[14] See also Rocke 1984.

[15] This has been observed before by Christie (Christie 1994).

[16] For a more detailed discussion of this problem see Klein 2001b and Klein 2002.

[17] Of course, theoreticians may make use of experimental results in a quite different way.

[18] From a pragmatic view intentionality is only one element of scientific investigation among many others. A scientific strategy or "method" may be established in form of a collectively shared habitus, and new intentions may emerge in the process of scientific research spurred by its unintended and unforeseen results and consequences.

[19] This has been taken into account to some extent in Duhem's and Quine's thesis of the theory-ladenness of observation; as a consequence it has been broadly acknowledged that it is not possible to separate clearly those elements of experimental marks (or "data") that are given to human experience and intellectual construction.

[20] For a more comprehensive overview on this kind of studies, see the introduction in Klein 2001c. "Representation" in these studies is a descriptive term which means signification and reference. It does not include any simultaneous commitment to the correspondence theory of truth and scientific realism.

[21] Analogously, if "representation" means all kinds of practices in experimental cultures from physical interventions and the production of experimental marks up to the last item in a chain of inscriptions, often termed a "model" and sometimes a "theory" by today's experimenters, we are deprived of any linguist resources that would allow us to analyze such practices. Consequently applied, we would end with a pretty monolithic picture of experimentation in experimental cultures rather than with Derridean differences. In this respect, it is interesting that Goodman uses the term "presentation" for the kind of reference at stake here, rather than "representation."

[22] For reasons of space, this aspect is not discussed in this paper. For a discussion see Klein 2002.

[23] For similar applications of the term "culture" with respect to science, see Latour and Woolgar 1986, 55; Latour 1987, 201; Jardine, Secord, and Spary 1996; Pickering 1995; Rheinberger 1995. Contemporary studies in the history of science which use a "cultural history" approach often apply the concept of scientific culture in the sense of collectively shared patterns of interpretations, value complexes, ideologies, mentalities, etc. Instead, I do not put the accent on the mentalistic connotations of "culture," but rather on the social, material and symbolic resources of a scientific community.

[24] "Mainly" is a restriction, which takes the role played by the broader social and cultural context into account. Insofar there is no absolute difference between the pluricentered style of experimental analysis and experimental cultures (see below).

[25] For the notion of "experimental system" see Rheinberger 1997.

REFERENCES

Bacon, F. 1879. *Works.* J. Spedding, R. L. Ellis, and D. D. Heath (eds.), vol. 4. London: Longmans & Co.

Boerhaave, H. 1741. *New Method of Chemistry. Including the History, Theory, and Practice of the Art. Translated from the Original Latin of Dr. Boerhaave's Elementa Chemiae, as published by himself. To which are added notes, and an appendix, showing the necessity and utility of enlarging the bounds of chemistry* (P. Shaw, Trans.). London: T. Longman.

Christie, M. 1994. "Philosophers versus Chemists Concerning 'Laws of Nature' ". *Studies in History and Philosophy of Science* 25: 613–629.

Galison, P. 1987. *How Experiments End.* Chicago: University of Chicago Press.

Galison, P. 1997. *Image and Logic: A Material Culture of Microphysics.* Chicago: University of Chicago Press.

Geoffroy, E. F. 1718. "Table des differentes rapports observés en Chimie entre differentes substances". *Histoire de l'Académie Royale des Sciences: Avec des Mémoires de Mathématique & de Physique pour la même Année, Mémoires,* pp. 202–212.

Goodman, N. 1997. "Sense and certainty". In C. Z. Elgin (ed.), *Nominalism, Constructivism, and Relativism in the Work of Nelson Goodman.* Vol. 1. New York: Garland Publishers, pp. 120–127.

Hacking, I. 1983. *Representing and Intervening: Introductory Topics in the Philosophy of Natural Science.* Cambridge: Cambridge University Press.

Hacking, I. 1992a. "The Self-Vindication of the Laboratory Sciences". In A. Pickering (ed.), *Science as Practice and Culture.* Chicago: University of Chicago Press, pp. 29–64.

Hacking, I. 1992b. "'Style' for Historians and Philosophers". *Studies in History and Philosophy of Science,* 23: 1–20.

Hacking, I. 1992c. "Statistical language, statistical truth, and statistical reason: The self-authentification of a style of scientific reasoning". In E. McMullin (ed.), *The Social Dimensions of Science.* Vol. 3. Notre Dame, Indiana: University of Notre Dame Press, pp. 130–157.

Heidelberger, M., & Steinle, F. (eds.) 1998. *Experimental Essays – Versuche zum Experiment.* Baden-Baden: Nomos Verlagsgesellschaft.

Holmes, F. L. 1989. *Eighteenth-Century Chemistry as an Investigative Enterprise.* Berkeley: University of California Press.

Jardine, N., Secord, J. A., & Spary, E. C. (eds.) 1996. *Cultures of Natural History.* Cambridge: Cambridge University Press.

Klein, U. 1994a. *Verbindung und Affinität: Die Grundlegung der neuzeitlichen Chemie an der Wende vom 17. zum 18. Jahrhundert.* Basel: Birkhäuser.

Klein, U. 1994b. "Origin of the Concept of Chemical Compound". *Science in Context* 7: 163–204.

Klein, U. 1995. "E. F. Geoffroy's Table of Different 'Rapports' Observed between Different Chemical Substances – A Reinterpretation". *Ambix* 42: 79–100.

Klein, U. 1996. "The Chemical Workshop Tradition and the Experimental Practice – Discontinuities within Continuities". *Science in Context* 9: 251–287.

Klein, U. 2001a. "Paper Tools in Experimental Cultures". *Studies in History and Philosophy of Science* 32: 265–302.

Klein, U. 2001b. "Berzelian Formulas as Paper Tools in Early Nineteenth-Century Chemistry". *Foundations of Chemistry* 3: 7–32.

Klein, U. (ed.) 2001c. *Tools and Modes of Representation in the Laboratory Sciences* (Vol. 222). Dordrecht: Kluwer Academic Publishers.

Klein, U. 2002. *Experiments, Models, Paper Tools: Cultures of Organic Chemistry in the Nineteenth Century*. Stanford: Stanford University Press.

Latour, B. 1987. *Science in Action: How to Follow Scientists and Engineers through Society*. Cambridge, Mass.: Harvard University Press.

Latour, B. 1990. "Drawing Things Together". In M. Lynch & S. Woolgar (eds.), *Representation in Scientific Practice*. Cambridge, Mass.: MIT Press, pp. 19–68.

Latour, B. 1999. *Pandora's Hope: Essays on the Reality of Science Studies*. Cambridge, Mass.: Harvard University Press.

Latour, B. & Woolgar, S. 1986. *Laboratory Life: The Construction of Scientific Facts*. Princeton: Princeton University Press.

Pickering, A. 1995. *The Mangle of Practice: Time, Agency, and Science*. Chicago: University of Chicago Press.

Rheinberger, H.-J. 1995. "From Experimental Systems to Cultures of Experimentation". In G. Wolters & J. G. Lennox (eds.), *Concepts, Theories and Rationality in the Biological Sciences: The Second Pittsburgh-Konstanz Colloquium in the Philosophy of Science; University of Pittsburgh, October 1–4, 1993*. Pittsburgh: University of Pittsburgh Press, pp. 107–121.

Rheinberger, H.-J. 1997. *Toward a History of Epistemic Things: Synthesizing Proteins in the Test Tube*. Stanford: Standford University Press.

Rocke, A. J. 1984. *Chemical Atomism in the Nineteenth Century: From Dalton to Cannizzaro*. Columbus: Ohio State University Press.

Shapin, S. & Schaffer, S. 1985. *Leviathan and the Air-Pump: Hobbes, Boyle, and the Experimental Life*. Princeton: Princeton University Press.

Steinle, F. 1997. "Entering New Fields: Exploratory Uses of Experimentation". *Philosophy of Science* 64 (Supplement): S65–S74.

Steinle, F. 1998. "Exploratives vs. theoriebestimmtes Experimentieren: Ampères erste Arbeiten zum Elektromagnetismus". In M. Heidelberger & F. Steinle (eds.),

Experimental Essays – Versuche zum Experiment. Baden-Baden: Nomos-Verlag, pp. 272–297.

Steinle, F. 2001. " 'Das Nächste ans Nächste reihen': Goethe, Newton und das Experiment". *Preprint series*. Vol. 169. Berlin: Max-Planck-Institut für Wissenschaftsgeschichte.

Stroup, A. 1990. *A Company of Scientists: Botany, Patronage, and Community at the Seventeenth-Century Parisian Royal Academy of Science*. Berkeley: University of California Press.

ARISTIDES BALTAS

ON FRENCH CONCEPTS AND OBJECTS
COMMENTS ON URSULA KLEIN

The task of the commentator is at times not only ungrateful but also very difficult. Custom has it that the work I am supposed to do here should aim at bringing out the shortcomings of Klein's paper and zoom in on its failings, its inconsistencies, its flaws and its mistakes. It is presumed that this is how the discussion to follow my comments will gain its fuel. Heat will hopefully be generated, the kind of heat taken as akin to intellectual excitement. My audience will be satisfied, that is, if what I come up with gives everybody the opportunity to vent her passion of thought and for thought. Nevertheless, I am almost sorry to say, I am today unable to discharge this task and contribute in generating this kind of heat. The reason is simple: at least as far as I am able to control its content, I find Klein's paper excellent.

As I cannot very well end up at this point to leave the floor free for you, I am obliged to look for another option for proceeding with my assigned task. The most "natural", the one everybody here should have certainly experienced at one time or another, would be that of my royally ignoring Klein's paper itself and my just taking it as an opportune platform for airing my own deep thoughts on its subject matter; deep by definition, since they would be precisely mine. However this option too is closed for me and not only because, in following it, I would inevitably appear to all of you as ridiculously narcissistic; this option is closed simply because I, most certainly, am neither a historian of science nor in any way related to chemistry, and hence I have neither credentials nor justification to cogitate on the object of Klein's concerns.

All standard options thus appearing as closed off, my assuming the responsibility for forwarding the following remarks can only stand, if at all, on a very flimsy basis indeed. To start what I am about to say, allow me then to try laying out this basis as succinctly as I can.

For a number of years now, my work has mainly concentrated, not on chemistry, but on the workings of physics, in terms which, I should admit, are quite idiosyncratic. Specifically, I have been trying to tackle more or less together the conceptual, the structural, the methodological, the cognitive, the historical and the social dimensions of my discipline (I am a physicist by training) in order to come to understand the ways in which they all hang together, or at least appear as hanging together, thereby endowing physics with the status (or the semblance) of, precisely, a unified discipline[1]. I am, of course, well aware that such an effort looks today not only as hopelessly outmoded but also as fully misguided. What is nowadays emphasized is the *disunity* of science[2] with the help of – I have no difficulty in admitting – telling, if not fully compelling, arguments. However, I still believe that this is not necessarily the end of the matter. At this stage of the discussion, the pragmatic argument I can forward in support of such a belief is very simple. Most if not all of the actual practitioners of the discipline, the physicists themselves, do take the unity of their discipline for granted despite the arguments to the contrary brought forth by philosophers, historians and sociologists of science. Why do they persist in believing this? Is it only because physicists are uneducated philosophically, historically and sociologically? Even if the answer to this last question is yes, I am persuaded that this cannot be the whole answer. Hence I feel free to persevere with my undertaking.

Given this, the basis for what I am about to say can amount only to the fact that Klein's chemistry, with all its own dimensions, is a scientific discipline that does not lie very far away, after all, from my physics, with all its own dimensions. The kind of issues I have been trying to face in respect to physics may thus not be very different from the kind of issues the study of chemistry brings forth. That I have characterized Klein's paper as excellent intends, among other things on which I will say a few words below, to point precisely at its value for my own undertaking as well as, I hope, for the discussion of us all here today. I am sorry, but my narcissism cannot be harnessed more effectively than that.

However here, as in most other places, things are more complicated. What I mean is that what Klein's paper actually says appears to lie at the antipodes of my concerns as I have just tried to outline them. Klein emphasizes not unity but disunity and not just that of the discipline of chemistry in its entirety but specifically that of its particular experimental dimension. Klein talks about different styles of experimental analysis; she pinpoints the many, relatively disjoint, centers, conceptual or otherwise, such styles were based on in the late 17th and early 18th centuries; and she distinguishes experimental styles of this sort from the more encompassing experimental cultures that succeeded them, the latter being viewed as incorporating various traditions and practices and as providing more comprehensive spaces for treating the corresponding objects of inquiry. How can Klein's approach help then my endeavor to locate what makes up the unity of physics as a discipline?

The answer is that Klein proceeds to her study with all the required theoretical tact and with a remarkable theoretical flair. This is to say that she does not seize the experimental dimension in itself, embark on it and take off at a tangent. My main reason for characterizing her paper as excellent is that she discusses the various aspects of the experimental dimension of chemistry not only by marshalling judiciously in an impressive amount of historical evidence but also by fully respecting the theoretical dimension of chemistry itself. In addition, her paper makes manifest the fact that Klein is fully aware that *her own endeavor* should be theoretically informed as well as theoretically articulated.

Klein talks of both scientific concepts and scientific objects. Regarding concepts, she distinguishes four different layers at work in experimental analysis and the loose relations those bear to one another. These are the layers first, of plant species as linked to plant anatomy and botanical classification; second, that of the more specifically chemical concepts of composition and analysis (affinity, immediate constituent, reaction, substance, etc); third, the more metaphysical concept of "principle" or "element"; and fourth the concepts related to the applications of chemistry for medical and pharmaceutical purposes. Regarding objects, Klein distinguishes carefully the two different meanings the term can acquire in the context of chemistry and, I presume, in that of experimental science in general. On the one hand, "scientific object" is a general "metaphysical" term referring to the subject matter of a particular scientific discipline or subdiscipline as a whole, that is the object whose knowledge the discipline in question intends to come up with. In this sense, the layers of concepts just mentioned go together with their corresponding specific objects of inquiry. On the other hand, "scientific object" may refer to material objects not found as such in nature but which are produced in the laboratory through the work of the scientists themselves. In that sense, scientific objects, "reify" (Bachelard's term) the conceptual elements and relations that guided the production and/or isolation of such objects.

The conceptual framework Klein relies on to tackle together all these levels, aspects and characteristics of the concepts and the objects of chemistry, is heavily indebted to notions such as "mark" and "system of signs" while she distinguishes carefully between the "signifiers" laboratory practice deploys and the material "signified" that corresponds to them in one way or another[3]. This vocabulary is characteristic of a whole tradition.

To formulate her theoretical framework, Klein appeals to the work of French speaking thinkers such as Gaston Bachelard and Jacques Derrida and, at least by implication, Ferdinand de Saussure. In addition, the figures of Lévi-Strauss, Foucault, or Althusser, although not mentioned by name, are almost discernible in the shadows. My considering her paper as excellent has then also to do with my viewing it as offering further evidence[4] that the French thinkers in question, misnamed collectively "structuralists" and/or "post-structuralists", have important things to say regarding the various aspects of the scientific endeavor[5]. Going one step further, I would maintain that the paper under discussion offers further evidence that the analytic/continental divide in philosophy not only cannot be tenable anymore for strong internal reasons (which, obviously, I cannot elaborate here) but also that it impedes our efforts to come to grips with the multiple and extremely varied facets making up the scientific endeavour. As my own work on physics has been trying to assimilate precisely this lesson, I cannot help finding Klein's approach particularly congenial.

However, once again, things are not that simple. The reception of the French thought of the 1960's (particularly in the US but also – and there from – to most other parts of the world, including, paradoxically, today's France) has been delivering an image which, I strongly believe, has very little to do with the fundamental concerns of the authors in question. They have all been conceived as being more or less theorists of literature, centering their work on what "texts" are and on how they function. Their being misnamed "structuralists" and/or "post-structuralists", despite the vehement denial of such an appellation by almost all of them, is an indicator of precisely this. Now, if that is correct, our applying more effectively their thought on our work as philosophers, historians or sociologists of science, depends upon our trying to become clear on what these thinkers were actually about The remainder of my talk will be devoted precisely to this attempt at clarification[6].

Everybody acknowledges that French thought of the 1960's has been strongly influenced by the structural linguistics of Ferdinand de Saussure, whose *Cours de Linguistique Générale* was published posthumously by his students in 1916. What most are misinterpreting, however, are the reasons why the work of Saussure became adopted the way it was by the French authors in question. It is first on this that we should try to become clear[7].

Without going into the details necessary for an adequate treatment, something neither time nor space allows, I claim that these reasons had very little to do with language as such, and even less with literature or with the "text". Saussure became important in the 1960's because he was perceived as proceeding paradigmatically to a kind of move which Klein mentions in her paper without, however, highlighting its fundamental importance. I mean that Saussure was perceived as he who had first come to endow linguistics with *its proper scientific status* through his having carved out its *proper object of study* with the help of a particular, systematically knit together, *array of concepts* established for the purpose, among which that of "structure" did indeed play a preponderant role. To put it very roughly, the object in question is made up from the structure of *langue,* as distinguished from the workings of *parole* that covers all the various acts of actual or possible enunciation. *Langue* itself, split into the levels of phonology, morphology and syntax, can then be studied by the system of concepts which establish in the first place these distinctions themselves as well as their consequences. This *co-constitutive relation* between the object of linguistics and the system of concepts offering its knowledge was considered by the French authors of the 1960's as the paradigmatic move that all disciplines have either already gone through or should go through, if they were to be accorded the status of proper sciences.

This general idea had more or less been driven home by what the *direct teachers* of the French authors of the 1960's appeared as having established. Those teachers – a fact seldom mentioned – were *philosophers and historians of the natural and mathematical sciences*, who, in that guise, had very little to do either with literature or with the "text". Four such teachers can be singled out. First and most important, at least as regards Klein's paper, comes Gaston Bachelard, whose work on physics and chemistry analyses the autonomous character of scientific development, a development which, by its very nature, goes against the grain of allegedly inescapable empirical or philosophical requirements. The work of Bachelard ties indissolubly together both the philosophical and historical aspects of the study of science, coming up with his particular brand of historicist philosophy of science which, at least in France, bears the name "*épistémologie historique*". We should mention, second, Alexandre Koyré, whose work highlighted the importance of the Scientific Revolution, making him thereby become one of the founders of the discipline of internal history of science. The influence of Koyré on Thomas Kuhn's ideas has been frankly admitted by Kuhn himself[8]. Third, there is Georges Canguilhem, the successor of Bachelard at the University of Paris, who developed Bachelard's ideas in respect to the biological and medical sciences. And last, there is Jean Cavaillès, a philosopher of mathematics executed by the Nazis and close friend of Canguilhem's, whose work can be summed up by the motto *"Autrement dit, il y a une objectivité, fondée mathématiquement, du devenir mathématique"*[9]. Cavaillès's last work, *Sur la Logique et la Théorie de la Science*, written in prison, was published posthumously by Canguilhem with a preface by Bachelard. Although Cavaillès was not the direct teacher of the 1960-generation, the strong endorsement of his views by both Bachelard and Canguilhem, to say nothing of his heroic death, made his few surviving works very important to all the authors in question.

It is clear, I hope, in what way Saussure's work relates to the guiding ideas of those four philosophers *cum* historians of the physical and mathematical sciences. To put it extremely schematically, the autonomy of scientific development that Koyré and Bachelard were claiming, each in his own particular way, for the case of either or both physics and chemistry, Canguilhem for the case of the biological and medical sciences and Cavaillès for mathematics, could be based only on the co-constitutive relation between the object and the conceptual system of these disciplines[10]. Saussure's work can thus be perceived as repeating for linguistics the move that guaranteed the scientific status of these older and more venerable disciplines. Accordingly, once Saussure's work becomes perceived under such a light, the way becomes open for trying to repeat an analogous move for all the social and human disciplines.

Although to establish it would require extensive arguments, we can nevertheless forward the idea that, in a sense, it was only natural that the task the French authors of the 1960's had set themselves was more or less that of doing for the various social and human disciplines what their teachers had highlighted as securing the scientific status of the physical and mathematical ones. Saussure's work had already established that this is indeed possible and, in any case, having one's name attached to the establishment of the scientific status of a whole discipline would be no mean achievement.

Given this, Lévi-Strauss was the first to take the lead and present his structural anthropology as founding scientifically the discipline of social anthropology. More or less at the same time, the *Annales* historians claimed that their way of practicing history was scientific while Althusser presented Marx's *Das Kapital* as founding no less than the "scientific continent" of history, that is as forming the basis for the scientific study of everything social and everything historical. Lacan, on his part, claimed that, through his reading, Freudian psychoanalysis could assert fully the status of the science of subjectivity while we can see under a similar light what Foucault tried to do in respect to the history of ideas and institutions or what Derrida was after with his "science" of the letter or "grammatology". Roland Barthes came up with a theory of literature and of general semiology also claiming scientific status while theoreticians of the cinema tried to achieve something analogous in respect to their own object of study. We see then that the authors in question may have worked on different existing disciplines or on none in particular (in that respect, Derrida as well as Gilles Deleuze are cases in point), they may not have appeared as agreeing much with one another, they, or at least some of them, would not perhaps accept to place their work under such a banner, but there is enough textual and inter-textual evidence that can buttress what we have said: if Spinoza was intoxicated by the idea of God, the French authors of the 1960 generation were intoxicated by the idea of science[11].

We should perhaps add that one of the reasons that their work has not, on the main, been perceived in this way is the fact that within the Parisian intellectual atmosphere everybody is supposed to know everything and hence there is no need for long explications as to what each author is after. Any of them could take for granted that everybody understood that their approach to the human and social disciplines was analogous to what they considered as established for the case of the natural and mathematical sciences and thus their vehement denial of "structuralism" can be easily explained.

To be fair, however, we should recognize that there were indeed reasons leading somebody unfamiliar with the intricacies of Parisian intellectual life to consider the authors in question as concentrating their work on literature and on the "text". For one, the importance accorded in the 1960's to Saussure's linguistics as well as the common knowledge that Lévi-Strauss, the initiator of "structuralism", owed a lot to the poetics of his friend Roman Jacobson, already lay the ground for such a view. Moreover, if we take into account the fact that, at the same period, Althusser was urging everybody to *read* Marx, that Lacan was expressly presenting his own work as a "simple" *re-reading* of Freudian psychoanalysis, a re-reading based, moreover, on Saussurean *linguistics*, that Foucault was talking a lot about *texts*, of what they allow to be formulated and of what they exclude, that Derrida was distinguishing sharply *writing* from *speech*, reversing, or rather deconstructing, all the relevant hierarchies, that Barthes was promoting a new "scientific" way to approach literature and *all systems of signs*, that cinema theorists were talking about *reading films as texts*, it was only natural that those who tried to understand what was happening without knowing the tradition these authors were in fact pursuing would indeed have perceived all the major figures of French thought of the 1960's as literary theorists and not as, certainly idiosyncratic, philosophers of science.

Be that as it may, where does all this leave us in respect to scientific disciplines? In a nutshell, my own answer is that we should consider very seriously the fundamental idea shared by practically all the French authors of the 1960's, namely that it is indeed the co-constitutive relation between a scientific object and the system of concepts that provide its knowledge that endow a discipline with its scientific status. Nevertheless, we should be careful to stress that there is always a third leading actor interposed between these two[12]. This actor, which the French authors of the 1960's unfortunately barely mention, constitutes the set of *experimental transactions* that tie together object and conceptual system in ways assuring that the claims formulated in or by the conceptual system can and do become effectively tested in respect to the object while, conversely, whatever experimental research on the object comes up with will eventually find a well defined place within the conceptual system. If we failed considering this as an additional defining characteristic of a scientific discipline, we would be obviously unable to speak of *objective* knowledge.

Conceived in this way, experimental transactions may be as much varied, uneven, multiform and multi-centered as our historical and/or sociological research can describe; they can include as many experimental styles, experimental cultures, practices and traditions as our work is able to reliably discern and to distinguish; and they may very well preserve their relative autonomy in respect to theory without necessarily offering either logical or historical precedence to it. As Klein stresses, the function of experimental work is not merely that of testing an already formulated theoretical claim; experiments might have a plethora of other functions, even to the point of coming up with radical scientific discoveries on terms practically all their own. Accordingly, if while talking about scientific disciplines we do indeed talk about three fundamental actors and not just two, three fundamental actors bearing *inter-constitutive relations with one another*, then the picture of science that emerges is wide enough to include all kinds of "disunities" in science without smashing scientific disciplines to the disparate bits and pieces that postmodern "thought" wants to force upon us.

Admittedly, this picture of science has not yet had a fair hearing within philosophy of science. Logical empiricism had hidden its possibility behind logic and method; the historicist approaches have hidden the same behind overarching historical schemes; and the social constructivist program has hidden everything behind the interplay of social forces, undeniable *per se*.

It is the case, however, that this picture forms a framework wide enough to cover all kinds of sober science studies, without straightjacketing ruthlessly science to some narrow positivistic, historicist or social constructivist pseudo-image and without making science appear as a hopeless muddle of disjoint undertakings. In this sense, philosophy of science, once it has learned the lessons of what the serious study of science at many levels and with various aims has brought forth, can find again the means to regroup so as to play the indispensable role that it has recently appeared as having more or less given up.

If I were allowed a concluding thought, I would say that this picture of science brings with it a whole array of questions that, to my knowledge at least, have not been given the attention they deserve. For example, one such question might be: do the three inter-constitutive elements of a scientific discipline that we have just tried to identify hang together in the same way *in respect to all disciplines* or does each discipline carry its own particular configuration? Concomitantly, if the latter is the case, what are the conditions *assuring the identity* of each scientific discipline and of its ways of developing? What exactly does distinguish any scientific discipline from others? Even if I have been trying to work on such issues for a number of years[13], I admit that I do not yet possess an answer. But I do believe that Klein's paper helps us a lot in going precisely in this direction. This is my final reason for considering it excellent.

National Technical University
Athens, Greece

NOTES

[1] See, for example, my "Physics as a Mode of Production", *Science in Context* 6 (2), 1993, 569–616.
[2] See, for example, P. Galison and D. J. Stump (eds.), *The Disunity of Science, Boundaries, Contexts, and Power*, Stanford University Press, 1996.
[3] Klein emphasizes that her work should not be considered as vindicating social constructivism. On the contrary, she stresses that marks and signifiers do not add up to mere systems of inscription but imply the existence of material referents. For the way she sees things, our ordinary notion of material reality does not evaporate.
[4] I have particularly in mind here Rheinberger's work, especially his 1997, wherein he employs Derrida's ideas for understanding important aspects of the history of biology. Klein herself takes this work seriously into account.

[5] I should add that Klein refers also to the work of Ian Hacking and Nelson Goodman, who are not at all hostile to the French authors in question.

[6] Much of what follows borrows heavily from my "HPS and STS: the Links", presented at a workshop organized by the Institute for Advanced Studies on Science, Technology and Society, in Graz, Austria, in June 2001, and which will be published in the yearbook of the Institute.

[7] For interesting appraisals of French thought of the 1960's, see Vincent Descombes, *Le même et l'autre*, Les Éditions de Minuit, 1979, translated by L. Scott-Fox and J. M. Harding as *Modern French Philosophy*, Cambridge University Press, 1980 as well as François Dosse, *History of Structuralism*, translated by Deborah Glassman, 2 Volumes, University of Minnesota Press, 1997.

[8] See, for example Kuhn's interview published in James Conant and John Haugeland (eds.) *The Road since Structure, Thomas S. Kuhn, Philosophical Essays, 1970-1993, with an Autobiographical Interview*, The University of Chicago Press, 2000.

[9] "In other words, there is an objectivity, mathematically founded of the development of mathematics". From his *Philosophie Mathématique*, Hermann, 1962, p. 28.

[10] For the case of mathematics, this co-constitutive relation needs serious spelling out. I cannot undertake this here but see, for example, Pierre Raymond, *L'histoire et les sciences*, Maspéro, 1977.

[11] For a general appraisal following more or less this line of interpretation see Quentin Skinner (ed.) *The Return of Grand Theories in the Human Sciences*, Cambridge University Press, 1985.

[12] See my "The Structure of Physics as a Science", in Diderik Batens and Jean Paul van Bendegem (eds.) *Theory and Experiment, Recent Insights and New Perspectives on their Relation*, D. Reidel Publishing Co, 1988, pp. 220–226.

[13] See, for example, my still in progress "Physics as Self-Historiography *in Actu*: Identity Conditions for the Discipline".

REFERENCES

Baltas, A. 1988. "The Structure of Physics as a Science". In D. Batens and J. P. van Bendegem (eds.), *Theory and Experiment, Recent Insights and New Perspectives on their Relation*. Dordrecht: D. Reidel Publishing Co., pp. 220–226.

Baltas, A. 1993. "Physics as a Mode of Production". *Science in Context* 6 (2): 569–616.

Baltas, A. Forthcoming. "HPS and STS: the Links".

Baltas, A. Forthcoming. "Physics as Self-Historiography *in Actu*: Identity Conditions for the Discipline".

Cavaillès, J. 1962. *Philosophie Mathématique*. Paris: Hermann.

Conant, J. and Haugeland, J. (eds.) 2000. *The Road since Structure, Thomas S. Kuhn, Philosophical Essays, 1970-1993, with an Autobiographical Interview*. Chicago: The University of Chicago Press.

Descombes, V. 1979. *Le même et l'autre*. Paris: Les Éditions de Minuit. Translated by Scott-Fox, L. and Harding, J. M. 1980. *Modern French Philosophy*. Cambridge: Cambridge University Press.

Dosse, F. 1997. *History of Structuralism*. Translated by D. Glassman, 2 Volumes. Minneapolis and London: University of Minnesota Press.

Galison, P. and Stump, D. J. (eds.) 1996. *The Disunity of Science, Boundaries, Contexts, and Power*. Stanford: Stanford University Press.

Raymond, P. 1977. *L'histoire et les Sciences*. Paris: Maspéro.

Rheinberger, H.-J. 1997. *Toward a History of Epistemic Things: Synthesizing Proteins in the Test Tube*. Stanford: Stanford University Press.

Skinner, Q. (ed.) 1985. *The Return of Grand Theories in the Human Sciences*. Cambridge: Cambridge University Press.

DONALD GILLIES

SOME COMMENTS ON "STYLES OF EXPERIMENTATION" BY URSULA KLEIN

This paper makes a distinction between "experimental analysis", which is illustrated by the example of the chemical analysis of plant and animal tissues in the eighteenth and early nineteenth centuries, and "experimental culture", which is illustrated by the example of organic or carbon chemistry from the 1820's. I found this distinction a useful one, and think that Klein has made a valuable contribution in characterising these two styles of experimentation. I will divide my comments on the paper into three sections. In (1) I will make a couple of points about the notion of experimental analysis. In (2) I will comment on Klein's account of the "conceptual system of chemical compound and reaction", which was developed in an earlier paper and is used in the present paper. In (3) I will consider a point of Klein's treatment with which I agree, and will try to add an argument in support of her position.

(1) Klein comments on p. 160: "'Experimental cultures' are the dominant style of experimenting in present sciences ..." Yet she adds on p. 168 that "... experimental analysis ... still exists in some present scientific disciplines and sub-disciplines", mentioning ecology in this context. Perhaps some parts of genetics also constitute an example of present-day experimental analysis. I mean systematic attempts to elucidate the genetic structure of a whole range of plants and animals. This does seem quite like the experimental analysis in Klein's own example of the attempts of eighteenth and early nineteenth century chemists to find out the chemical constituents of plants and animals. In footnote 5, Klein relates experimental analysis to Bacon's "'chemical history of vegetables' that he included in his natural history." Indeed experimental analysis seems to me rather Baconian in character. It involves systematic data gathering by a whole community, where there does also seem to be the hope of discovering what Bacon called a "secret of excellent use" in the form of a substance of great practical value.

(2) In both Klein's historical examples, use is made of what she calls (p. 162) "the conceptual system of chemical compound and reaction." She begins by describing (p. 163) "the conceptual network of chemical compound and reaction, built in the decades before and after 1700 ... in the domain of chemistry which investigated purified natural or artificially synthesized salts, acids, alkalis and metals." She says that (p. 164): "During the eighteenth century, the conceptual system of compound and reaction, or aspects of it, became a constitutive element of any chemical experimentation, including the experimental analysis of plant tissues. Yet it was never tested or explicitly justified by chemists. Rather, it functioned as an indispensable internal prerequisite of experimentation ..." She further says that (p. 172): " ... the conceptual system of chemical compound and reaction was a long-term cognitive framework, built in the decades before and after 1700 and changed only in the twentieth century with the emergence and acceptance of new physical theories and quantum chemistry." These ideas were developed by Klein in her earlier 1994 paper, where she sees the emergence of the concepts of chemical compound and reaction as part of a transformation of Paracelsian chemistry, which could be referred to as (1994, p. 201) "a scientific and chemical revolution." This suggestion of an earlier chemical revolution around 1700 seems to me a very valuable one, but I think it may lead Klein to exaggerate a little the continuity in the concepts of compound and reaction between 1700 and the twentieth century. I would argue that the more familiar chemical revolution at the end of the eighteenth century involved debates which led to some change in the concepts of compound and reaction. For example Lavoisier and his followers would have spoken of the combination of hydrogen and oxygen to produce water, while the phlogistonites would have spoke of the phlogistication of dephologisticated air to produce water. This issue was certainly debated vigorously, and the concepts of compound and reaction used by the two parties to the debate do not seem to be identical.

As regards the second example: experimental culture of carbon chemistry that developed from the late 1820s onward, Klein argues (p. 174) that the chemists involved "used this Berzelian formula as an unquestioned prerequisite for constructing an interpretative model of the reaction under experimental investigation ..." Historically she is quite right here, but I would regard this as somewhat fortuitous. I can see no a priori reason why some surprising experimental finding could not have brought about some questioning of the concept of Berzelian chemical formula.

(3) On pp. 176 – 178, Klein has an interesting discussion of the distinction between (i) entirely culturally constructed sign representations of an object

e.g. e which stands for electron, and (ii) experimental marks e.g. a trail in a cloud chamber produced by an electron. This discussion is in response to the following remark of Rheinberger's which is quoted on p. 176: "It is unnecessary to distinguish between machines that 'transform matter between one state and another' and apparatuses or 'inscription devices' that 'transform pieces of matter into written documents'." Klein thinks that this (*ibid.*) "... is justified to the extent that representation of a scientific object by means of culturally available sign systems has a referential structure that is analogous to that between experimental marks and the scientific objects they refer to." However she insists nevertheless that (p. 177): "... in our historical-epistemological reconstructions of scientists' representational activities it is necessary to distinguish between 'experimental marks' that also comprise elements that are given to humans and sign systems that are intellectually and culturally constructed in their entirety. If this differentiation is given up in the history and philosophy of science, the notions of scientific empirical 'experience' and of 'experiments' no longer have any distinctive meaning at all, but become identical to writing, modeling and theory formation tied to scientists' work with sign systems." Klein is quite right in my view to insist on the distinction which Rheinberger seems to deny. Although there is indeed some analogy between e and a trail in a cloud chamber, there are profound differences. To produce a trail in a cloud chamber requires carrying out an elaborate procedure which is not always successful, and which requires apparatus which is not easy to obtain. By contrast I can write "let e stand for an arbitrary electron" at any time. Assimilating these two different ways in which an electron can be "represented" seems to me quite misleading.

Department of Philosophy, King's College London
London, United Kingdom

REFERENCES

Klein, U. 1994. "Origin of the Concept of Chemical Compound". *Science in Context* 7: 163–204.

GÜROL IRZIK

IMPROVING "STYLES OF EXPERIMENTATION"
A COMMENT ON URSULA KLEIN

Until recently, philosophers and historians of science were guilty of "theory chauvinism". Experimentation in its own terms, context and complexity was rarely discussed, and when it was, it was discussed only in relation to theory and even then it was unjustly subsumed under "observation". In the last two decades the situation has changed radically. A new field of experiment studies with a whole new discourse flourished, and just like experimentation itself, it has acquired a life of its own – indeed, not just one but "many lives of its own". Now, we commonly talk of "a new philosophy of experiment", "a sociology of experiment", "an anthropology of experiment", and so on.

Ursula Klein's paper explores the sense in which experimentation has many lives. If the aim of experiment is not the testing and justification of theories, she asks, what is it and what are the preconditions that make experimentation possible? To answer these questions, Klein presents two historical case studies from chemistry in the eighteenth and early nineteenth centuries. She shows with admirable historical detail and clarity that there are at least two styles of experimentation. One of them is what she calls "the pluricentered style of experimental analysis", which was practiced widely in plant chemistry in the eighteenth century and gradually faded away in the first half of the nineteenth century. Since experimental analysis has been largely ignored by philosophers and historians of science, Klein devotes the first part of her paper to a meticulous discussion of this topic. The other one is "the experimental cultures" which emerged from the end of the first quarter of the nineteenth century and constitutes the familiar style of experimentation in most sciences today. The second part of her paper discusses this development, which overshadowed experimental analysis.

The style of experimental analysis can be described, in terms of its objects, goals, concepts and, finally, assumptions and postulates. Although Klein does not mention it explicitly, we may add its methods to this list as well. The objects of experimental analysis are plant species. Its goals are primarily knowledge of the particular plant species, especially, knowledge of their composition from organic substance components, the purity of the extracted materials, the classification of these species and the discovery of the fundamental ontological principles that constitute them. Its methods are distillation and extraction by solvents. Its central concepts are "analysis", "synthesis", "component", "composition", "compound", "affinity", "elements" and "principles", which form a network. Finally, experimental analysis is guided by a number of both metaphysical and methodological postulates and assumptions: (1) the whole consists of, and is determined by, its parts; (2) the substance components of chemical compounds are simpler substances (later called "principles" or "elements") (3) of which there are only a handful, and (4) they can be obtained by the methods of experimental analysis; (5) all chemical transformations are the result of various ways in which the components are combined by chemical affinity, etc.

Described in this way, the pluricentered style of experimental analysis comes out as an extremely rich and surprisingly structured activity despite the fact that its success was limited. It provided a qualitative but rather comprehensive conceptual framework within which the community of chemists in Europe carried out their scientific activity. Indeed, it seems that the style of experimental analysis was much more than mere analysis (note that it contained its own network of concepts and assumptions), so it would not be an exaggeration to say that it functioned something like a (qualitative) paradigm. Klein hints at this when she says that there is a "profound conceptual rupture" between experimental analysis and the kind of earlier activities displayed by, for instance, Paracelsian alchemy. This emerges more clearly when she points out that the conceptual system of experimental analysis "is never tested or explicitly justified by chemists. Rather, it functioned as an indispensable internal prerequisite of experimentation" (this volume, p. 164).

It is worth noting that Klein's use of the term "conceptual system" covers *both* concepts *and* statements, though she does not always distinguish between them carefully. Similarly, the latter variously includes assumptions and postulates (both of which are Klein's own terms), but we are never explicitly told what distinguishes them from each other. Of the five statements above that characterize the style of experimental analysis, the first one is referred to as a postulate. Does this mean that the rest are assumptions? How do they differ from postulates? Were any of them thought to be testable despite the fact that they were not subjected to any tests as Klein claims?

I raise these questions because they are directly relevant to the central theses of Klein's paper, namely that experiments rarely aim at theory testing, justification, clarification and problem solving left open by a theory, and more specifically that experimental analysis has had totally different goals than these. Despite the centrality of these theses, the notion of theory curiously remains unexplained by Klein. What is a theory? Is it a set of empirical postulates only? Does it include assumptions as well? Is the conceptual system part of the theory? More specifically, what was the theory behind the pluricentered style of experimental analysis? Was there just one or several? Did the five assumptions and postulates above comprise the theory? What exactly was the relationship between them and the concepts listed above? To make a stronger case for Klein's theses, these questions need to be answered explicitly.

I do not deny that most philosophers of science had a rather narrow conception of experimentation and thought that the major and perhaps the only function of experiments was the testing of theories. It is well known, for example, that Popperians turned testability by observations and experiments into the hallmark of science. But it is equally true that well before the new studies of experimentation, some of the early post-positivist philosophers of science, especially Thomas Kuhn, attributed a much larger role to experimentation than Klein acknowledges. Indeed, this is the very point of the notion of paradigm articulation that takes place during normal science. Kuhn told us emphatically that during normal science the purpose of experiments was never paradigm testing, but rather paradigm articulation, which involved, among other things, fact-gathering, using observations and experiments that were "particularly revealing of the nature of things" (Kuhn 1970, p. 25): the computation of physical constants, measurement of specific gravities of materials, boiling and melting points of elements and so on. It seems to me that normal scientific activity accommodates the style of experimental analysis that Klein has described so masterfully.

Similarly, in *The Scientific Image*, van Fraassen, as an eloquent defender of the constructive empiricist philosophy of science, argued that the relationship between experiment and theory is more complex than mere testing and justifying: experiments are also functional in the construction and completion of theories understood as models, and conversely, theories are used in the design of experiments (van Fraassen 1980, pp. 73–77). Thus, when Klein says in the second part of her paper that the goal in using Berzelian formulas in the experimental culture that was emerging during the first half of the nineteenth century was not the testing of the theory of chemical portions, but rather the construction of models that fit the empirical phenomena such as chemical reactions, she is simply repeating a point made by van Fraassen before the new philosophy of experiment arose (unless of course Klein has in mind a sharp distinction between models and theories, in which case she again has to tell us what she means by models and how they differ from theories).

So I am wondering whether Klein's characterization of the "old" conception of experimentation before the "new" philosophy emerged makes it appear narrower than it was. In fairness, however, I must add that neither Kuhn nor van Fraassen (nor anybody else for that matter before the eighties) developed a conception of experiment detached from theory. The new philosophical and historical studies on experiment have enriched and widened our understanding precisely by developing such a conception.

I have said very little about the second part of Klein's paper, where she talks about the emergence and nature of the experimental culture in chemistry, but that is because I agree with much of it. I especially applaud her healthy realism, which uncompromisingly maintains the distinction between mind-independent "experimental marks" and human-made sign systems. We should never lose sight of the fact that however much construction goes into the designing of apparatuses and experiments, once the experiment is allowed to run, it is nature, not we human beings, that decides the fate of the pointers, traces, or the computer print out. There is therefore a strong sense in which marks themselves, their properties and the facts about them, are not human constructions. Otherwise, there is ultimately no point in carrying out an experiment, no matter how open-ended it may be.

Let me finish with two minor quibbles. On p. 170 (this volume) Klein writes that "the new objects of inquiry *must* first be stabilized, identified and demarcated *before any theory about them can be thought of*" (my emphasis). It seems to me that this is an overstatement. It is certainly *possible* that a new object (and its various properties) is first theoretically predicted and then experimentally discovered. Neutrinos are a case in point. The second quibble I have concerns Klein's point that experimental marks need to be interpreted by the experimenter (this volume, p. 170) and that "it is not possible to separate clearly those elements of experimental marks (or 'data') that are given to human experience and intellectual construction" (this volume, fn. 19). But this has nothing to do with the "Duhem's and Quine's thesis of the theory ladennes of observation" as Klein thinks (*ibid.*). What is known as the Duhem–Quine thesis in the philosophical literature is that no hypothesis or theory T can be tested in isolation from other hypotheses, called auxiliary assumptions A; what is tested is always the conjunction T and A. The Duhem–Quine thesis is therefore about the holistic nature of testing, not the theory-ladennes of observation. The latter was championed originally by Norwood Hanson and later by Thomas Kuhn and many others. Now, to say that experimental marks or data need to be interpreted is certainly true and not seriously disputed in contemporary philosophy. The other characterization in terms of nonseparability, on the other hand, has been disputed. Since eventually Klein herself rejects it, it is best to drop footnote 19 altogether.

Philosophy Department, Bogazici University
Istanbul, Turkey

REFERENCES

Kuhn, T. 1970. *The Structure of Scientific Revolutions*. 2nd ed. Chicago: Chicago University Press.

van Fraassen, B. 1980. *The Scientific Image*. Oxford: Clarendon Press.

DAVID ATKINSON

EXPERIMENTS AND THOUGHT EXPERIMENTS IN NATURAL SCIENCE

1. Falling bodies

Let us begin with Aristotle and Galileo Galilei. In some places, Aristotle claims merely that heavier bodies fall more quickly than lighter ones:

> The mistake common to all those who postulate a single element only is that they allow for only one natural motion shared by everything. ... But in fact there are many things which move faster downward the more there is of them. (Aristotle, *De Caelo,* Book III/v/304b)

We shall call this the weak Aristotelian dogma: it is the qualitative statement that heavier bodies fall faster than lighter ones. Moreover, in other places Aristotle maintains that the natural motion of a body is proportional to its weight. Here we have to understand what he meant by *natural motion,* or perhaps which property of a falling body comes closest to the ancient notion of natural motion, before we can reasonably consider the question of inconsistency. Be that as it may, here is another passage that is less susceptible of interpretational uncertainty:

> If a certain weight move a certain distance in a certain time, a greater weight will move a same distance in a shorter time, and the proportion which the weights bear to one another, the times too will have to one another, e.g. if the half weight cover the distance in x, the whole weight will cover it in x/2. (Aristotle, *De Caelo*, Book I/vi/274a)

We shall call this the strong Aristotelian dogma: it is the quantitative statement that times of fall, from a given point to a lower point, of bodies of differing weights, that are alike in other ways, are inversely proportional to their weights.

The great contributions made by Galileo to physics have little to do with his claims, via his spokesman Salviati, that the Aristotelian dogma, whether in its weak or its strong version, implies a logical inconsistency. It is sufficient to point to a physical situation in which Aristotle's dogma, even in its strong form, is empirically correct. Since an inconsistent argument points at nothing at all, but Aristotle's argument does in fact indicate a realizable configuration, it follows that the dogma cannot be internally inconsistent. The case which gives Galileo the lie is that of bodies falling in a fluid (such as air or water) at their terminal velocities in the case of laminar fluid flow. Consider this quotation:

> We see that bodies which have a greater impulse either of weight or lightness, if they are alike in other respects, move faster over an equal space, and in the ratio which their magnitudes bear to one another. ... In moving through *plena* it must be so; for the greater divides them faster by its force. For a moving thing cleaves the medium either by its shape, or by the impulse which the body that is carried along or is projected possesses. (Aristotle, *Physica,* Book IV/viii/216a)

Under the restriction of laminar flow, and for bodies of identical size and shape, as specified by the Stagyrite, the viscous forces are proportional to the velocities, and so the terminal rates of fall are proportional, the times of fall inversely proportional, to the weights. It is not part of my thesis that Aristotle espoused, or could have espoused, this detailed interpretation, nor that Galileo excluded, or might have excluded, the particular case of terminal motion with laminar flow. I maintain simply that the possibility of a realizable model of the strong Aristotelian dogma frees it of any possible logical inconsistency.

Galileo's own resolution of the imagined inconsistency in the doctrine that different bodies fall at different rates (as implied by the weak dogma) was that all bodies must fall at the same rate (Galilei 1638). Moreover, he presents this as a truth about the world that is accessible to reason, rendering experiment unnecessary.

> *Salviati:* But, even without further experiment, it is possible to prove clearly, by means of a short and conclusive argument, that a heavier body does not move more rapidly than a lighter one, provided both bodies are of the same material, and in short are such as those mentioned by Aristotle. ... If then we take two bodies whose natural speeds are different, it is clear that on uniting the two, the more rapid one will be partly retarded by the slower, and the slower will be somewhat hastened by the swifter. Do you not agree with me in this opinion?
> *Simplicio:* You are unquestionably right.
> *Salviati:* But, if this is true, and if a large stone moves with a speed of, say, eight, while a smaller stone moves with a speed of four, then when they are united, the system will move with a speed less than eight; but the two stones when tied together make a stone larger than that which before moved with a speed of eight. Hence the heavier body moves with less speed than the lighter; an effect which is contrary to your supposition. Thus you see how,

from your assumption that the heavier body moves more rapidly than the light one, I infer that the heavier body moves more slowly. ... We infer therefore that large and small bodies move with the same speed, provided they are of the same specific gravity. (Galilei 1638, p.108)

Galileo was in imagination arguing with Aristotelians and the above quotation may be regarded as being largely a polemical device. As an argument, it is disappointing in both its destructive and constructive aims: not only must the criticism of Aristotle's dogma be defective, as we have just seen, but the new Galilean dogma concerning free fall is itself a *non sequitur*. Again we will not immediately demonstrate this, but we rather draw attention to the fact that a physical model exists in which different bodies fall at different rates, even in vacuo. In a nonuniform gravitational field, as in the terrestrial situation, the rate of fall is a function of the distance from the centre-of-mass of the earth: a body at a higher elevation falls less quickly than one at a lower elevation. Moreover, the rate of fall can depend, in special circumstances, on other parameters too, such those defining magnetic or electric fields, etc.

Thus the insufficiency of Galileo's reasoning has been shown indirectly by means of physical models in which both the destructive and the constructive aspects of his reasoning fail. More directly, the reason for the failure is Galileo's insertion into the argument, as if it were self-evident and not in need of empirical testing, the following supposition:

(S1) Natural speeds are mediative.

By this it is meant that if two bodies with different natural speeds are bound together, the natural speed of the composite lies between those of the constituents. Whether this is true or not depends of course on the meaning attached to the notion of natural speed. For Aristotle it had a significance bound up with the notion of the natural places of earth, fire and so on. If we reinterpret natural speed as acceleration, following Galileo's lead, then indeed it is true that two bodies, with different accelerations, if bound together, will thenceforth have an acceleration lying at an intermediate value: accelerations are mediative, as we shall prove by using Newton's laws. However, if, as indicated above, we interpret natural speeds as terminal velocities in a fluid with laminar flow, the situation is more complicated. In the first place, if we consider simply binding together two bodies that have different terminal velocities, the terminal velocities will generally be mediative. Other possibilities are open, according to which the terminal velocities, even of composite objects, could be proportional to the weights and thus be additive rather than mediative. This would be the case if the bodies were placed one at a time, or both together, in an impermeable container of negligible weight. The situation is even more complicated when the terminal velocity is reached in a condition of turbulent fluid flow, as is often the case in practice.

In short, Galileo's thesis [S1] is anything but self-evident, being in fact in some circumstances empirically true and in others empirically false. Any claim that his argument offers a glimpse into a Platonic realm of truth, rendering empirical testing unnecessary, is demonstrably false.

Such a claim is indeed made by J.R. Brown (1991). According to Brown, Galileo's thought experiment is a classic in the field, since, so it is claimed, an old belief has been destroyed by pure thought, and replaced by new knowledge concerning the world, without the need for a real experiment, that is, without extra empirical input. For Brown, the landscape of empirico-theoretical truths (and also that of purely mathematical truths) is something that is there to be observed by a sufficiently cultivated inner eye. Brown's world view is denied by J.D. Norton (1996), for whom thought experiments are disguised arguments. On the basis of a careful study of the *epistemology* of thought experiments (as opposed to, for instance, their impact on the scientific community), Norton concludes that thought experiments "can do no more than can ordinary thinking with its standard tools of assumption and argument". T.S. Gendler (1998), on the other hand, opposes both Brown and Norton. She argues that thought experiments are "guided contemplations", i.e. arguments with a particularly strong persuasive power, their "justificatory force". Apparently taking her inspiration from Mach, Gendler ascribes this justificatory force to the fact that, in a thought experiment, instinctive and hitherto unarticulated empirical knowledge suddenly becomes organised and manifest. For all her astute analysis, I do not agree with Gendler, if she means to suggest that Galileo did not need to make real experiments, but merely to articulate what before his time was inarticulate. Galileo performed, and needed to perform, real experiments. In the appendix we provide further technical analysis of the mechanics of falling bodies.

2. EPR thought experiment

In quantum mechanical theory, an *observable*, like the position or velocity of a particle, is represented by an operator, not just by a number. One may like to think of this as a sort of table of all the possible results of measurements of the observable. What the result of such a measurement will be in a given case depends on the state of the system to be measured. For example, suppose that the system is an excited calcium atom, which decays into its ground state, emitting two light quanta, or photons, in the process. Quantum mechanics is silent about the direction of propagation of the photons, and thus about the result of a measurement of the position, at a given time, of one of them. However, if a definite result of a position measurement of one of the photons has been obtained, that of the other can be inferred from theory – successive measurement results are correlated, that is the prediction. The classical, or common sense interpretation of this fact would be that the positions of both photons, as functions of time, exist, and are indeed correlated, but that one has insufficient information about the detailed nature of the decaying calcium atom to make unconditional predictions. Such a situation is common enough in classical theory. The Copenhagen dogma asserts, however, that the positions of the photons *do not exist* prior to a measurement of one of them, and that after a measurement of one photon's position, and its consequent entification, the state of the other photon is immediately changed, its position being entified by the act of measurement performed on the other.

A similar story may be told about the momenta of the photons: no unconditional prediction of the outcome of a momentum measurement is possible, but if the momentum of one photon is measured, that of the other is thereby also known. But there is more. A measurement of the momentum of one photon gives knowledge about the momentum of both photons, but it gives no information about the position, either of the photon that is subjected to the measurement, nor about that of the other. The Copenhagen interpretation claims that a particle, whose momentum has just been measured, *does not have a position,* that is, it is not simply that we do not know what the position is, but there is no position to know. Moreover, this applies equally for the second photon, the one that has not been subjected to measurement. The idea that a measurement of one property of a localized system might well change the value of another property of the same system is easily understandable, but that this other property should, as a result of the invasive nature of the measurement, fail to exist, is counterintuitive, to say the least. That the ontological status of the second photon, with respect to the existence of its position or its momentum, should depend on, and be changed by, what is done to the first photon, which is out of contact with, and indeed may be greatly separated from, the first photon, seemed to Einstein, Podolsky and Rosen (EPR 1935) to be an absurdity. In Einstein's view, such a holistic interdependence of different parts of reality would nullify the physicist's profession. There is a *sous-entendu* that, given the manifest empirical successes and technical applications of the physicist's trade to date, such a nullification would be intolerable. At any rate Einstein, to the end of his days, would not tolerate it.

For Bohr and Heisenberg, the new quantum ontology seemed to be enforced by the disjunction of

1 The mathematics of Heisenberg's matrix mechanics, and
2 The evangelical conviction that the new quantum mechanics is, in its domain of application, complete.

The latter conviction was strengthened by von Neumann's flawed proof (Neumann 1932) that hidden variables, which might 'complete' quantum mechanics and restore classical ontology, are logically impossible. The reason that simultaneous existence was not accorded to incompatible observables like position and momentum is that the Copenhagen dogma proclaims an observable, represented by an operator, to be well-defined in a given state, and by an insidious epistemic slide, *to be*, only when the vector that represents this state is an eigenvector of the operator. Since noncommuting operators like those representing position and momentum do not possess common eigenvectors, it follows, given the epistemic slide, that a photon, or any other system, cannot possess both a position and a momentum.

Einstein and Bohr agreed that the quantum ontology, as sketched above, was inescapable if one assumes quantum mechanics to be complete. For Einstein, this was tantamount to a proof that quantum mechanics cannot be complete, and this is the burden of the EPR paper. Bohr accepted, nay created and exulted in the new ontology; and he welcomed any proof, defective or not, that hidden variables could not be tacked on to the fledgling discipline. The stand-off between Einstein and Bohr was complete, and it continued to their deaths, seemingly unaffected by unfolding events.

The first significant event was theoretical, when David Bohm created a new version of quantum mechanics that was empirically indistinguishable from the Copenhagen canon, but which implemented the unimplementable (Bohm 1952). Hidden variables, in the form of his quantum potential, were inserted into the interpretation of the wave-function of quantum mechanics. It was left to John Bell to put his finger on the weak point of von Neumann's no-go theorem, namely the assumption that the measurement outcome of an observable that is represented as the sum of two operators is necessarily the sum of the separate outcomes of measurements of the observables corresponding to those operators (Bell 1987).

It was also David Bohm who initiated the second important step in the development by retooling the EPR effect in terms of the components of spin of the two particles, instead of their positions and momenta. This reformulation paved the way for John Bell, who derived an inequality that must be satisfied by correlation coefficients between spin measurements performed on the two particles in the EPR configuration, on condition that local, noncontextual hidden variables exist, and that stochastic independence in the sense of Kolmogorov is equivalent to physical independence. Eighteen years were to elapse between the derivation of the inequality and the demonstration of its experimental violation by Alain Aspect et al. at Orsay (Aspect 1982).

Some exegeses of the EPR paper speak of a "paradox", others of a thought experiment or theorem, but its true significance for the development of physics lies in its development from the stand-off of a thought experiment and of two competing world views (Einstein versus Bohr) via theoretical insights (Bohm and Bell) to a genuine experiment (Aspect). Had this genuine experiment not been performed, the EPR thought experiment would have remained a fruitless stand-off.

I once asked Bell whether he had thought, after he had derived his inequality, that it would be violated by nature. "Well", he said, "I expected it would be, but I hoped it would not be!" He added with a smile, "then I would have been famous." In the period that experiments were being planned to test the inequality, and, by extension, quantum mechanics itself, the emotions of the community of physics were divided between the aspirations of the realists and the confidence of those who practiced quantum methods on a day-to-day basis. The impact of Aspect's work was not merely negative, in destroying the hidden-variable assumptions of Bell, but also constructive, for it verified in detail the numerical predictions of quantum mechanics.

For our purposes, the lesson to be learned from the ascent

1 EPR thought experiment
2 Bohr-Einstein stand-off
3 Bohm reformulation of EPR
4 Bell inequality
5 Aspect experiment

is that a thought experiment which supports two contradictory intuitions can fruitfully point the way forward, by stimulating theoretical development and suggesting a real experiment, to a resolution of the dilemma. As a thought experiment alone, the value of EPR was limited to the challenge it made to Bohr to sharpen his distinction between quantum object and classical measuring instrument, but he was never able to convince Einstein that his interpretation of quantum mechanics was viable.

3. String theory

The modern candidate for a Theory of Everything is string theory, according to which the known fundamental particles, photons, electrons, quarks and so on, with their associated fields, are nothing more nor less than different frequencies of vibration of the postulated string. In other words, the building blocks of our universe are notes on a cosmic string, rather than autonomous elementary particles. The theory aims at a definitive unification of all known forces, including gravitation. According to string theory, the gravitational attractive force between two particles should increase more rapidly, as the particles approach one another, than do the other forces, until gravity is as strong as all these other forces. However, to test string theory experimentally, one would have to penetrate to impracticably tiny distances, or equivalently to accelerate particles to impossibly high energies before allowing them to collide with one another. In this sense, string theory looks more like a recipe for a thought experiment than a genuinely physical theory that can be put to the test of experiment.

To give a rough idea how impractical it would be to perform a real experiment to test string theory, let us consider briefly the history of particle accelerators at CERN, in Geneva. The first machine, a proton synchrotron (PS), came on-line in 1959 and accelerated protons to an energy of 28 GeV (the unit, the giga-electron-volt, corresponds roughly to the energy that would be produced if one proton were to be annihilated, turning all its rest-mass into energy). In 1980 the SPS (S for "super") produced protons and antiprotons at 170 GeV per particle, and in 2005 the Large Hadron Collider (LHC), a machine with a diameter of 27 km, is expected to produce protons of energy 7000 GeV. An American project to build an even bigger machine in Texas was killed by the former president Clinton; and there are no plans anywhere to build bigger accelerators than LHC. The economic limit seems to have been reached. In forty-five years the maximum attainable energy per particle has risen by a factor of 250. To test string theory adequately one would have to produce energies that are ten to the power sixteen (ten thousand million million) *times higher* than those that LHC will produce in 2005. It seems safe to say that we will never be able to produce energies anywhere near this value, and that string theory can never be confronted with the crucial test of experiment.

Is string theory truly a scientific theory? String theory could be tested *in principle*, and in this it differs from unscientific world systems. But it will never be testable *in practice*, and in this it differs radically from Newton's or Einstein's theories of gravitation. Suppose that string theory can be completed in a consistent manner; and suppose that it accommodates the Standard Theory of elementary particles, as well as Einstein's General Theory of Relativity, as a low-energy approximation. This future Theory of Everything would postulate certain new properties of gravity at very high energies, where these new properties *de facto* cannot be tested.

Edward Witten, the most prominent proponent of string theory, once said, on being asked about experimental support for string theory: "Things fall." The chain of implication may be reconstructed as follows: string theory contains Einstein's general relativistic theory of gravitation, Einstein's theory contains Newton's theory as an approximation, and Newton's theory describes quantitatively the falling of 'things', like planets, moons and apples. The weakness of the answer is apparent if one reverses the order, and considers the nested inductions: Newton's explanatory theory is subsumed in Einstein's curved space-time explication-explanation, and this is further seen as a property of multidimensional strings. As for Einstein's inductive leap, it predicts not only the precession of Mercury's quasi-elliptical orbit, but does so with good numerical accuracy. This and other successful predictions support the realist's conviction that Einstein's leap was at least in the right direction. But what now of the string theorists' specific claims, as for example that space-time has ten dimensions (of which we can perceive only four)? Perhaps all that could be said, in the most favourable case, is that string theory is, or may come to be, one of the possible unifying logical systems relating General Relativity and the Standard Model. In this sense one might then claim it to be a scientific theory.

Ed Witten once said that, while General Relativity and presumably the Standard Model are included in string theory, they can hardly be claimed as predictions of that theory. At best one might call them *postdictions* (Witten 1999). But even that claim would be too generous. For although string theory predicts the particles of the Standard Model, it predicts also the existence of a host of other particles, of which there is no sign.

On the other hand, he did say that supersymmetry is a genuine prediction of string theory, and suggested that, if supersymmetric partners of ordinary particles were to be found in future high-energy experiments, that would constitute a confirmation of string theory. This claim is however overly enthusiastic. In the first place, supersymmetry antedates string theory, and its experimental observation, while being good news for string theorists, who need supersymmetry to avoid causal paradoxes involving *tachyons*, would not specifically favour string theory above the earlier supersymmetric descriptive programs. In the second place, since string theory does not place any bounds on the masses of the postulated supersymmetric partners, a failure to detect any of them in experiments at LHC, for example, could always be shrugged off with the claim that the masses must then be higher, out of reach of the new machine. In short, string theory's prediction that supersymmetry exists is not falsifiable.

String theorists do not think of themselves as mathematicians, and certainly not as philosophers (it is still the case that, for many scientists, "philosophizing" is put on a par with daydreaming or sloppy reasoning). No, string theorists think of themselves as full-blooded physicists: they really want to say something about the furniture of the world, and they hope that their mathematics provides a glimpse of the world as it is. Could we honour such aspirations; and, if so, does this not imply that string theory differs somewhat from an explicatory program? The latter can explicate a theory, for example the Copenhagen Interpretation is an explicatory program that aims at expounding the notion of complementarity, and clarifying the distinction between classical observer and quantum object. On this view string theory is not concerned with explication, in the sense of making a certain intuition in an explanatory program *clear* (*cf.* Kuipers 2001). Rather, it aims at *unification*, that is the bringing together of apparently different explanations into one coherent logical or mathematical framework. If no new predictions can in practice be empirically tested, however, then we have to do here with a new sort of metaphysics: different empirically confirmed explanations could be underpinned by a mathematical theory whose essentially new ontological claims cannot be tested in the crucible of experiment.

4. Envoi

To summarize, we have considered three theoretical systems, from three different historical periods, with three very different kinds of status. Galileo's thought experiment, designed to destroy the Aristotelian dogma and instate his own in its place, has been shown to be logically deficient in both its negative and in its positive aspects. As a thought experiment, it leaves something to be desired, to say the least; that its major conclusion is in the case of free fall approximately correct, however, is not an accident: it is the result of real experiments performed by Galileo with steel balls and inclined planes. The EPR thought experiment threw Einstein's *a priori* convictions about the nature of physical reality into sharp relief, but they made no impression at all on Bohr's equally dogmatic espousal of the new natural order of things. However, unlike Galileo's thought experiment, which merely pretended to take the place of, and to render superfluous, real experimentation, the idea of Einstein, Podolsky and Rosen, transformed by Bohm and Bell, led to a real experiment, the result of which would have disturbed Einstein[1], but which has had ramifications which have not yet been exhausted. Finally, the string theory of everything purports to be scientific, but it seems not to be susceptible to any serious experimental test, although it arrogates to itself all the empirical successes of the partial theories that it subsumes.

Appendix

Gendler analyzes Galileo's thought experiment with especial acumen. A generalized version of Gendler's analysis, couched in modern language, is that the following three statements are not consistent with one another:

[S1] Accelerations of falling bodies are mediative.
[S2] Weight is additive.
[S3] Accelerations of falling bodies are proportional to their weights.

Here [S3] is an interpretation of what we called the strong form of the Aristotelian dogma. It implies, for instance, that if two falling bodies $B(1)$ and $B(2)$ have weights $W(1)$ and $W(2)$, with $W(1) < W(2)$, then $B(1)$ accelerates less than does $B(2)$: $a(1) < a(2)$. To show the inconsistency of [S1] – [S3], suppose that [S2] and [S3] are true. Then

$$a(1) / W(1) = a(2) / W(2) = a(12) / W(12)$$

and hence

$$a(12) = a(1) W(12) / W(1) = a(1) [W(1) + W(2)] / W(1) > a(1)$$

$$a(12) = a(2) W(12) / W(2) = a(2) [W(1) + W(2)] / W(2) > a(2)$$

which is inconsistent with [S1], that is,

$$[S2] \& [S3] \rightarrow \neg [S1]$$

By rejecting [S3], and by stating that the acceleration of all falling bodies of the same material is the same (thereby making [S1] trivially true), Galileo succeeds in avoiding a contradiction.

As Gendler explains at length, an Aristotelian who wishes to parry the destructive force of Galileo's argument, while retaining [S3], would have to deny [S1] or [S2], or both. He might for instance, against all reasonable common sense, postulate an essential difference in mechanical behaviour between bodies that are merely united (i.e. tied, or glued together) and bodies that are unified (i.e. truly one, whatever that means). He could for instance claim that weights *and* accelerations are additive for unified bodies, but that both are mediative for composite bodies whose pieces are merely tied together. Gendler's view is that even a dyed-in-the-wool Aristotelian, on being confronted with these exotic ways of saving the Master's theory, would recant and deny any reality to such a distinction between union and unification.

Thus Galileo is right in accepting [S1] and [S2] and rejecting [S3]. However, the fact is that [S1] and [S2] are true, that [S3] is false, but that nevertheless Galileo's conclusion is also false. The statement that objects of the same material always fall at the same rate, Galileo's escape from the Aristotelian contradiction, is not true. In other words, with the caveat that *natural speed* has been replaced by acceleration, the destructive part of Galileo's thesis is correct, but the constructive part is incorrect. Such, at least, is the verdict given by Newton's laws of motion and gravitation. To find out where exactly the flaw is situated, we will first scrutinize, from a Newtonian point of view, [S1] and [S2] successively.

The gravitational forces acting on bodies $B(1)$ and $B(2)$, i.e. their weights, are $W(1)$ and $W(2)$. Let their inertial masses be $m(1)$ and $m(2)$, respectively. From Newton's second law of motion, $F = ma$, the accelerations are thus given by $a(1) = W(1)/m(1)$ and $a(2) = W(2)/m(2)$. It is part of Newton's theory that:

[N1a] Forces, and hence in particular gravitational forces, are additive.[2]
[N1b] Inertial masses are additive.

According to [N1a] and [N1b], the force acting on the composite made by tying bodies $B(1)$ and $B(2)$ together is $W(1) + W(2)$, while the inertial mass of the composite is $m(1) + m(2)$. The acceleration of the composite is

$$a(12) = [W(1)+W(2)] / [m(1)+m(2)] = [m(1) a(1)+m(2) a(2)] / [m(1)+m(2)]$$

If we suppose that $a(1) < a(2)$, then the expression on the right is increased if $a(1)$ is replaced by $a(2)$, whereas it is decreased if $a(2)$ is replaced by $a(1)$. It follows therefore immediately that $a(1) < a(12) < a(2)$. In other words, Newton's laws [N1a] and [N1b] imply [S1], i.e. accelerations are indeed mediative.

As far as [S2] is concerned, the additivity of weights, one might at first think that it is equivalent to [N1b]. But this is not so, for in Newton's system weight is a function of a body's *gravitational mass*[3] and the local gravitational field. However, since weight is a force, [S2] is implied by [N1a]. Conclusion: Newton underwrites both [S1] and [S2]. Thus, on pain of falling into an Aristotelian contradiction, [S3] must be wrong.

This is all of course exactly in accordance with Galileo's reasoning. So where did Galileo go wrong? It is precisely in the constructive part of his thought experiment, where he replaces [S3] by the statement that all weights fall with the same acceleration. For this statement is neither the only possibility to escape from the Aristotelian contradiction, nor is it correct for bodies falling to the surface of the earth. The reason is that the earth's gravitational field is not uniform, since the earth is a spheroid and the weight of a body depends not only on its gravitational mass, but also on how high it is above the surface of the earth.

The acceleration of a falling body at two earth-radii from the centre of the earth is only one quarter what it is when the body is close to the earth's surface. As a matter of fact, in this case the acceleration is precisely proportional the weight, but the weight is a function of the distance between the centre of mass of the earth and that of the body. The rate at which the body's velocity increases is independent of its gravitational mass, but dependent on its position. Galileo's solution is only correct in a uniform gravitational field, and that the earth does *not* have such a uniform field is a brute empirical fact.

Of course, the differences in accelerations are very small for a difference in elevation of a few metres. Be that as it may, the size of the effect is not at issue here. In the presence of a homogeneous gravitational field (and in the absence of air), different bodies would indeed fall at the same rate. However, this is not an *a priori* statement about the way bodies fall; indeed, given that the earth's gravitational field is inhomogeneous, it is not even an accurate statement. Logically speaking, Galileo's claim that the failure of Aristotle's dogma implies that all bodies fall at the same rate is a *non sequitur;* and moreover empirically it is in fact not the case that all bodies fall at the same rate.

At this juncture, a modern apologist for Galileo might remark that the inhomogeneity of the gravitational field could be seen as a disturbing factor, on a par with air friction. It requires after all little effort to postulate a homogeneous gravitational field in order to reinstate Galileo's thought experiment in all its pristine splendour. Our answer to this apologist would be that one can only postulate such a field within the framework of a theory of gravitation. For one can only identify the inhomogeneity of a gravitational field as a disturbing factor if one knows enough about it. Galileo lacked such knowledge. He did not have a theory of gravitation, at least not one in which gravitational forces (i.e. weights) drop off as the inverse square of the distance from the centre of the earth. Such a theory was only invented a generation later by Newton, who was able to test it quantitatively with the help of his calculus; he was able to compare the motions of falling apples with that of the moon, as it 'falls' endlessly in its month-long orbit around the earth.

The same goes of course for any other putative disturbing factor: one needs a theory to be able coherently to postulate conditions under which it would be absent. It would be circular to require all unspecified disturbing factors to be absent, so that Galileo's law of falling bodies be correct. Since physical laws are tested by their empirical implications, it is not only that thought experiments are subordinate to the theories which they inspire (and by which they are inspired), but it is also the case that disturbing factors must likewise be considered, controlled and rendered manageable within a theoretical framework.

In short, Galileo's conclusion that all bodies fall from the same height at the same rate, if air friction is negligible, and there are no appreciable electrostatic or magnetic forces at work, is approximately correct; but this fact is not a consequence of pure logic, despite Salviati's ease in discomfiting the Aristotelian straw-man, Simplicio. What is more, a good case can be made that Aristotle was interested in falling bodies in situations where fluid viscous forces are important:

> ... in the ratio which their magnitudes bear to one another. ... In moving through *plena* it must be so. (Aristotle, *Physica,* Book IV/viii/216a, *vide supra*)

In this case Newton's law of motion reads

$$ma = mg - \kappa v$$

m being the mass (strictly speaking, the inertial mass on the left and the gravitational mass on the right), g the acceleration due to gravity (assumed constant), v the instantaneous velocity, and κ the frictional coefficient due to viscous drag. This equation has the following solution for the velocity:

$$v = [\,1 - \exp(-\kappa t/m)\,]\, mg/\kappa$$

and this yields, for the terminal velocity,

$$v(\text{term}) = W/\kappa$$

where W = mg is the weight. Evidently, if we interpret Aristotle's natural speed as terminal velocity in a medium *(in plena)*, then Aristotle is right that the speed of a body is proportional to its weight! But what has happened then to the Galilean *contradictio?* The three statements [S1] – [S3] are replaced by

[S1'] Terminal velocities of falling bodies are additive *(not mediative).*

[S2'] Weights are additive.

[S3'] Terminal velocities of falling bodies are proportional to their weights.

Due to the change [S1] → [S1'], there is now no inconsistency. Twin sisters suspended from one parachute fall twice as quickly as one sister. The Stagyrite is vindicated!

It is not my contention that Aristotle had the above interpretation in mind when enunciating what I have called the strong dogma, only that such an interpretation is possible, and it serves, among other things, to throw further doubt on the worth of Galileo's thought experiment. The conclusion, of course, applies only in very special circumstances, namely for two bodies of the same shape and size (and then only if the fluid motion is laminar rather than turbulent). Aristotle *did* write

> We see that bodies which have a greater impulse either of weight or lightness, *if they are alike in other respects,* move faster over an equal space, and in the ratio which their magnitudes bear to one another. (Aristotle, *Physica,* Book IV/viii/216a, *vide supra*, my italics)

But the situation can be more complicated. Consider, for example, skydiving. In this sport one jumps from an aeroplane but does not immediately open a parachute. Now the twins fall at the same terminal speed as does one sister alone! And how fares the child of one of the sisters, if he is pushed out of the plane too? He does not fall as quickly, of course. But if he is fastened with a rope to his mother? Will not the mother, the fast body, be partially braked by her son, the slow body? Do you not agree with me in this opinion?

Simplicio: You are unquestionably right.

How do we explain this? Clearly we have to replace v(term) = W/ κ by

$$V(1) = W(1) / \kappa(1) \quad V(2) = W(2) / \kappa(2) \quad V(12) = W(12) / \kappa(12)$$

where V, W and κ are the terminal velocities, the weights and the friction coefficients, respectively, for the bodies (1) (the mother), (2) (the son) and (12) (the system comprising the mother and son, fastened together). The weights are proportional to the volumes of the bodies, but the frictional coefficients are proportional to their effective cross-sections, which increase approximately as the power 2/3 of the weights. In this more complicated scenario, we have in place of the statements [S1] – [S3] the following

[S1"] Terminal velocities of falling bodies are *(again)* mediative.

[S2"] Weights are additive.

[S3"] Terminal velocities of falling bodies are functions of weights and of cross-sections.

The last form, vague as it is, can be extended to the case of bodies falling under turbulent conditions.

University of Groningen
Groningen, The Netherlands.

NOTES

[1] Bell added a characteristic footnote to Einstein's famous dictum, "Raffiniert ist der Herrgott, aber boshaft ist er nicht", to the effect that, had he been still alive in 1982, the year of Aspect's experiment, Einstein might well have had reason to suspect malice on the part of the Almighty.
[2] In general, forces obey the rules of *vector addition,* but in the present case of forces that are parallel to one another, vector reduces to scalar addition.
[3] The gravitational mass of a body may be defined as the coefficient of proportionality between the body's weight and the gravitational field in which it is situated. This is conceptually different from the inertial mass of the same body, which is the coefficient of proportionality between a force acting on the body (for example its weight) and its resulting acceleration. That the two kinds of mass are numerically equal has been experimentally tested to high accuracy. This equality was built into the very foundations of Einstein's general theory of relativity. Because of the equality, two bodies of different gravitational masses, if placed in the same gravitational field (with no retarding forces), will experience the same acceleration, precisely because the ratio of the two bodies' inertial masses is the same as the ratio of their gravitational masses. (Atkinson and Peijnenburg 2002)

REFERENCES

Aristotle, 1960. *De Caelo,* with an English translation by W.K.C. Guthrie, and *Physica,* with an English translation by P.H. Wicksteed and F.M. Cornford. Harvard: Heinemann.

Aspect, A., Dalibard, J., Roger, G. 1982. "Experimental Tests of Bell's Inequalities Using Time-Varying Analyzers". *Physical Review Letters* 49: 1804–1807.

Atkinson, D., and Peijnenburg, J. Forthcoming. "Galileo and Prior Philosophy in Natural Science".

Bell, J.S. 1987. *Speakable and Unspeakable in Quantum Mechanics.* Cambridge: Cambridge University Press.

Bohm, D. 1952. "A Suggested Interpretation of the Quantum Theory in Terms of 'Hidden' Variables", I and II. *Physical Review* 85: 155–193.

Brown, J.R. 1991. *The Laboratory of the Mind: Thought Experiments in the Natural Sciences.* London-New York: Routledge. Paperback edition: 1993.

Einstein, A., Podolsky, B., and Rosen, N. 1935. "Can Quantum-Mechanical Description of Physical Reality Be Considered Complete?". *Physical Review* 47: 777–780.

Galilei, G. 1638. *Discorsi e dimostrazioni matematiche, intorno a due nuove scienze,* translated into English by Henry Crew and Alfonso de Salvio under the title: *Dialogues Concerning Two New Sciences.* Dover Publications (1954).

Gendler, T.S. 1998. "Galileo and the Indispensability of Scientific Thought Experiment". *British Journal for the Philosophy of Science* 49: 397–424.

Kuipers, T.A.F. 2001. *Structures in Science.* Dordrecht: Synthese Library.

Neumann, J. von 1932. *Mathematische Grundlagen der Quantenmechanik.* Berlin: Springer. Translated into English under the title: *Mathematical Foundations of Quantum Mechanics.* Princeton University Press (1955).

Norton, J.D. 1996. "Are Thought Experiments Just What You Thought?". *Canadian Journal of Philosophy* 26: 333–366.

Witten, E. 1999. *Duality, Spacetime and Quantum Mechanics.* Seminar, http://doug-pc.itp.ucsb.edu/online/plecture/witten/

DANIEL ANDLER

THE ADVANTAGES OF THEFT OVER HONEST TOIL
COMMENTS ON DAVID ATKINSON

David Atkinson asks whether nonempirical constructions can lead to genuine knowledge in science, and answers in the negative. Thought experiments, in his view, are to be commended only insofar as they eventually lead to real experiments. The claim does not rely on a general study, conceptual or historical, of thought experiments as such: the range of the paper is at once narrower and broader. Atkinson views thought experiments as commonly understood as just one kind of episode in the development of physics in which real experimentation is bypassed, and he believes that such episodes are justified only inasmuch as they are transitory stages on the way to genuine empirical inquiry.

Atkinson wants to kill with one arrow what is usually regarded as two different birds: the notion that thought experiments proper can be persuasive in themselves, and the thesis that theories which cannot be brought to the tribunal of experience can nevertheless belong to science. He thus implicitly opposes *three* views which are commonly, albeit not universally, held: (i) some thought experiments are conclusive; (ii) some theories belong to science despite not being evidently and concretely amenable to empirical corroboration; (iii) the two issues are largely independent.

The standpoint from which Atkinson operates is a rather strict form of empiricism, one which relies on a fairly sharp distinction between the conceptual and empirical dimensions of inquiry. My outlook is rather different: the conceptual and empirical seem to me to be intertwined, both conceptually, as suggested by Quine's critique of logical positivism, and empirically, as revealed by the evidence provided by science itself in its daily and historical reality.

Rather than take the high road, I propose to focus first on the critique to which Atkinson subjects Galileo's thought experiment, and the lessons he draws from his analysis. The other two case studies I will not examine individually, due first to insufficient expertise, second to the dialectics of the situation. I have no quarrel with the conclusions which Atkinson draws from the history of EPR: I certainly don't doubt that such happy *dénouements* are productive for science and deeply satisfying from both a historical and an aesthetic perspective.

As for string theory, it seems to me to raise a separate issue. High-level theories are notoriously hard to confront with experience. Evolutionary theory is perhaps the most familiar example; Popper categorized it as a metaphysical research program for precisely that reason. Most philosophers of biology today would be reluctant to set it so sharply apart from more typical scientific theories, such as Newtonian mechanics or molecular genetics. But the important question which Atkinson raises, as I understand it, is whether mathematical physics, which does not deal with one-time only sequences of events and in that sense is, contrary to evolutionary theory, fully theoretical, is not committed to stricter standards of verifiability. It certainly seems that verification procedures which are in principle possible but call for impossible feats of engineering, though not unrelated to the situation of forever lost historical or paleontological archives, create a novel predicament: the impossibility goes deeper, one is inclined to think. However I do not see exactly how one could rule out theoretical developments which would bring the higher-level theory closer to a, surely quite indirect, confrontation with empirical data. I will not attempt to go further into the matter.

In the second part of this comment, I shall attempt to gain, from the consideration of the set of case studies which Atkinson picks, as it were, as his data base, a perspective on his methodological doctrine.

1. Galileo's "conclusive argument"

Atkinson faults Galileo for claiming that logic alone shows the Aristotelian dogma (in either strong or weak form) to be inconsistent (the "destructive aim" of Galileo's argument), and the Galilean doctrine about free fall *ipso facto* correct (the "constructive aim").

Atkinson's argument takes the following form:

(i) Galileo offers the following argument P:

Let B be the common ground accepted by both thinkers, including the hypotheses concerning the two bodies whose rate of fall is being compared, and A the weak Aristotelian dogma. Then there is an assertion Z such that from B and A jointly follow both Z and not-Z. Therefore, granted the consistency of B, A must be false. Hence not-A.

(ii) There are conditions under which both B and A+, the strong Aristotelian dogma, are true.

(iii) Hence it cannot be the case that A (which is logically weaker than A+) is "internally inconsistent", as Galileo claims: his "destructive aim" is defeated.

(iv) The "new Galilean dogma" G – all bodies fall at the same rate – follows from not-A, according to Galileo. But the inference is unsound: in fact, there are conditions C' under which A is indeed false, but so is G. Thus his "constructive aim" is equally defeated.

(v) The crucial part of P is (S1), the assumption that (given B), natural speeds are mediative. As a matter of fact, there are conditions D under which they are, and conditions D' under which they aren't. Hence, it cannot be the case that S1 is a logical (conceptual) consequence of B, as Galileo maintains.

The complete analysis of Atkinson's argument would require a lot of work, especially since it is divided into a straightforward, non-technical part in the body of the paper, and an "interpretative", technical part in the Appendix. I grant the physics, of course, and find Atkinson's discussion of Gendler's reconstruction of Galileo's argument against Aristotle highly illuminating in its own right.

I do on the other hand have a pair of objections.

There is first a logical point. Atkinson insists on the logical structure of the argument. But in *strictly* logical terms, the new Galilean dogma G *is* (by the double-negation rule) logically equivalent to not-A. So, if, as Atkinson argues, it is not the case that not-A is proved, Galileo cannot be faulted for making the *further* logical mistake of positing G on the basis of the rejection of A..

Readers familiar with Gendler 1998 may think that Atkinson has the following issue in mind: rephrasing A as "Rates of fall vary (nontrivially) with weight", not-A comes to "Rates of fall remain constant when weight varies" *but* Galileo's dogma *further* asserts that rates of fall are insensitive to everything else, such as position with respect to the center of the Earth. *Pace* Gendler, Norton and others, I think this is a red herring: the disagreement between Aristotle and Galileo regards rate of fall as a function of weight, *all else remaining constant*. No doubt there are circumstances which we can imagine under which *color* of the bodies matters, circumstances which neither Aristotle nor Galileo could have imagined. But this presumed fact is neither here nor there: they would have been happy to take this issue on had they come to suspect it had potential interest; as it is, they were concerned with dependency with respect to *weight* not *color*.

What Atkinson actually proves is something different. He shows that (a) there is an *interpretation* J of Galileo's argument, and circumstances C such that A, under J, is actually true; (b) there is an *interpretation* J', and circumstances C' such that although A+, under J', is indeed false, nevertheless not-A, under J', is also false. (As I understand the arguments, J involves terminal rates of fall in non-vacuous media, while J' involves accelerations *in vacuo*.) This is logically possible because A does not entail A+. But surely Galileo cannot be faulted for (implicitly) choosing one fixed interpretation in the short excerpt of the dialogue quoted; in fact, had he shifted interpretations half-way in the argument, he could have rightly been faulted for disingenuousness.

Still, one might think, the case against Galileo remains basically sound: Galileo has failed (i) to prove by logical means alone that terminal velocities in non-vacuous media are constant (hence that the weak dogma, thus construed, is false) *and* (ii) to prove by logical means alone that accelerations *in vacuo* are constant (hence in establishing his own "dogma").

However, as "grammar" suggests, one cannot fail unless one has tried. My second and main objection to Atkinson's overall argument is this: Galileo does *not* attempt to provide a *logical proof* of either the falsity of the Aristotelian dogma(s) or the truth of his own theory. The reason is simply that no concept of logical truth in a sense commensurate to our own, and to the one Atkinson relies on, is available to Galileo and his contemporaries. What Galileo is after is an *argument*, which he hopes will be conclusive. Nor is this a mere verbal issue: as his extended discussion in the *Discorsi* shows, he is quite aware of the fact that his conclusions result not from the sole application of logical rules, but implicate a variety of considerations, ranging from interpretation of the terms involved, to acceptability of idealizations, common sense, convergence of arguments, rejection of putative objections, and so forth[1]. There is even an unarticulated consideration which plays a role in the central point's persuasiveness, for Galileo as well as many of his readers, *viz.* the very straightforwardness of the sub-argument that natural speeds are mediative. Finally, Atkinson's hard work is clearly of a nonlogical nature: even if, contrary to fact, Galileo were committing a logical fallacy, Atkinson would not be meeting him on logical grounds, but on the customary grounds where physicists appraise arguments from *physics*.

Once this is accepted, it remains to get clear on the following two related issues. First, is Atkinson right in faulting Galileo's argument form? Second, what, if anything, is wrong about the central sub-argument?

The first issue raises a host of problems. I will briefly focus on just two, which are related and both concern the vexing phenomenon of defeasability. Arguments, however tight and apparently conclusive, regularly turn out to be defeasible (though thankfully not always defeated): airtightness is not of this world, it appears, whether in science or in lay reasoning. Atkinson, with laudable fairness to Galileo, does not blame him for not having thought of some of the particular ways in which his arguments could go wrong; but he does blame him for not realizing that this could well happen. What he faults in Galileo is what he sees as his conceptual dogmatism. But is Galileo really dogmatic? Does he really think of his arguments as indefeasible? I have my doubts, nor do I think that we are offered evidence to that effect. Further, does Galileo base his alleged dogmatism on a misplaced trust in conceptual reason, as opposed from appraisal of empirical evidence? Again, I do not presume to decide; but I do have a suspicion that Galileo might have less confidence than Atkinson in the virtues of empirical results, and (as I stated earlier), if this is the case I side with Galileo against Atkinson.

The second problem regards rigor: shouldn't we simply admit that Galileo, with all due reverence, is insufficiently rigorous? After all, he must be, since his argument contains loopholes. But *all* arguments do, and standards of rigor, no matter how stringent, cannot provide complete protection. They constrain only the conditions under which subjective evaluations of the soundness of the ideally completed argument are communicated among scientists. The tighter the standards, the more confident one can be that the *actual* (perforce enthymematic) argument under scrutiny is *essentially* sound. However, rigor has a cost which needs to be justified by the subjectively evaluated risk that a trap may have been laid out by nature. So the standards of rigor change over time, tending towards greater strictness as we find out, by falling into them over and over again, that traps abound. That this tightening of standards occurs even in the history of mathematics tends to show that the conceptual/empirical distinction is irrelevant.

About the sub-argument, Atkinson is correct in pointing out that Galileo's way of settling the question of what would be the rate of fall of the composite object, were the rates of fall of its components different, can strike the modern reader as flippant. However, suppose for a moment that, like Galileo, you believe that in fact natural speeds are invariant in our world (neglecting air resistance). A world in which they are not is therefore, as you see it, imaginary. Reasoning on imaginary worlds is, of course, what thought experiments are all about, and it is a notoriously dangerous exercise, because it involves proceeding with counterfactual premises as if they were true in the real world, or, more precisely, it means operating in a possible world closest to our own in which the premises are true. Galileo had no choice but reason as he did; he used his common sense, i.e. his wordly knowledge (his "naive physics"), to speculate about an otherwordly event. But what else could he go by? Common sense is the guide to follow until reasons to doubt its conclusions come to light: defeasibility again. As Atkinson implies, other worlds need to be kept under the control of a pre-existing theory. But Galileo has such a theory, albeit in a state of less than perfect scientific crystallization. Maybe we have reason to put in question, with hindsight, the choice of this particular counterfactual situation (for example, because we suspect that it is under-described). But exactly the same fate awaits *any* experiment, real as well as imaginary.

2. Conceptual truths, defeasibility, and corroboration

Seen as a whole, Atkinson's paper appears to rest on three broad assumptions: (i) There is a sharp separation between conclusions based on experience and conclusions based on conceptual analysis; (ii) The essential weakness of the latter is that they are vulnerable to new facts; (iii) Conceptual analysis too easily evades the tribunal of experience. I concede that Atkinson does not take an explicit stand on those general issues, and might well want to deny that his case logically depends on the assumptions as I just stated them. The observations that follow are based on my possibly overly schematic reconstruction of his background doctrine. I will take up the three posits in turn.

(i) Matters of fact *vs* matters of meaning. What Atkinson means by "(pure) logic", I take it, is not really what goes under that label in contemporary philosophy, but rather conceptual analysis. He argues that the resources of conceptual analysis, buttressed, when a non-Platonic realm is being investigated (free fall *vs* numbers for example), by common sense, are too weak to bring about results enjoying the robustness of those procured by empirical means.

The difficulty here is familiar: the conceptual/empirical distinction is unclear. There are two well-trodden ways to see this. The (broadly) conceptual way is to follow Quine's rejection of the analytic/synthetic distinction: whether or not we should take a step back and allow for more of a principled distinction where Quine sees none at all, it remains that we have for now no stable notion of a conceptual truth. The (broadly) empirical way is to consider the actual practice of scientific inquiry, where conceptual and empirical considerations are inextricably intertwined. This is not to deny that there is *some* distinction, which can be usefully drawn, on the fly, in a rough and ready way. In fact, I have intentionally worded my objection so that it involves the very distinction whose status I question. What I challenge is the idea that purely empirical facts, plus logic (in the strict sense), in the absence of any conceptual ingredient, yield robust non trivial results. If one considers the process which includes not only the actual performance of a real experiment, but the work which goes into setting it up and drawing the moral which the scientist draws, one immediately realizes that some thought-experimental procedures are brought in at nearly every step. Although they do not amount to thought experiments standing on their own, they do involve the consideration of counterfactual mini-scenarios, the outcome of which is usually regarded as too obvious to call for a separate checking procedure. Whether or not this shows that the blurring of the analytic/synthetic distinction spills over and messes up the borderline between thought and real experiments seems to me a serious possibility, which I will not explore further. For the purpose at hand, the consequence seems to be that thought experiments, and, more broadly, concept-intensive (or largely armchair-based) inquiries in science are not distinctively more fragile than the more typical empirical inquiries for logical reasons connected with their high conceptual content.

(ii) Defeasibility. Are thought experiments, and other concept-intensive inquiries, vulnerable to defeat by unexpected factors in a sense or in a way in which real experiments are not? Atkinson's argument certainly points in that direction: what's wrong with Galileo's thought experiment, in Atkinson's eyes, is that its conclusions, both destructive and constructive, are defeated in circumstances which Galileo could not imagine. By implication, real experiments are not exposed to the same danger.

As I've stated above, I believe this is wrong: conclusions reached by way of real experiments are just as defeasible. Let's see why. Whenever we draw inferences based on some fact about the real world, we draw on default assumptions; in other words, we operate not in the real world W, but in the world W' which we can imagine to be like the real world in every relevant respect. We perform on W' a thought experiment, using our beliefs about the real world. This is not different from the case of an actual thought experiment, except for the fact that the imaginary world W" there is explicitly posited from the beginning and presumed to be distinct from W: it obeys some condition A which is known, or presumed, to be false in W. This in turn explains why we resort to thought experiments at all: W does not satisfy the preconditions of the experiment.

So when we do a thought experiment, we operate on a world which we know (or believe) to be different from our own, yet not to such a degree that we cannot rely on our real-world knowledge to navigate: Superman operates pretty much exactly like Batman, except for the flying part. When we do a real experiment, we navigate in the real world in much the same way as the absent-minded and short-sighted Mr. Magoo makes his way among crevices which are temporarily filled by backs of hippos or heads of giraffes; the difference is of degree, not of nature: what makes real experiments work (and the inferences we draw from them sound), when all goes well, are the high probabilities of the default assumptions they unknowingly rely on.

This is not to say, of course, that in respects other than in-principle defeasibility, there are no serious conceptual differences between (typical) real experiments and (typical) thought experiments. It matters a lot that the world W" of a thought experiment is known (or presumed) to differ from the real world W in some respect relevant to the question under investigation. First, as we have just reminded ourselves, it accounts for the fact that we are motivated, and often have no choice but, to perform the thought experiment rather than a real one. Second, the value of the thought experiment resides in the contrast it allows us to discern and make explicit between our world and some neighboring unreal worlds. One family of cases involves imaginary worlds which are *crucially* different from our own: they satisfy a condition A which our purpose is precisely to *prove* that it is false in W: these sorts of thought experiments are the quasi-empirical analogue of formal proofs by *reductio*. Another family involves worlds which are *inessentially* different from our own: this is the case of idealization. The ethereal beauty of Galileo's thought experiment is that it combines both cases, while remaining utterly simple: four worlds are involved, the real one, the one which is like ours only frictionless, the one which is like ours only Aristotelian, and the one in which the experiment is conducted, which is like ours except for being both Aristotelian and frictionless.

Finally, Atkinson's intuition that real experiments are, as a general rule, more reliable than thought experiments, is partly vindicated by the fact that, as science progresses, the evidential network in which a real experiment is performed becomes tighter, and leaves less room for default assumptions being falsified in the real world. By contrast, in a thought experiment involving an imaginary world, the network is by definition pulled apart to make room for the assumption A which is false in the real world, issuing in a less than perfectly controllable loosening of the connections surrounding A.

(iii) *Corroboration*. Galileo's thought experiment (as opposed to the real experiments he performed with inclined planes and so forth), and string theory (as opposed to conventional theories in fundamental physics), cannot be confronted with experience, and this, Atkinson tells us, is a fault. But what is the confrontation supposed to achieve? Atkinson writes as if the scientists' goal were to identify observable consequences of a theory under assessment such that, if the consequences are actually observed, the theory is thereby vindicated. What's wrong with thought experiments, he thinks, is that there are no facts at all to be had in an imaginary world; while string theory, according to him, does not proprietarily entail facts which it would in the real world be feasible to check.

But as we all know no amount of corroborated consequences can establish a theory: data underdetermine theories. In fact, the more ambitious the theory, the more it goes beyond its observable consequences. So the difference between acceptable and unacceptable candidates to scientific theoryhood cannot reside in corroborability, in this all-or-nothing construal. It is a matter of domain-specific wisdom. Atkinson is not alone among physicists, it seems, to wonder whether the added intelligibility which string theory may provide balances the increase in indirectness which affects its contact with observable facts. But this, surely, is not a matter of logic nor even methodology in a broad sense.

While we can grant Atkinson that thought experiments are less than fully conclusive; that some theories seem so distant from empirical corroboration as to bring doubts regarding their place in science; and lastly that there is a connection between the two issues, *viz.* they both engage more armchair than lab or field work, we must resist, I suggest, his overall picture, with speculation on one side, together with fragile and temporary results, and possibly a hint of cheapness, and on the other side experiment and confrontation with the real world, hand in hand with robustness, stability, scientific honor and earnestness. Atkinson's distinctions, or so I have argued, are not aligned in the way which would warrant this picture. Science is more of a blend in which the conceptual and empirical dimensions are intertwined, and owe their identity more to their interrelations and to the historical and local context than to any durable, intrinsic properties.

Université de Paris-Sorbonne (Paris IV)
Paris, France

NOTES

[1] I am grateful to Marta Spranzi for providing me with expert explication of Galileo's text.

REFERENCES

Sorensen, R. A. 1992. *Thought Experiments.* New York: Oxford University Press.

MIKLÓS RÉDEI

THINKING ABOUT THOUGHT EXPERIMENTS IN PHYSICS
COMMENT ON
"EXPERIMENTS AND THOUGHT EXPERIMENTS IN NATURAL SCIENCE"

The main claim of Atkinson's paper is that "... thought experiments in physics are of value only when they are related to or inspire real scientific experiments". The paper illustrates its thesis by analyzing three examples: Galileo's thought experiment intended to show that, contrary to Aristotle's view, bodies of different weight do not fall with different speed, the EPR experiment designed to prove incompleteness of quantum mechanics and, as a third example, string theory.

The claim's tacit assumption is that there is a well defined and sharp difference between thought experiments and "real" scientific experiments. I call this assumption tacit because no attempt is made in the paper to spell out systematically the conceptual difference between thought as opposed to "real" experiments. Lack of an epistemic specification of this difference takes much of the bite out of the claim, and as a result the claim remains somewhat weak and vague. Rather than trying to make Atkinson's claim sharper, in what follows I comment on the three examples Atkinson analyzes.

In Atkinson's evaluation Galileo's thought experiment does not disprove either the weak (qualitative) or the strong (quantitative) Aristotelian dogma concerning the rate of fall of heavy bodies. Why? Because in Atkinson's interpretation Galileo's thought experiment is an attempt to give an a priori argument against the Aristotelian dogma, which Atkinson takes to be a statement that can only be verified (or falsified) empirically. On such an interpretation of Aristotle's dogma and of Galileo's thought experiment Galileo's attempt is futile indeed, and there can hardly be any disagreement about this. But the details of Atkinson's analysis of the Aristotelian dogma and of Galileo's thought experiment raise a few questions. Atkinson argues that the Aristotelian dogma cannot be proved to be logically inconsistent (as Galileo thought to have succeeded in showing) because under certain physically realizable conditions (bodies falling in a viscous medium such as water) the "dogma" is in fact empirically true. While it is indeed true that bodies of different shape fall at different speed in a medium, this is true so obviously that one wonders whether referring to this situation has any significance. It is difficult to think that Galileo would have denied that bodies of different shape would fall at different speed in a viscous medium. Is it not the case that the disagreement between Aristotle and Galileo was about the rate of falling of bodies in the absence of any medium, i.e. in vacuum?

Atkinson takes the position that even if we answer this question in the positive, Galileo's conclusion that bodies with different weight fall at the same speed is no less a dogma than Aristotle's because "... a physical model exists in which different bodies fall at different rates, even in vacuo. In a nonuniform gravitational field, as in the terrestrial situation, the rate of fall is a function of the distance from the centre-of-mass of the earth: a body at a higher elevation falls less quickly than one at lower elevation." (p. 211) Again, true of course; however, it is clear from Galileo's thought experiment that Galileo was thinking of the rate of fall of different bodies at the same location: a body at the foot of the Himalayan mountains cannot be "united" even in thought with one on the peak, and Galileo's thought experiment assumes that the two bodies can be united. If so, what significance does the fact have for Galileo's thought experiment that the gravitational field of the earth is nonuniform?

In short: it seems that what Aristotle and Galileo were arguing about was the rate of speed of different bodies at a given location in the gravitational field in vacuum. Any reference to circumstances and situations that place the argument in a different context seems to create unnecessary and misleading complications in the evaluation of the debate between Aristotle and Galileo.

The above is not to be taken as an attempt to defend Galileo's thought experiment against Atkinson's charge that the thought experiment is ineffectual if interpreted as an a priori argument. I am in complete agreement with Atkinson on his pointing out the major tacit assumption in Galileo's argument: that natural speeds are mediative (p. 211). This assumption is indeed not self-evident and is in need of empirical testing, as Atkinson emphasizes. And this assumption is in fact used by Galileo in his argument. Following Gendler, Atkinson shows that if natural speed is taken to be acceleration, then natural speed can indeed be shown mediative within the framework of classical Newtonian mechanics. Since Newtonian mechanics is not a priori true, I find myself in agreement also with Atkinson's main evaluation of Galileo's thought experiment: "Galileo's conclusion that all bodies fall from the same height at the same rate, if air friction is negligible, and there is no aprreciable electrostatic or magnetic forces at work, is approximately correct; but this fact is not a consequence of pure logic, despite Salvati's ease in discomforting the Aristotelian straw-man, Simplicio." (p. 222)

The EPR thought experiment was proposed by Einstein, Podolsky and Rosen in a relatively short but extremely influential Physical Review paper published in 1935. Atkinson's evaluation of this thought experiment is that "... its true significance for the development of physics lies in its development from the standoff of a Gedankenexperiment and of two competing Weltanschauungen (Einstein versus Bohr) via theoretical insights (Bohm and Bell) to a genuine experiment (Aspect). Had this genuine experiment not been performed, the EPR thought experiment would have remained a fruitless stand-off." (p. 215).

There is no denying the significance of the fact that EPR-type experiments have actually been performed in the early eighties by Aspect and his co-workers. Yet, it seems to me that emphasizing only this aspect of the Aspect experiment does not exhaust the significance of the EPR thought experiment. If it were true without qualification that if Aspect's genuine experiment "had not been performed, the EPR thought experiment would have remained a fruitless standoff", then one would expect that after the measurement had actually been done, the standoff was resolved fruitfully. But this is not the case. The debate about whether quantum mechanics is a complete theory did not stop after the Aspect experiment had been done. Quite on the contrary, the Aspect experiments marked the beginning of a new wave of debates. Also, if the EPR thought experiment's significance did lie exclusively in the fact that about 50 years later a similar experiment was performed, then it is difficult to explain L. Rosenfeld's reaction to the EPR paper: "It was an onslaught [that] came down upon us as a bolt from the blue" (Jammer 1995, p. 129).

That the debate about completeness of quantum mechanics raged on after Aspect's experiment had been performed is partly due to the fact that the separation of "thought" experiments and "real" or "genuine" experiments does not seem to be as neat as one would like to think. There is far too much "thought" involved in a "real" experiment to say without any further qualification that the significance of a thought experiment lies exclusively in what it is realizable from it. Aspect's experiment and the debate about what it confirms or falsifies is a nice example of the well known theory ladenness of "real" experiments.

It also is remarkable that EPR-type correlation experiments were designed after the Aspect experiment, and they are still being made. Once I asked a leading experimental physicist whether he designs those experiments because he expects to find measurement results that contradict quantum mechanical predictions. His reply: "Of course not – I expect what quantum theory predicts". So I am wondering whether J. Bell's utterance (quoted by Atkinson (p. 215)) namely that he expected empirical violation of Bell's inequalities could be considered typical in the physics community.

Atkinson points out that in an EPR situation "... successive measurement results are correlated, that is the prediction" of quantum theory (p. 213), and those correlations are confirmed by experiments. Atkinson also indicates that observing correlations is not at all something exceptional or unique as long as one can conceive an explanation of them. The problem with the EPR correlations is that we do not possess a satisfactory explanation of these correlations. One has in principle two options to explain a correlation: by assuming a causal influence between the correlated events or by assuming some other factor (a common cause) that is responsible for the correlation. Formulating the need to explain correlations and identifying only these two options for an explanation is the content of what became called Reichenbach's Common Cause Principle. This principle is not formulated in Atkinson's paper, but its apparent violation by EPR correlations lies at the heart of the difficulty related to the EPR correlations. In the case of EPR correlations the option of explaining the correlation by assuming causal influence between the correlated events is available only at the expense of violating relativity theory since the correlated events are spacelike. But the other option also seems to be blocked by No-go theorems that spell out the impossibility of a common cause explanation of EPR correlations. The status of these No-go theorems is a controversial issue which is currently debated in the foundational literature (see (Rédei Forthcoming) for a review of the recent developments concerning the status of Reichenbach's Common Cause Principle from the perspective of quantum correlations, and (Rédei 2002a) for partial results about spacelike correlations prediicted by relativistic quantum theory).

Atkinson's somewhat skeptical remarks about the empirical nature (hence usefulness) of string theory are based on a real concern. If a theory such as string theory aspires to be a physical theory but it "... will never be testable in *practice*" (p. 217), then it is indeed *practically metaphysics*, even if it is testable in principle. That string theory will *never* be testable in practice, is however too strong a statement. While Atkinson's argument in favor of this assertion should make one very skeptical about the possibility of an empirical test of string theory because the energy per particle required for a test seems to be out of reach if one extrapolates the rate by which the available energy per particle has been increasing, this does not entail that the energies will in the future remain unattainable. If in assessing the status of string theory one takes a Popperian position about empirical testability as a demarcation criterion, like Atkinson does, then why not accept another idea of Popper regarding the unpredictability of future as well? If one does this, the energies required for testing might be available one day. If history of science is any indication it is likely however that string theory will have long been superseded by other, perhaps even more outlandish theories by then.

Department of History and Philosophy of Science, Faculty of Science
Loránd Eötvös University, Budapest, Hungary

REFERENCES

Jammer, M. 1985. "The EPR Problem in Its Historical Development". In P. Lahti and P. Mittelstaedt (eds.), *Symposium on the Foundations of Modern Physics. 50 Years of the Einstein-Podolsky-Rosen Gedankenexperiment*. Singapore: World Scientific, pp. 129–149.

Rédei, M., Summers, S.J. 2002a. "Local Primitive Causality and the Common Cause Principle in Quantum Field Theory". *Foundations of Physics* 32: 335–355.

Rédei, M. Forthcoming. "Reichenbach's Common Cause Principle and Quantum Correlations". In J. Butterfield and T. Placek (eds.), *Modality, Probability and Bell's Theorems*. Dordrecht: Kluwer Academic Publishers.

MICHAEL STÖLTZNER*

THE DYNAMICS OF THOUGHT EXPERIMENTS
A COMMENT ON DAVID ATKINSON

The issue of thought experiment, or *gedanken* experiment, is a classic in the philosophy of science. However, the literature has until recently been surprisingly sparse. There is, nevertheless, substantial disagreement as to what a thought experiment is. It was the Danish physicist Hans Christian Ørsted who coined the term within the context of German *Naturphilosophie* (Witt-Hansen 1976, Kühne 2002). More prominent is today the usage of *gedanken* experiments by the staunchest critic of such metaphysics, Ernst Mach; not least in virtue of their paradigmatic role for Albert Einstein whose railway embankments and freely falling elevators had an enormous influence on modern theoretical physics. Interestingly, Mach held that the purest thought experiments occur in mathematics which, on his account, was economically ordered experience. A similar connection was introduced into modern philosophy of mathematics by Imre Lakatos who contraposed the informal mathematical thought experiment to the formal Euclidean proof. "Thought-experiment (*deiknymi*) was the most ancient pattern of mathematical proofs." (1976, p. 9 fn.1) The terminological parallel, to be sure, was drawn by Lakatos because the cited book of Árpád Szabó interprets *deiknymi* as "to make the truth or falsity of a mathematical statement visible in some way;" (1978, p. 189) with the progress of Greek mathematics *deiknymi* developed into the technical term for formal proof. As I shall argue, Lakatos's identification of informal proof and thought experiment makes sense only when assuming the Machian notions of thought experiment and intuition. As both philosophies tend to blur the boundary between formal arguments and experiences and are oriented at reconstructing the history of a thought experiment, they will prove helpful in the context of the present comment.

On different grounds and without referring to Lakatos or Szabó, Nicholas Rescher has equally located the origin of thought experiment in the earliest epoch of Greek thinking. Identifying largely thought experimentation in philosophy with hypothetico-deductive reasoning, he concludes that its use "in philosophy is as old as the subject itself." (1991, p. 32)

Most contemporary interpreters, however, connect thought experiment to modern scientific theorizing and experimentation, or to a philosophy striving to be scientific in the sense of Newton and Einstein. Roughly speaking, three systematic dimensions prevail in the present literature. First, the debate between John Norton (1991, 1996) and James R. Brown (1991) concerning the epistemological status of thought experiments has attracted considerable attention. Are thought experiments merely arguments couched in a somewhat pictorial form or do some of them permit us to reach genuine knowledge about Platonic truths governing nature? There are attempts to overcome such an alternative. Tamar Szabó Gendler, for instance, considers thought experiment as "a fulcrum for the reorganization of conceptual commitments." (1998, p. 415)

By introducing novel categories by which we make sense of the world, this reconfiguration allows us to recognize the significance of certain previously unsystematized beliefs. ... Thus the thought experiment brings us to new knowledge of the world, and it does so by means of non-argumentative, non-platonic, guided contemplation of a particular scenario (*ibid.*, p. 420).

This already leads back to the dynamic (or historical) account emphasized by Kuhn (1977) and Lakatos (1976, 1978), albeit within a starkly diverging conception of scientific progress.[1]

Second, where are thought experiments located on the scale between theory and experiment? On Norton's account, they are closer to theory, or at least to the argumentative analysis of an experiment, and they can accommodate rather general philosophical principles into a scientific argument. Andrew D. Irvine, however, holds that "the parallel between physical experiments and thought experiments is a strong one." (1991, p. 150) All assumptions of a thought experiment must be supported by independently confirmed observations, and the thought experiment typically has repercussions on a certain background theory. On Irvine's account, the fact that "many thought experiments are meant to precede real experiments in which the original thought experiment's premises are actually instantiated" (*ibid.*, p. 151) and the fact that some elements of a thought experiment are assumed to be true, proves that it typically contains some but not only counterfactual elements. Ronald Laymon proposes to render benign the counterfactual character of thought experiments involving frictionless surfaces and the like by treating them as "ideal limits of real experimentation." (1991, p. 167)

[This requires] to (1) show that there exists a series of experimental refinements such that real experiments can be made to approach the postulated idealized thought experiment, and (2) show that there exists a series of theoretical corrections that can be made to apply to real experiments such that once corrected real experiments look increasingly like the original thought experiment. (*ibid.*, p. 174)

Paul Humphreys places himself on the other end of the spectrum.

The function of real experiments that is simulated by thought experiments is the isolation of those features of the world that are represented in a theoretical model and the approximation, as closely as possible, of the idealizations that are employed therein. This function of real experiments, which is to narrow the gap between theory and the world as it naturally presents itself to us, is nowadays almost exclusively guided by theory, and it indicates that thought experiments lie much closer to theory than to the world, or even its experimental refinements. (1993, p. 218f.)

Instead Humphreys detects clear parallels between thought experiments and computer simulations. First, both "require adjustments to bring them into conformity with existing empirical data," by picking appropriate boundary conditions, parameter values, etc. Second, both numerical experiments on a computer, e.g., Monte Carlo simulations, and thought experiments provide new knowledge insofar as they "allow us to explore properties of the theoretical model lying behind the simulation such as its robustness under changes in the idealizations and parameter values." Third,

simulations often deliberately alter the parameters involved in law schemata to produce laws that are not descriptive of our real world. Thus, one can simulate the trajectory of a planet under an inverse 1.99 law, rather than an inverse square law, to explore the effects this would have on the stability of N-body orbital motion, a process which is clearly the same as that involved in a thought experiment. (*ibid.*, p. 219)

On this basis, Humphreys considers thought experiments as "explorations and refinements of theoretical models" which are best described by "something like Lakatos's account of mathematical progress, applied to scientific models." (*ibid.*, p. 220, italics dropped)

The first and second dimensions combine in a proposal by Borsboom, Mellenbergh, and van Heerden. They claim that there exists a class which cannot be reformulated as an argument. Such a functional thought experiment "is not aimed at refuting or supporting a theory, but has a specific function within a theory. In the case of frequentist statistics, it functions as a semantic bridge, providing a real world interpretation for the abstract syntax of probability." (2002, p. 384) As an example they cite a brainwashing procedure invoked to apply the frequentist conception of probability in psychological testing. Contrary to the authors' assumption, there exist functional thought experiments within modern physics. General relativity and quantum mechanics have taught us that the theory of measurement can be part and parcel of the theory itself; already measuring a magnetic field by a charged test body can count as a functional thought experiment – though one of restricted interest.

Third, one may ask whether a thought experiment can succeed or fail in the same (non-trivial) sense as a real experiment. As a positive conclusion (Janis 1991) obviously depends on how the thought experiment is embedded into a set of background hypotheses or the theory under investigation, the question of success is linked to the informative content and semantical depth of the thought experiment. As Gendler rightly argues, the effect of a deep conceptual reorganization is not exhausted by reconstructing it as an argument. Even many mathematicians diagnose a substantial difference here; e.g. William Thurston: "We are not trying to meet some abstract production quota of definitions, theorems and proofs. The measure of our success is whether what we do enables *people* to understand ... mathematics." (1994, p. 163) To be sure, such a view would follow from a Platonist theory of intuition as proposed by Brown (1991); but to my mind also a psychological theory of scientific intuition à la Mach suffices to require a certain semantical depth from thought experiments as compared to other counterfactual arguments.

After this (incomplete) sketch of present debates, let me turn to Atkinson's paper. To my mind, it represents an important contribution to the second dimension and reasonably investigates thought experiments as a driving force of a research program. Unfortunately, this question is not sufficiently distinguished from a value judgment that proves problematic given the intimate relations of modern physics and mathematics. Atkinson holds that thought experiments that do not lead to theorizing and to a real experiment are generally of much less value that those that do so. Empiricists can hardly deny that triggering a real experiment strongly increases the epistemological value of the initial thought experiment. Yet this is not its only value as the following passage suggests.

Some exegeses of the EPR paper speak of a 'paradox', others of a thought experiment or theorem, but its true significance for the development of physics lies in its development from the stand-off of a thought experiment and of two competing world views (Einstein versus Bohr) via theoretical insights (Bohm and Bell) to a genuine experiment (Aspect). Had this genuine experiment not been performed, the EPR thought experiment would have remained a fruitless stand-off. (This volume, p. 215)

Here I disagree. The inability to resolve the EPR stand-off would have told us something important about the conceptual structure of quantum mechanics. In the same vein today the existence of different interpretations of quantum mechanics which are empirically equivalent up to very special situations, provides important insights into the significant leeway within this theory. As I shall argue below, historically the EPR-case was not so linear as Atkinson's scheme suggests and it built upon purely theoretical and mathematical progress as well. On the other hand, Atkinson's proposal deliberately spoils the rigid distinction between thought experiments and real experiments that are pre-conceived in thought. This yields a problem with his third example. As in general relativity and cosmology, there exist thought experiments whose sole purpose is to test the coherence of string theory or to justify its applicability in principle, but which are hardly accessible by real experiments that arise from them. Such thought experiments are very close to mathematical thought experiments which Atkinson's one-dimensional experiment-oriented equally fails to appraise. Let me take this example first.

String theory: mathematical experimenters

String theory's aspirations at a unified theory of physical interactions are strongly at odds with the practical impossibility of serious experimental tests. Experimenters would have to restore the conditions prevailing in the first spilt second of our universe at the expense of incredibly high energies. Atkinson rightly recapitulates the new metaphysical aspect of string theory: "different empirically confirmed explanations could be underpinned by a mathematical theory whose essentially new ontological claims cannot be tested in the crucible of experiment." (This volume, p. 218) However, his criticism concerns the issue of testability and does not imply that there is little use for thought experiments.

String theory is a unificatory research program that purports to provide a Theory of Everything. This claim, to my mind, makes possible to formulate some instructive thought experiments which are not directed at a subsequent real experiment. Nevertheless, they provide the only available opportunity to obtain independent confirmations of this theory. Here is an example. Steven Weinberg believes that "string theory has provided our first plausible candidate for a final theory." (1993, p. 169) Finality, to his mind, can be verified by stability against thought experiments: "the final theory [is] one that is so rigid that it cannot be warped into some slightly different theory without introducing logical absurdities like infinite energies." (*ibid.*, p. 12) In such a "logically isolated theory every constant of nature could be calculated from first principles; a small change in the value of any constant would destroy the consistency of the theory." (*ibid.*, p. 189) This is not just a counterfactual argument because in string theory the values of the constants of nature are fixed by certain phase transitions in the early universe.[2] Thus there exists a cosmological background theory.

Weinberg's variation of fundamental constants and the reference to a general principle fulfill Mach's conditions for thought experiments. In the first place, "the basic method of thought experiments, as with physical experiments, is variation." (Mach 1976, p. 139) In the latter we vary really existing physical conditions, in the former we vary the facts in thought. "Physical experiment ... [is] the natural sequel to thought experiment, and it occurs wherever the latter cannot readily decide the issue, or not completely, or not at all." (*ibid.*, p. 148) On the other hand, "[d]eliberate, autonomous extension of experience and systematic observation are thus always guided by thought and cannot be sharply limited and cut off from thought experiment." (*ibid.*, p. 149)

The second element of Machian thought experiment is "adaptation of isolated ideas to more general modes of thought developed through experience and the search for agreement (permanence, unique determination), the ordering of ideas in sequence." (*ibid.*, p. 125f.)[3] Mach's analysis of Stevin's famous thought experiment emphasizes the principle of unique determination; on pain of obtaining a perpetuum mobile – which would contradict our most basic daily experiences – there is no reason for the chain to turn right or left. Weinberg's employment of the principle of unique determination in variation, accordingly, characterizes his counterfactual reasoning as a thought experiment. Yet the comparison with Stevin's argument also shows a poverty of the string thought experiment. The values of the constants of nature are not immediately accessible to daily experience.

Another purely theoretical problem of present string theory is the fact that there are many different but mutually dual string theories. One can devise thought experiments showing that the transition between them does not yield measurable effects.[4] Similar thought experiments about experimental indistinguishability exist in all gauge-invariant theories; in general relativity they enjoy quite a fundamental status. The objective of such thought experiments is typically the exploration and justification of the conceptual framework.

Mach moreover held that both the methods of thought and physical experiment first developed in mathematics and spread from there to the natural sciences because the experiences in mathematics were of a simpler kind. "The change and motion of figures, continuous deformation, vanishing and infinite increase of particular elements here too are the means that enliven enquiry." (*ibid.*, p. 145)

Even where the exposition of a science is purely deductive, we must not be deceived by the form. We are dealing with thought constructions which have replaced previous thought experiments, *after* the result had become completely known and familiar to the author. (*Ibid.*, p. 144, missing words inserted according to the German original)

This passage could well have been written by Lakatos who emphasized that each formal result in mathematics was reached by a long-winded informal development in which thought experiments, e.g. stretching polyhedra, are the intermediate step between naive conjectures and the systematic method of analysis and synthesis (see Glas 1999).

In the above-mentioned thought experiments, rigorous mathematicians act as the experimenters' substitute. Interestingly, the notorious unfeasibility of experimental verification combines with a spectacular success of those mathematical insights which string theorists had obtained in an informal manner. This interaction between two communities with different standards prompted a broad discussion among mathematicians how to appraise non-rigorous (informal) results. Arthur Jaffe and Frank Quinn, two eminent mathematical physicists, distinguish two stages of the mathematical research process.

First, intuitive insights are developed, conjectures are made, and speculative outlines of justifications are suggested. Then the conjectures and speculations are corrected; they are made reliable by proving them. We use the term *theoretical* mathematics for the speculative and intuitive work; we refer to the proof-oriented phase as *rigorous* mathematics. (1993, p. 1)

But they "are not suggesting that proofs should be called 'experimental' mathematics. There is already a well-established and appropriate use of that term, namely to refer to numerical simulations as tests of mathematical concepts." (*ibid.*, p. 2) Rather does the terminology express a functional analogy between rigorous proof and experimental physics. Both correct, refine and validate the claims of their theoretical counterparts.

As I have argued elsewhere (2002a), "theoretical mathematics" hardly has a reasonable ontological status within a logicist and syntax-oriented philosophy of mathematics. Platonism, of course, attributes an objective meaning to unproven theorems, such that perhaps Brown's Platonic theory of thought experiments performs best in mathematics. But a quasi-empirical ontology in the sense of Lakatos (1978) does the job equally well and avoids the notorious pitfalls of Platonism. While "Euclidean" theories are built upon indubitable axioms from which truth flows down through valid inferences, in quasi-empirical theories truth is injected at the bottom by virtue of a set of accepted basic statements. In the latter case, truth does not flow downward from the axioms, but falsity is retransmitted upward. Theoretical physics is, of course, quasi-empirical and empirical in the usual sense. Mathematical thought experiments are only quasi-empirical by virtue of the flow of truth. There are two types. A proof-thought experiment "leads to a decomposition of the original conjecture into subconjectures." (1976, p. 13f.) Like an experimental technique or a partial result, e.g., a lower bound on the measured quantity, it remains valid even if the proof does not prove; thought experimenters may not need a conjecture "to devise an analysis, i.e. a test thought-experiment." (*ibid.*, p. 78)

In my conception the problem is not to prove a proposition from lemmas or axioms but to discover a particularly severe, imaginative 'test-thought experiment' which creates the tools for a 'proof-thought experiment', which, however, instead of *proving* the conjecture *improves* it. The synthesis is an 'improof', not a 'proof', and may serve as a launching pad for a research programme. (1978, p. 96)

The thought experiment may enter the hard core of a research program after further steps of refinement by proofs and refutations. Refutations are suggested by counterexamples that either concern the conjecture (*global* counterexamples) or the lemmas (*local* counterexamples). There are three types: (i) Global, but not local counterexamples logically refute the conjecture. They are what mathematicians call a counterexample. (ii) If a global counterexample is also local, it does not refute the theorem, but confirms it. (iii) Local, but not global counterexamples show a poverty of the theorem, such that one has to search for modified lemmas. Cases (ii) and (iii) are not genuinely logical, but *heuristic* counterexamples. There are two different methods to deal with counterexamples. After a logical counterexample one restricts the domain of the guilty lemma to exclude the counterexamples as unintended interpretations. After a heuristic counterexample one tries to keep as much as possible from the initial thought experiment and its heuristics by stretching the basic concepts so as to reach also logical counterexamples. The method of proofs and refutations thus proceeds by combined overstatements and understatements. Lakatos's terminology is useful for Atkinson's next example.

Aristotle versus Galileo: on standard models

Galileo's alleged refutation of the Aristotelian theory of falling bodies is one of the most-discussed thought experiments in history. Atkinson provides a very helpful analysis couched in modern language and concludes against Gendler that rearranging articulate and inarticulate knowledge did not suffice. "Galileo performed, and needed to perform, real experiments" (this volume, p. 212). Against the Galilean claim that the Aristotelian dogma that heavier bodies fall faster than lighter ones is inconsistent, Atkinson argues by physical counterexamples.

It is sufficient to point to a physical situation in which Aristotle's dogma ... is empirically correct. Since an inconsistent argument points at nothing at all, but Aristotle's argument does in fact indicate a realizable configuration, it follows that the dogma cannot be internally inconsistent. The case which gives Galileo the lie is that of bodies falling in a fluid (such as air or water) at their terminal velocities in the case of laminar fluid flow. (*ibid.*, p. 210)

[There exists] a physical model ... in which different bodies fall at different rates, even in vacuo. In a nonuniform gravitational field, as in the terrestrial situation, the rate of fall is a function of the distance from the centre-of-mass of the earth: a body at a higher elevation falls less quickly than one at a lower elevation. Moreover, the rate of fall can depend, in special circumstances, on other parameters too, such those defining magnetic or electric fields, etc. (*ibid.*, p. 211)

Accordingly, "Galileo's solution is only correct in a uniform gravitational field, and that the earth does *not* have such a uniform field is a brute empirical fact." (*ibid.*, p. 221) All this is, of course, true. Atkinson's reasoning joins in with a long list of proposals how the Aristotelian historically could have defended himself and which particular interpretation of Aristotle Galileo set out to disprove (*cf.* Kühne 2002). But, to my mind, all this does not invalidate Gendler's point that there occurred an intuitive reorganization of conceptual commitments which is not exhausted by the argument view. This also touches upon the relation between thought experiment and physical experiment because, as Mach had emphasized, the latter is guided by the former and thus by theoretical assumptions and background commitments. Let me illustrate this by a little story.

According to the legend, Galileo experimented by dropping objects from the leaning tower at Pisa. Let me tell the legend of his Bolognese colleague Aldrovandi who experimented inside the even more leaning towers of Garisenda and Asinelli in order to prevent disturbance by rain fall. Since the Piazza dei Miracoli is situated outside the city center, on clear days Galileo could see the horizon and feel that it is safe to assume a uniform gravitational field. Although feeling the inclination of the tower, his Bolognese colleague had to take the lines of brickstones as a measure of the distance and finding that the orbits of falling bodies bend according to this measure he might well have concluded that they are also subject to a sort of Coriolis force like water in a sink.

Both Galileo's thought experiment and his real experiment presuppose a uniform gravitational field and the absence of friction while Aldrovandi and Aristotle do not even formulate these concepts. In Lakatosian terms, Galileo's inconsistency is a local but not global counterexample against the Aristotelian dogma; it becomes a global counterexample under these further assumptions. Galileo skillfully expanded the Aristotelian concepts in such a way that they blatantly contradicted everyday experience thus making the thought experiment so convincing. Yet at bottom the thought experiment yields only a heuristic counterexample showing that the Aristotelian theory fails to reflect the difference between – now in Newtonian terms which Galileo deliberately set aside – the cause of the acceleration and other factors like resistance of the medium. In this dialectic of counterexamples we find conceptual reorganization at work. Galileo reclassified the models to which the theory is applied and developed new basic concepts; Atkinson's model is a non-standard model for Galileo that with the benefit of hindsight favors the Aristotelian theory. But historically Aristotle's theory did not contain the distinction by virtue of which the model is non-standard for Galileo. This reclassification could have been launched by a mere thought experiment, as a theoretical limit case of the inclined plane experiments, but the situation was most likely to lead into an experimental investigation, even without the necessity of empirical justification. This tendency towards experimental resolution is even stronger if theorizing yields a stand-off between two interpretations.

EPR, Bohm, Bell, and all that

Atkinson studies the historical "ascent" from the (1) EPR thought experiment and the (2) Bohr-Einstein stand-off, to (3) Bohm's reformulation of EPR, (4) the Bell inequality until (5) the Aspect experiment finally decided the stand-off between the Copenhagen "evangelical conviction that the new quantum mechanics is, in its domain of application, complete ... [which] was strengthened by von Neumann's flawed proof that hidden variables, which might 'complete' quantum mechanics and restore classical ontology, are logically impossible" (this volume, p. 214) and Einstein's conviction that "a holistic interdependence of different parts of reality would nullify the physicist's profession." (*ibid.*, p. 213)

After 1935, Bohr and Einstein remained committed to their respective criteria of quantum reality. Many historical studies have shown that the Copenhagen camp was anything but homogeneous and Einstein's realism was complex enough to make him reject Bohm's causal interpretation (1952). Today there is a great variety of different interpretations on the market and despite Bohm's reformulation the EPR-thought experiment has to compete with the double-slit experiment for the situation which most blatantly violates our common sense ontology. Looking at the historical development I can assent to Atkinson's first lesson "that a thought experiment which supports two contradictory intuitions can fruitfully point the way forward, by stimulating theoretical development and suggesting a real experiment, to a resolution of the dilemma" (this volume, p. 215). Indeed dilemmas have a high motivational value. But as already mentioned I have problems with Atkinson's one-dimensional view according to which the import of EPR would have been very limited without becoming the first step towards the spectacularly precise experimental tests of quantum mechanics. To prove his thesis, Atkinson would have to show (a) that in the historical course between (1) and (5) the integrity of the original stand-off has not been violated or (b) that the final physical experiment actually covers the original philosophical intuitions of Bohr and Einstein to a sufficient extent. Let me express some doubts and claim that the way was much more thorny and also contained genuinely theoretical advances and progress in experimental technology.

To those who rejected Bohm's interpretation (1952), his reformulation of EPR alone did not signify any theoretical progress as the poor response during the 1950s showed. But within the program to experimentally investigate the puzzles of quantum mechanics launched in the 1960s it quickly became clear that Bohm's thought experiment was much easier to realize in practice, in particular with neutrons and photons. In order to set up the first convincing experiment, Alain Aspect spent a lot of time developing appropriate photon sources and reach sufficient efficiencies, technological achievements that were useful for other experiments in quantum optics. And he had to meet a requirement which John Bell kept emphasizing over the years, to wit, that for a profound experimental test of locality the polarizers must be set only during the flight of the photons.

Yet recognizing the pivotal role of locality, to my mind, was only possible on the basis of Bell's inequalities. But their historical position is quite complex. On the one hand, they substantially contributed to the ascent (1)–(5) by singling out the quantity that was actually measured by Aspect and the subsequent experiments. On the other hand, they represented a milestone within an exclusively theoretical ascent that also emerged from a thought experiment. To my mind, Bell's inequalities represent a generalization of von Neumann's No-hidden variable theorem (Stöltzner 2002b). Von Neumann's proof was not flawed, as Atkinson and Bell hold, but based on an inappropriate notion of hidden variables. The situation was quite similar to the case of Galileo. By a thought experiment von Neumann rigorously constructed hidden parameters by partition (1932, p. 161) and proved that this contradicted an axiom concerning the additive character of observables and operators which he held to be natural (*ibid.*, p. 167). Pointing to Bohm's interpretation which violated this axiom but recovered quantum mechanical predictions and by inventing an even simpler model, Bell showed that the axiom was not natural at all. Von Neumann's proof thought experiment was thus rendered inconclusive and supplanted by a better one. Thus in the Lakatosian sense, von Neumann's thought experiment did not fail. Is this really so?

According to Allen Janis, a thought experiment fails in a non-trivial sense "when its analysis leads to a correct conclusion, but not one that accomplishes the purpose that provided its motivation" (1991, p. 116). What the experimenter expects is irrelevant; although Millikan wanted to refute Einstein's theory of the photoelectric effect, his experiment was a successful confirmation of it. EPR, however, comes close to the failed experiments to measure the proton lifetime which only provided an upper bound. It "does not show that quantum mechanics is complete. It only fails to establish that quantum mechanics is incomplete." (*ibid.*, p. 117) Of course, one can learn from failures. Martin Carrier rightly rejects this interpretation of EPR: thought experiments like real experiments "sometimes exhibit a Duhemian uncertainty about which hypothesis should be held responsible for an anticipated result." (1993, p. 415) In the EPR case it was not evident whether locality or completeness had to go, and the decision was successfully reached only decades later. To my mind, Duhemian holism is both a presupposition of the acceptable part of Atkinson's analysis and the reason why its one-dimensional reading is not warranted. Unlike real experiments or computer simulations, in thought experiments not every background assumption outside the core of the argument has to be fixed in advance. This is the reason why an initial thought experiment like EPR can survive such a long-winded ascent nearly unscathed. Axiomatized thought experiments like von Neumann's fail in a rather precise sense. On the other hand, there are essential elements for such a historical ascent which were *per se* not directed at experimental verification of the initial thought experiment. Had there been an immediate opportunity to experimentally test the 1935 version of EPR, it might not even have resolved the philosophical stand-off. But also recognizing the failure of von Neumann's No-hidden variable theorem by means of a simple counterexample would have had little impact without Bohm's model and Bell's detailed analysis. It was above all Bell's theorem that showed scientists the alternative between locality and a description in terms of hidden variable once a certain inequality was experimentally confirmed. Moreover, Bell's theorem was more general than the concrete form of quantum mechanics.

As a matter of fact, still today the EPR thought experiment is a fertile incentive. Greenberger, Horne, and Zeilinger (1989) proved generalized Bell inequalities between three states instead of two. In this case the situation is intuitively evident because one obtains a contradiction rather than just a correlation. This generalization should be counted as a progress even before the attempted experimental verification. Should this experiment fail, the EPR story would take perhaps another fruitful turn.

Institute Vienna Circle, Vienna, Austria, and
IWT, University of Bielefeld, Germany

NOTES

* I am indebted to David Atkinson, Martin Carrier, Don Howard, Ulrich Kühne, and Jeanne Peijnenburg for valuable suggestions and comments.
[1] For a convincing criticism of Kuhn's account, see (Humphreys 1993); for an analysis of Lakatos's views, see (Glas 1999).
[2] To be sure, these aspirations are exaggerated; see (Stöltzner and Thirring 1994).
[3] John Norton (1991) shows that principles of equal generality figure prominently in Einstein's thought experiments.
[4] Richard Dawid, private communication.

REFERENCES

Atkinson, D. 2003. "Experiments and Thought Experiments in Natural Science". This volume, pp. 209–225.

Bell, J.S. 1987. *Speakable and Unspeakable in Quantum Mechanics.* Cambridge: Cambridge University Press.

Bohm, D. 1952. "A Suggested Interpretation of the Quantum Theory in Terms of 'Hidden' Variables", I and II. *Physical Review* 85: 155–193.

Borsboom, D., Mellenbergh, G.J. and van Heerden, J. 2002. "Functional Thought Experiments". *Synthese* 130: 379–387.

Brown, J.R. 1991. "Thought Experiments: A Platonic Account". In Horowitz and Massey (eds.), pp. 119–128.

Carrier, M. 1991. "Critical Discussion" of (Horowitz and Massey 1991). *Erkenntnis* 39: 413–419.

Gendler, T.S. 1998. "Galileo and the Indispensability of Scientific Thought Experiment". *British Journal for the Philosophy of Science* 49: 397–424.

Glas, E. 1999. "Thought-Experimentation and Mathematical Innovation". *Studies in History and Philosophy of Science* 30: 1–19.

Greenberger, D.M., Horne, M.A. and Zeilinger, A. 1989. "Going Beyond Bell's Theorem". In M. Kafatos (ed.), *Bell's Theorem, Quantum Theory, and Conceptions of the Universe*. Dordrecht: Kluwer, pp. 69–72.

Horowitz, T. and Massey, G.J. (eds.) 1991. *Thought Experiments in Science and Philosophy*. Savage, MD: Rowman and Littlefield.

Humphreys, P. 1993. "Seven Theses on Thought Experiments". In J. Earman et al. (eds.), *Philosophical Problems of the Internal and External Worlds. Essays on the Philosophy of Adolf Grünbaum*. Pittsburgh and Konstanz: University of Pittsburgh Press and Universitätsverlag Konstanz, pp. 205–227.

Irvine, A.D. 1991. "On the Nature of Thought Experiments in Scientific Reasoning". In Horowitz and Massey (eds.), pp. 149–166.

Jaffe, A. and Quinn, F. 1993. "'Theoretical Mathematics': Toward a Cultural Synthesis of Mathematics and Theoretical Physics." *Bulletin of the American Mathematical Society* 29: 1–13.

Janis, A.I. 1991. "Can Thought Experiments Fail?". In Horowitz and Massey (eds.), pp. 113–118.

Kühne, U. 2002. *Die Methode des Gedankenexperiments. Untersuchung zur Rationalität naturwissenschaftlicher Theorieformen.* Ph.D. dissertation. University of Bremen.

Kuhn, T.E. 1977. "A Function for Thought Experiments". In *The Essential Tension.* Chicago: University of Chicago Press, pp. 240–265.

Lakatos, I. 1976. *Proofs and Refutations. The Logic of Mathematical Discovery.* Cambridge: Cambridge University Press.

Lakatos, I. 1978. *Mathematics, Science and Epistemology. Philosophical Papers 2.* Cambridge: Cambridge University Press.

Laymon, R. 1991. "Thought Experiments by Stevin, Mach and Gouy: Thought Experiments as Ideal Limits and as Semantic Domains". In Horowitz and Massey (eds.), pp. 167–191.

Mach, E. 1976. *Knowledge and Error. Sketches on the Psychology of Enquiry.* Dordrecht: Reidel.

Neumann, J. von. 1932. *Mathematische Grundlagen der Quantenmechanik.* Berlin: Julius Springer.

Norton, J.D. 1991. "Thought Experiments in Einstein's Work". In Horowitz and Massey (eds.), pp. 129–148.

Norton, J.D. 1996. "Are Thought Experiments Just What You Thought?". *Canadian Journal of Philosophy* 26: 333–366.

Rescher, N. 1991. "Thought Experimentation in Presocratic Philosophy". In Horowitz and Massey (eds.), pp. 31–41.

Stöltzner, M. 2002a. "What Lakatos Could Teach the Mathematical Physicist". In G. Kampis, L. Kvasz, and M. Stöltzner (eds.), *Appraising Lakatos. Mathematics, Methodology, and the Man.* Dordrecht: Kluwer, pp. 157–187.

Stöltzner, M. 2002b. "Bohm, Bell and von Neumann. Some Philosophical Inequalities Concerning No-go Theorems". In T. Placek and J. Butterfield (eds.), *Modality, Probability, and Bell's Theorem.* Dordrecht: Kluwer (NATO series), pp. 35–36.

Stöltzner, M. and Thirring, W. 1994. "Entstehen neuer Gesetze in der Evolution der Welt". *Naturwissenschaften* 81: 243–249.

Szabó, Á. 1978. *The Beginnings of Greek Mathematics.* Budapest: Akadémai Kiadó.

Thurston, W. 1994. "'Theoretical Mathematics': Toward a Cultural Synthesis of Mathematics and Theoretical Physics". *Bulletin of the American Mathematical Society* 30: 161–177.

Weinberg, S. 1993. *Dreams of a Final Theory*. London: Vintage.

Witt-Hansen, J. 1976. "H.C .Örsted, Immanuel Kant, and the Thought Experiment". *Danish Yearbook of Philosophy* 13, 48–65.

GIORA HON

AN ATTEMPT AT A PHILOSOPHY OF EXPERIMENT

INTRODUCTION

"What exactly is an experiment in physics?" Pierre Duhem posed this question almost a century ago. Apparently, he was concerned not only with the content of the question but also with its reception, since he added forthwith: "This question will undoubtedly astonish more than one reader. Is there any need to raise it, and is not the answer self-evident?" (Duhem 1974, p. 144) Recent attention to the issue of experimentation illustrates that the answer to this question is not self-evident and that there is a philosophical and historical interest in raising the question: what exactly is an experiment in physics and indeed in science generally? The study of experiment has constituted one of the principal forces in reshaping history and philosophy of science during the last two decades. We therefore can assure Duhem that today his question has finally struck a responsive audience. We now consider experiment a central issue in history and philosophy of science, a concept that needs explication and elucidation.

However, the rich studies of experiment indicate that the philosophy of experiment is lagging behind the extensive historical studies of experimentation and the many facets that historians of experimentation have addressed, facets such as technological, cultural, sociological and anthropological. It appears indeed that a divide separates the historians of science from the philosophers of science as to experimental practice. The divide may be clearly discerned in the collection, *Experimental Essays – Versuche zum Experiment* (Heidelberger and Steinle 1998). It appears that a stronger case for the philosophy of experiment should have been made (Radder 1998). To be sure, there have been attempts at such philosophy and I shall outline a few of them shortly. However, these attempts have not cohered into a forceful and cohesive philosophical analysis of experiment, incisive at once for epistemology and for the historiography of experimentation. My objective then is to contribute towards a philosophy of experiment. I grope to bridge the divide between history and philosophy of scientific experimentation by developing a historically informed philosophy of experiment.

In this study I focus, as it were, on the inner working of experiment. There are quite a few philosophical discussions of experimentation that are concerned with the paramount relation between experiment and theory, that is, the role of experiment in the overall framework of the scientific method – in a word, all what is external to experiment be it analytical, logical or methodological. For example, the notion of error statistics in experimentation that concerns directly with the relation between theory and experimental results (Mayo 1996; but see Hon 1998b), or for another example, the experiment as an interrogative procedure that executes some kind of erotetic logic (Hintikka 1988) – these external issues do not constitute the theme of my paper. Rather, my interest lies in the internal elements that comprise experiment, their physical and logical interrelationships, their governing principles – in sum, the internal "working", as it were, of experiment which brings about a result, that is, a feature of the world we have come to know.

As a preliminary step, I shall identify and characterize what appear to me the principal obstacles to the construction of a philosophy of experiment, obstacles that have proved quite recalcitrant. An outline of the tension between history and philosophy of experiment will serve as a background.

HISTORY VS. PHILOSOPHY OF EXPERIMENT

The position of the historians of science may well be represented by the view of Jed Buchwald. He claims succinctly and bluntly that

> living sciences cannot be corralled with exact generalizations and definitions. Attempting to capture a vibrant science in a precise, logical structure produces much the same kind of information about it that dissection of corpses does about animal behavior; many things that depend upon activity are lost.

Indeed, according to Buchwald, "axiomatics and definitions are the logical mausoleums of physics" (Buchwald 1993, pp. 170–171). The position of the contemporary historian of science is then to regard science as an activity, not an end result but a process, a "living" and "vibrant" process. The historian's claim is that any generality in the form of, say, logical structure, simply kills this lively activity. The metaphor of the living and the dead appears to be crucial to Buchwald and to historians of science at large. They follow Kuhn's directive, which he formulated right at the beginning of his *Structure of Scientific Revolutions*. According to Kuhn, the aim of history of science "is a sketch of the... concept of science that can emerge from the historical record of the research activity itself." "Activity" appears to be the key feature as distinct from "finished scientific achievements" (Kuhn 1970, p. 1).

The historian may well be happy therefore with a detailed description and a thorough analysis of the activity – the "living" particular; but the philosopher must strive, as Hacking put it simply and directly, for "both the particular and the general" (Hacking 1992, p. 29). There is no escape. If we want to do philosophy, that is, if we believe that philosophy has a bearing on a certain kind of activity, we have then to seek its general features, its underlying principles. In other words, we have to uncover logical structures and characterize methodological principles that govern this activity, without however losing sight of its particulars, namely, its "living" execution. Now, as to experimentation, it is unquestionable that philosophy ought to have a bearing on this activity – it being one of the chief methods of obtaining scientific knowledge. We have then no choice but to analyze experiment *in vitro* as it were, keeping a wide eye on its features as an activity *in vivo*. Buchwald's claim should serve as a warning rather than a condemnation. We should give heed to this warning and follow Whitehead's cautious dictum: "Too large a generalisation leads to a mere barrenness. It is the large generalisation, limited by a happy particularity, which is the fruitful conception" (Whitehead 1929, p. 39).

Thus, a well-developed philosophy of experiment should bring together in a consistent fashion both the normative aspect of the experimental activity – its descriptive as well as prescriptive dimensions, and a comprehensive theoretical conception of experiment that throws light on its internal features, features that underwrite the reliability of the knowledge thus obtained. I propose the notion of experimental error as an efficient vehicle for attaining this objective.

A claim to knowledge in the form of a proposition may be found in time, by various means, to be either true or false. A conceptual system contains by its very nature such claims of which some are found, whatever the system, erroneous. It is commonly expected of the proponent of such system to address the problem of error and to explain the failed attempts at knowledge. The most habitual approach is to analyse errors in terms of the system itself. By doing so, the entire structure of the system – its constituting elements and governing principles – becomes exposed.

Consider for a very brief example Descartes' system of philosophy. As expected, he conceived of the notion of error in the very terms with which he constructed his philosophical system. In Descartes' system error is associated with the cleavage introduced between will and reason. When free will is not restrained and it cajoles successfully the intellect to assent to a proposition that is neither distinct nor clear then, according to Descartes, an error occurs – an indication that the God given faculty of free will has been misused (Hon 1995, pp. 5–6). Thus, a study of Descartes' conception of error reveals immediately the central elements of his philosophical system and its governing principles. In this vein, I present in this essay an outline of Bacon's theory of error; it serves as a background to the philosophical analysis of experiment that I develop (for further clarification of this method of inquiry see Hon 1998a, § 2, 3).

I consider experiment a philosophical system that aims at furnishing knowledge claims about the world, be the world physical or social. Like any philosophical system, experiment comprises elements and governing principles. Given the above method of inquiry, I propose that a study of sources of error arising in this system will throw light on its working. Thus I seek generalizations of the experimental activity that emerge through a study of the notion of experimental error. I claim that while capturing a central feature of the experimental activity, namely, seeking to minimize if not eliminate errors, the notion of experimental error also reflects, albeit negatively, principal conceptual features of experiment. To be more specific, the thesis exploits types of experimental errors as constraints by which one may uncover general features of experiment. It may be seen that the articulation of the notion of experimental error originates in the normative dimension – how to address, rectify and indeed avoid errors in the execution of experiment. However, this articulation reflects at the same time structures and governing principles of experimentation. The attempt then is to capture at once, via the notion of experimental error, both the normative aspect and the theoretical conception of experiment. I shall be concerned in this essay only with the theoretical conception of experiment.

SETTING THE PHILOSOPHICAL SCENE: TWO CLUSTERS OF PROBLEMS

To set the philosophical scene, it is useful first to identify the obstacles that obstruct the way to a viable philosophy of experiment. I discern two principal clusters of obstacles to the construction of such philosophy. Not surprisingly, both clusters have to do with the transition from the particular to the general. For reasons that will shortly become clear, I call the first cluster epistemological and the second methodological. As it happened, right at the beginning of the last century two physicists-cum-philosophers published pioneering, influential works that bear on these issues. Ernst Mach published in 1905 his *Knowledge and Error*. In this collection of essays he addresses problems pertaining, in his words, to "scientific methodology and the psychology of knowledge" (Mach 1976, pp. xxxii). Mach dedicated one essay to the analysis of physical experiment and to identifying its leading features (Ch. XII). A year later, in 1906, Pierre Duhem published his book entitled *The Aim and Structure of Physical Theory*. It is in this book that Duhem posed the question to which I referred at the outset of my talk: "What exactly is an experiment in physics?" (Duhem 1974, p. 144) While Duhem focuses on the epistemological problem, Mach is concerned with methodological issues.

1. The epistemological cluster: the transition from matter to argument

The first cluster of obstacles to a philosophy of experiment is in my view the transition from the material process, which is the very essence of experiment, to propositional knowledge – the very essence of scientific knowledge. As Duhem sees it, the experimental physicist is engaged in "the formulation of a judgment interrelating certain abstract and symbolic ideas which theories alone correlate with the facts really observed." The conclusions of any experiment in physics, and for that matter in science, are indeed "abstract propositions to which you can attach no meaning if you do not know the physical theories admitted by the author" (*ibid.*, pp. 147–148). The end result of an experiment is not, to refer once again to Duhem, "simply the observation of a group of facts but also the translation of these facts into a symbolic language with the aid of rules borrowed from physical theories" (*ibid.*, p. 156). In other words, the obstacle I wish to identify is the problematic passage from matter that is being manipulated and undergoes some processes, via observations to propositions – a language expressed in interrelated symbols – whose meaning is provided by some theory.

Andrew Pickering, to turn to a contemporary author, addresses this problem as a substantial element of the issue of realism. Pickering writes that he is concerned with the process of "finding out about" and "making sense of"; that is, he inquires into the relation between articulated scientific knowledge and its object – the material world (Pickering 1989, p. 275). He conceives of a three-stage development in the production of any experimental fact: a material procedure, an instrumental model and a phenomenal model (*ibid.*, pp. 276–277). These three stages span according to Pickering the material and conceptual dimensions of the experimental practice. It is in the arching of these two dimensions that the passage from matter to knowledge should be forged. Pickering is of the opinion that this passage "is one of made coherence, not natural correspondence." In other words, the coherence between material procedures and conceptual models is an artificial product due to actors' successful achievements in accommodating the resistances arising in the material world (*ibid.*, p. 279).

In a different vein, I have had recourse elsewhere to a concept that I called "material argument" (Hon 1998a, especially §4). I was trying with this concept to bring together in a philosophical context all the elements which are involved in experimentation: the theoretical context and the scheme of manipulation, the material processes and the resulting scientific knowledge which is essentially propositional. I introduced the notion of "material argument" precisely for the purpose of rendering intelligible the transition from the process of manipulating matter to the process of inferring propositions that characterize experimental knowledge, namely, the declared end result of experiment. Experiment, I claimed, is a procedure, a physical process, which can be cast into an argument of a formal nature (*ibid.*, p. 233). But this discussion should not detain us further. Suffice it to remark that the transition from matter to proposition presents the first set of difficulties for a philosophy of experiment. I call this cluster of obstacles the epistemological issue.

2. The methodological cluster: transcending the list of strategies, methods, procedures, etc.

The second cluster of obstacles is at the level of manipulation of matter – the very essence of physical experiment; I refer to this cluster as methodological. Here we are concerned with the transition from the myriad of strategies, methods, procedures, conceptions, styles and so on, to some general, cohesive and coherent view of experiment as a method of extracting knowledge from nature. From a philosophical perspective it would have been fruitful had we obtained a general yet fundamental scheme of experiment that captures in a tight economic fashion this myriad of facets and features. This goal may be anathema to Buchwald's historical view of experiment, but in my opinion it is crucial for a philosophical understanding.

A convincing historical account that exhibits the enormous variety of facets and features which experiment possesses is Darrigol's notion of "transverse principles" which he applied to nineteenth century electrodynamics. These principles are not general rules of scientific method; they are rather methodological precepts that regulate at once theory and experiment, hence "transverse principles". Guided by tradition or one's own ingenuity, the physicist follows a transverse principle that links one's theoretical conception of the physics which one studies, to actual experimentation. Clearly, the application of the principle contributes much to the formation and definition of the physicist's methodology (Darrigol 1999, pp. 308, 335).

Consider Faraday for an example. According to Darrigol, Faraday's theories "were rules for the distribution and the interplay of various kinds of forces." Faraday dispensed with the Newtonian distinction between force and its agent. In Faraday's view, "an agent could only be known through actions emanating from it" (*ibid.*, pp. 310–311). Thus, the best course to take in the study of body acting on anther body consisted in mapping the various positions and configurations of the body acted upon. This position called for a principle of contiguity. It is this principle that regulated, according to Darrigol, both the theoretical and experimental practice of Faraday:

On the theoretical side, this principle entailed his concept of the lines of forces as chains of contiguous actions and his rejection of the dichotomy between force and agent. On the experimental side, it determined the emphasis on the intermediate space between sources and the exploratory, open charactèr of his investigations (*ibid.*, p. 312).

When Darrigol juxtaposes this approach of Faraday to the studies of other nineteenth-century electrodynamicists, the variety and richness of conceptions of theory and experimental practices become apparent.

Darrigol argues persuasively for a close connection between theory and experiment in nineteenth-century electrodynamics. As it is so tightly connected to theory, the conception of experiment and its actual procedures become, at least in this historical episode, enormously varied and complex. The question immediately presents itself as to how should one, as a philosopher, capture in general terms this enormous variety of conceptions of experiment and the concomitant practices of material procedures?

To take another example, Rom Harré analyses experiments by their assigned goals: spelling out the formal aspects of the method involved (e.g., finding the form of a law inductively); developing the content of a theory (e.g., finding the hidden mechanism of a known effect) and technique (e.g., the power and versatility of apparatus). Like Darrigol's "transverse principles", Harré's principle of organization of kinds of experiment according to their goals also demonstrates the enormous variety of facets and features which experiment possesses (Harré 1983).

In his essay on the leading features of physical experiment, Mach realizes that these features may not be exhausted. It seems then that a generalization may not be attained. The formative features of experiment, which Mach describes, have been abstracted, so he writes,

> from experiments actually carried out. The list is not complete, for ingenious enquirers go on adding new items to it; neither is it a classification, since different features do not in general exclude one another, so that several of them may be united in the experiment (Mach 1976, p. 157).

Is the list indeed open or is it in fact in the final analysis constrained? If no constraints were to be imposed on this method of inquiry, then no classification and indeed no generalization would be obtained. The approach, in a word, would be eclectic and *ad hoc*.

A good illustration of a detailed list which goes beyond Mach's preliminary list and yet remains *ad hoc*, is Allan Franklin's list of "epistemological strategies" which he convincingly buttresses with elaborated case studies. Here is the list of strategies which Franklin has drawn:

1. Experimental checks and calibration, in which the apparatus reproduces known phenomena.
2. Reproducing artifacts that are known in advance to be present.
3. Intervention, in which the experimenter manipulates the object under observation.
4. Independent confirmation using different experiments.
5. Elimination of plausible sources of error and alternative explanations of the result.
6. Using the results themselves to argue for their validity.
7. Using an independently well-corroborated theory of the phenomena to explain the results.
8. Using an apparatus based on a well-corroborated theory.
9. Using statistical arguments (Franklin 1990, p. 104; *cf.* also 1986, chs. 6, 7, and 1989).

Franklin argues that these strategies have been designed to convince experimenters that experimental results are reliable and reflect genuine features of nature. The list of strategies demonstrates according to Franklin the different ways experiments gain credibility. Practising scientists pursue such strategies to provide grounds for rational belief in experimental results

(Franklin 1989, pp. 437, 458). For Franklin the use of these strategies has then the "hallmark of rationality" (Gooding et al. 1989, p. 23) and in that sense he is seeking to contribute to a philosophy of experiment.

However elaborated and complex, the list of strategies which Franklin puts forward, is essentially similar to the list which Mach presents in his essay on the leading features of experiment. Like Mach, Franklin is aware of the limitation of this approach – the account is *ad hoc*. Franklin indeed states that the strategies he documented are neither exclusive nor exhaustive. Furthermore, these strategies or any subset of them do not provide necessary or sufficient conditions for rational belief. "I do not believe", he states, that "such a general method exists" (Franklin 1989, p.459). Nevertheless, Franklin is convinced that scientists act rationally. According to the unfailing optimism of Franklin, scientists use, as Gooding, Pinch and Schaffer aptly put it, "epistemological rules which can be applied straight-forwardly in the field to separate the wheat of a genuine result from the chaff of error" (Gooding et al. 1989, pp. 22–23).

Franklin is much concerned with the working scientist, or rather the practising experimenter, and it appears that the strategies he lists have been in fact abstracted from actual experiments, precisely as Mach did a century earlier. As such his list, although rich and varied, remains eclectic and *ad hoc*. While each item on the list provides a thorough and detailed illustration of an experimental procedure that is designed to give grounds for rational belief, there appears to be no overall guiding principle to govern the list itself. Such a list cannot be completed since no constraint is being imposed. A coherent generalisation appears therefore impossible.

This is then another problem that is posed to the philosopher of experiment, namely, how to transcend "the list"? How to generalize the various items that comprise the list? In attempting an answer to this question we should give heed to Hacking's warning and be careful not "to slip back into the old ways and suppose there are just a few kinds of things, theory, data, or whatever" (Hacking 1992, p. 32; *cf.*, p. 43).

THE "'ETC.' LIST"

Following Hacking, I call this problem the "'etc.' list". In his "Self-Vindication" paper, Hacking refers to several authors and in particular to Pickering and Gooding, identifying in their writings lists of items. So, for example, what Pickering calls "pragmatic realism" is the co-production of: "facts, phenomena, material procedures, interpretations, theories, social relations etc." (Hacking 1992, p. 31). Similarly, Hacking portrays Gooding as having another "'etc.' list." According to Hacking, Gooding "speaks of an 'experimental sequence' which appears as the 'production of models, phenomena, bits of apparatus, and representations of these things'" (*ibid.*, p. 32). We agree, Hacking continues, "that the interplay of items in such a list brings about the stability of laboratory science" (*ibid.*). On his part, Hacking gives the *matériel* of an experiment a crucial role to play in the stabilization process of experimental science. By the *matériel* he means

the apparatus, the instruments, the substances or objects investigated. The matériel is flanked on the one side by ideas (theories, questions, hypotheses, intellectual models of apparatus) and on the other by marks and manipulations of marks (inscriptions, data, calculations, data reduction, interpretation) (*ibid.*).

It looks then as if Hacking presents us with an "'etc.' list" of his own. Hacking however is not content with "lists and etc.'s" (*ibid.*), and he ventures a taxonomy of elements of experiment which takes him further afield, beyond Mach and Franklin.

The conception that in experiment the *matériel* is flanked on one side by *ideas* and on the other by *marks* is the clue to Hacking's proposal for making the open list converge onto three groups of elements of experiment, namely, "ideas, things, and marks" (*ibid.*, p. 44). "Ideas" are the intellectual components of experiment; "things" represent the instruments and apparatus, and finally "marks" comprise the recording of the outcomes of experiment. Apparently, Hacking is not worried by Mach's claim that classification will not do, "since different features do not in general exclude one another, so that several of them may be united in the experiment" (Mach 1976, p. 157). In fact, Hacking delights in constructing a flexible taxonomy, since in his view the stability of experimental results arises from precisely the very interplay of elements – whatever the case may be the taxonomy should *not* be rigid (Hacking 1992, p. 44). With this taxonomy Hacking seeks at once to demonstrate, in his words, the "motley of experimental science", and to contribute towards a philosophy of experiment so that one would not meander, as he puts it, "from fascinating case to fascinating case" (*ibid.*, pp. 31–32).

In what follows, I wish to address this second cluster of problems, that is, the methodological issue – the "etc." list. My objective is to transcend the list much in the spirit of Hacking but based on a different line of argumentation, then reach the taxonomic stage and aim beyond it to experimental principles.

THE GUIDING IDEA: APPROACHING KNOWLEDGE FROM THE PERSPECTIVE OF ERROR

As I have indicated, my guiding idea is to study experiment by the nature of its possible faults. I suggest that light may be shed on experimentation by examining and ordering possible sources of error in experiment. My approach takes then a different route altogether from that of Franklin. I am not seeking epistemological strategies that are designed to secure reliable outcomes that may in turn provide basis for rational belief. As I have argued, this approach results in an open, *ad hoc* list. I am looking rather for general characterizations of classes of possible sources of error. We shall see that in many respects the emerging typology of classes of experimental error reflects, albeit from a negative perspective, Hacking's typology. There will be however some crucial differences. It is hoped further that the resultant typology would serve as a framework for developing a theory of experiment out of which general principles may emerge.

By way of clarification, here is a brief account of how sources of experimental error may be broached. Consider the standard approach to experimental error, that is, the dichotomy of systematic and random error. Clearly, this dichotomy reflects an interest in the mathematical aspect of error: does a deterministic law govern the error? Or is it a statistical law? In the former case, as is well known, the error is systematic and in the latter it is random. The dichotomy is very useful and much in use in the practice of experimentation, especially in the analysis of the results by introducing correction terms and reducing the data. The dichotomy could therefore be included in the list of strategies. However, the distinction throws no light on the source of the error; in other words, philosophically it is not useful. Error that may originate in the presupposition of incorrect background theory is classified together with an error that has originated in a faulty calibration – both being systematic. For another example, small error in judgment on the part of the observer in estimating the scale division and unpredictable fluctuations in conditions such as temperature or mechanical vibrations of the apparatus, are classified together since these errors are all random in nature (for a detailed analysis see Hon 1989b, pp. 474–479).

I maintain that for philosophical purposes analysis should be focused on the source of error while clear distinctions should be drawn among different kinds of possible sources. From an epistemological perspective, one is interested in the source of error and not so much in the mathematical features of the error and the means of calculating it away – the causal feature being of a higher interest than the pragmatic one. Thus, for example, errors that have originated in the use of the apparatus should be set apart from errors that pertain to the interpretation of the data. It is hoped that once distinctions among the different kinds of source of error are being introduced, retained and elaborated, the structure of the method at stake would come to light. Specifically, as we shall see, the features of the different kinds of source of error reflect the various elements that are involved in experimentation.

The approach to knowledge from its negative perspective, that is, from errors and faults, is not new. In fact, "the first and almost last philosopher of experiments" – to use Hacking's characterization of Francis Bacon (Hacking 1984, p. 159) – employed a similar methodology. Bacon was philosophically aware of the problem of error and explicitly addressed it. Indeed, he deployed the notion of error as a lever with which he hoisted his new program for the sciences. As expounded in the *Novum Organum* (Bacon [1620] 1859; 1960; 1989; 2000), his programmatic philosophy consists of two principal moves: first, the recognition of error and its rebuke if not elimination, and then the commencing anew of the true science based on experiment and induction. I shall presently argue that Bacon's conception is found wanting especially when experiment, the very instrument of his research, is in question. The shortcomings of his approach would be the key to my move. So here is a précis of Bacon's theory of error.

BACON'S TYPOLOGY OF ERRORS: THE FOUR IDOLS

Bacon argues in his celebrated *Novum Organum*, that Aristotle "has corrupted Natural Philosophy with his Logic; ... he has made the Universe out of Categories" (Bacon 1859, p. 39 (I, lxiii)). In Bacon's view, the application of Aristotle's doctrine has rather the effect of confirming and rendering permanent errors which are founded on vulgar conceptions, than of promoting the investigation of truth (Bacon 1859, pp. 13–14 (I, xii); *cf.*, Bacon 2000, p. 10 (Preface to "The Great Renewal") and p. 28 (Preface)).

Bacon builds his program on the doctrine that truth is manifest through plain facts, but for this claim to be valid the student of nature has to get rid of all prejudices and preconceived ideas – "freed from obstacles and mistaken notions" (Bacon 2000, p. 13 (Preface to "The Great Renewal")). As Bacon instructs, "the whole work of the mind should be recommenced anew (*ut opus mentis universum de integro resumatur*)" (Bacon 1859, p. 4 (Preface); 1989, p. 152); only then would the student experience things as they are. "Our plan", he explains, "consists in laying down degrees of certainty, in guarding the sense from error by a process of correction ... and then in opening and constructing a new and certain way for the mind from the very perceptions of the senses" (Bacon 1859, p. 3 (Preface)). In this way, Bacon concludes, "we are building in the human Intellect a copy of the universe such as it is discovered to be, and not as man's own reason would have ordered it (*Etenim verum exemplar mundi in intellectu humano fundamus; quale invenitur, non quale cuipiam sua propria ratio dictaverit*)" (*ibid.*, p. 120 (I, cxxiv)). Thus the first task of the scientist is to eliminate errors from his or her cognition by the "expiation and purgation of the mind (*expiationibus et expurgationibus mentis*)", and only then can the scientist enter "the true way of interpreting Nature (*veram interpretandæ naturæ*)" (*ibid.*, p. 51 (I, lxix)). Bacon states explicitly this objective in the full Latin title of the book: *Novum Organum, sive indicia vera de interpretatione naturæ*, that is, *The New Instrument, or True Directions for the Interpretation of Nature* (Bacon 2000, p. 11, fn 8). The project then is to put an end to an unending error – *infiniti erroris finis* (*ibid.*, p. 13 (Preface to "The Great Renewal"); 1989, p. 133) and to seek "a true and lawful marriage between the empirical and the rational faculties (*Atque hoc modo inter empiricam et rationalem facultatem ... conjugium verum et legitimum in perpetuum nos firmasse existemamus*)" (Bacon 2000, pp. 11–12 (Preface to "The Great Renewal"); 1989, p. 131).

Bacon therefore finds it necessary to expound in considerable detail the subject of the obstacles to the true interpretation of nature, before proceeding to unfold his positive program: the method of inductive inquiry based on experimentation. He devotes nearly the whole of the first book of *Novum Organum* – "the destructive part" (Bacon 2000, p. 89 (I, cxv)) – to the examination of these obstacles: "the signs and causes of error (*signis et causis errorum*) and of the prevailing inertia and ignorance" (*ibid.*, p. 89 (I, cxv); 1989, p. 210) which he calls idols, idols of the mind. The term "idol" conveys at once the meaning of the Platonic concept of *eidolon* – fleeting, transient, image of reality as well as religious undertones. *Eidolon* stands as an antithesis to the concept of *idea*: "*humanae mentis* idola" vs. "*divinae mentis ideas*" (Bacon 1859, pp. 16–17, fn (I, xxiii)).

Although Bacon claims that "to draw out conceptions and axioms by a true induction is certainly the proper remedy for repelling and removing idola" (*ibid.*, p. 21 (I, xl)), he still finds it of great advantage to explicitly indicate the idols and expound them in detail. For, as he explains, "the doctrine of idola holds the same position in the interpretation of Nature, as that of the confutation of sophisms does in common Logic" (*ibid.*, p. 21 (I, xl)). In other words, to use Jardine's formulation, "the idols ... bear a relation to the inductive method analogous to that which cautionary lists of fallacious arguments bear to syllogistic" (Jardine 1974, p. 83). As I have indicated, I wish to advance further from mere "cautionary lists" and to obtain a conceptual scheme of experiment based on a typology of sources of error. Bacon's theory of error, his typology of idols and its critique, serves as a philosophical illustration of the approach I am taking.

Bacon classifies four types of idol that, as he puts it, "block men's minds (*mentes humanas obsident*)": idols of the tribe (*tribus*), the cave (*specus*), the marketplace (*fori*) and the theatre (*theatri*) (Bacon 2000, p. 40 (I, xxxix)).

I) Idols of the tribe

The first type of idols consists of idols of the tribe; that is, errors incidental to human nature in general. The most prominent of these errors are the tendency to support a preconceived opinion by affirmative instances, whilst neglecting all counter examples; the tendency to generalize from a few observations, and to consider mere abstractions as reality. Errors of this type may also originate in the weakness of the senses, which affords scope for mere conjectures (Bacon 1859, pp. 21, 24–29 (I, xli, xlv-lii)). Bacon warns the student of Natural Philosophy against the belief that the human sense is the measure of things. For Bacon, "the human intellect is like an uneven mirror (*speculi inæqualis*) on which the rays of objects fall, and which mixes up its own nature with that of the object, and distorts and destroys it" (*ibid.*, p. 21 (I, xli)). To obtain the true interpretation of nature, the human mind should function, according to Bacon, like an even mirror.

II) Idols of the cave

The second kind of idols consists of idols of the cave. These errors are incidental to the peculiar mental and bodily constitution of each individual (the cave is a direct reference to Plato's simile in the *Republic*). These errors may be either of internal origin, arising from the peculiar physiology of the individual, or of external origin, arising from the social circumstances in which one is placed by education, custom and society in general (*ibid.*, pp. 22, 29–30, 32–33 (I, xlii, liii, lviii)).

III) Idols of the marketplace

The third class of idols comprises idols of the marketplace, that is, errors arising from the nature of language – the vehicle, as Bacon puts it, for the association of men, their commerce and consort (*ibid.*, pp. 22–23, 33–35 (I, xliii, lix, lx)). Language, according to Bacon, introduces two fallacious modes of observing the world. First, there are some words that are merely "the names of things which have no existence (as there are things without names through want of observation, so there are also names without things through fanciful supposition)." Secondly, there are "names of things which do exist, but are confused and ill defined" (*ibid.*, p. 34 (I, lx)). Bacon is aware of the opaqueness of language to nature and that may lead the researcher astray. He therefore cautions the researcher of the faults of language.

IV) Idols of the theatre

Finally, the fourth class of idols consists of idols of the theatre. These are errors which arise from received "dogmas of philosophical systems, and even from perverted laws of demonstrations" (*ibid.*, p. 23 (I, xliv); *cf.*, pp.35–49 (I, liv, lxi-lxvii)). Here Bacon refers mainly to three kinds of error: sophistical, empirical and superstitious. The first error corresponds to Aristotle who has, according to Bacon, "made his Natural Philosophy so completely subservient to his Logic as to render it nearly useless, and a mere vehicle for controversy" (*ibid.*, p. 30 (I, liv; *cf.*, lxiii)). The second error, the empirical, refers to leaping from "narrow and obscure experiments" to general conclusions. Bacon has in mind particularly the chemists of his time and Gilbert and his experiments on the magnet (*ibid.*, pp. 41–42 (I, liv, lxiv; *cf.*, lxx)). The third error, the superstitious, represents the corruption of philosophy by the introduction of poetical and theological notions, as is the case according to Bacon with the Pythagorean system (*ibid.*, pp. 42–44 (I, lxv)).

Concluding his discussion of the idols, Bacon demands that all of them "must be renounced and abjured with a constant and solemn determination" (*ibid.*, p. 49 (I, lxviii)). He insists upon purging (*expurgandus*) and freeing (*omnino liberandus est*) the intellect from the idols, so that "the approach to the Kingdom of Man (*regnum hominis*), which", as Bacon conceived of his quest, "is founded on the Sciences, may be like that to the Kingdom of Heaven (*regnum cœlorum*)" (*ibid.*). Thus, having performed these "expiations and purgations of the mind", one "may come to set forth the true way of interpreting Nature" (*ibid.*, p. 51 (I, lxix)). The religious connotation is explicit and should be underlined.

Clearly, Bacon's doctrine of the idols is systematic and methodical if somewhat contrived. He neatly classifies the idols as "either adventitious or innate. The adventitious," Bacon explains, come into the mind from without – namely, either from the doctrines and sects of philosophers or from perverse rules of demonstration. But the innate are inherent in the very nature of the intellect, which is far more prone to error than the sense is. (Bacon 1960, p. 22 (The Plan of the Great Instauration); on the history of Bacon's scheme see Spedding, Note C, in Bacon 1989, pp. 113–117, and p. 98 fn 1).

The classes of idols proceed progressively from the innate to the adventitious, from the most persistent to the easiest to discard. They reflect as much as "they are separable or inseparable from our nature and condition in life", to use Spedding's formulation (Bacon 1989, p. 91 fn 4; 98 fn 1 and Note C, pp. 113–117). The idols commence with the general character of human beings – the tribe – move on through the features of individuals that comprise the tribe – that is, the cave – further on to the daily intercourse of common life: negotiations and commerce between individuals – the marketplace – and reach finally the doctrines that individuals conceive and believe in – the theatre. Bacon is aware of the fact that the innate features are hard to eradicate, so that these idols cannot be eliminated. "All that can be done", he instructs, "is to point them out, so that this insidious action of the mind may be marked and reproved (else… we shall have but a change of errors, and not clearance)…" (Bacon 1960, p. 23 (The Plan of the Great Instauration)). By contrast, the adventitious idols, principally those of the theatre, could and should be eliminated (Bacon 2000, p. 49 (I, lxi)). Having undergone these epistemological ablutions, and "clarified the part played by the nature of things (*rerum natura*) and the part played by the nature of the mind (*mentis natura*)", one is ready according to Bacon to commence anew the true interpretation of nature (*ibid.*, p. 19 (Plan of "The Great Renewal"); 1989, pp. 139–140. *Cf.* I, cxv).

Bacon designed the typology to shed light on the nature of sources and causes of error (*causas errorum*) (Bacon 1989, p. 186 (I, lxxviii)). The scheme of idols presents a systematic and methodical view of the elements involved in the obstruction of knowledge: the interplay of sources of error pertaining to the nature of the mind in general, to individuals and their community, to language and doctrines. The scheme may appear somewhat artificial, but it constitutes an essential element of Bacon's comprehensive conception of the emergence of new knowledge and its impediments. In many respects the scheme of idols anticipated new disciplines, namely, the study of anthropology, ethnology, psychology, linguistic and cultural, political and religious ideologies (Coquillette 1992, pp. 233–234; for references see p. 300, fn 24).

A CRITIQUE OF BACON'S SCHEME

The question naturally arises whether or not this all-embracing typology of sources of error is applicable to the very method of research that Bacon advocates for use, that is, experimentation. "It will doubtless occur to some", Bacon acknowledges the question, that

> there is in the Experiments themselves some uncertainty or error; and it will therefore, perhaps, be thought that our discoveries rest on false and doubtful principles for their foundation (Bacon 1859, pp. 111–112 (I, cxviii)).

This appears to be a surprising remark. Could it be that Bacon's proposed method of research is open to objections and that all the cleansing and ablutions were for nothing? No! Bacon dismisses the threat right away; "this is nothing", he exclaims, "for it is necessary that such should be the case in the beginning." By way of an analogy he explains that

> it is just as if, in writing or printing, one or two letters should be wrongly separated or combined, which does not usually hinder the reader much, since the errors are easily corrected from the sense itself. And so men should reflect that many Experiments may erroneously be believed and received in Natural History, which are soon afterwards easily expunged and rejected by the discovery of Causes and Axioms (*ibid.*, p. 112 (I, cxviii)).

Bacon assures us that we should not be disturbed by these objections and he reiterates this confidence in his outline for experimental history (Bacon 1960, p. 280 (viii)). However, he admits that

> it is true, that if the mistakes made in Natural History and in Experiments be important, frequent, and continuous, no felicity of wit or Art can avail to correct or amend them (Bacon 1859, p. 112 (I, cxviii)).

Thus, if there lurked at times "something false or erroneous" in Bacon's Natural History which have been proved with "so great diligence, strictness, and", Bacon adds, "religious care", what then must be said, he asks rhetorically, "of the ordinary Natural History, which, compared with ours, is so careless and slipshod?, or of the Philosophy and Sciences built on ... quicksands?" (*ibid.*)

Notwithstanding Bacon's resolute assurance, the objections are disturbing. Bacon appears to be waving his hands, so to speak, rather than providing convincing arguments in defence of his position. He would have us believe that the analogy between a printer's error and an experimental error is a faithful one. However, it is precisely the sense of the context – the meaning which is given according to Bacon's analogy – that the experimental sciences lack and in fact seek to discover. The two types of error, namely, the printer's and the experimental, are categorically different. (I distinguish elsewhere between these two possible faults. I call the former mistake and the latter error, see Hon 1995.)

Surprisingly, it appears that Bacon did not apply consistently his critical scheme of errors to the very instrument of his inquiry – experiment. Admittedly, he was concerned with errors that beset the mind: once one had purged one's mind from the idols and, to use Bacon's mirror metaphor, smoothed away with religious fervour every protrusion and cavity in one's intellect so that it became an even surface reflecting genuinely the rays of things (Bacon 1960, p. 22), one was then ready to embark on the true way of interpreting nature. At issue here is not whether this instruction to cleanse one's mind is practicable or not, but rather can the instrument of one's inquiry be itself an object of critical scrutiny. Indeed, as we have seen, it had taken some time before the question: "What exactly is an experiment in physics?" was explicitly raised and addressed (Duhem 1974, p. 144).

The persistent impediment that the occurrence of errors poses knowledge resulting from experimentation is not covered by Bacon's scheme of idols of the mind. Bacon's trust in his method of inquiry, which he expressed with his off-hand dismissal of experimental errors, is objectionable. I follow up this criticism and propose to examine the different idols that beset experiment.

THE IDOLS OF EXPERIMENT:
SCRIPT, STAGE, SPECTATOR AND MORAL

The construction of a scheme of idols that beset experiment has a similar objective to Bacon's scheme, but the analysis goes further in that it explicitly argues that the scheme reflects underlying principles of experimentation, that is, the principle of classification reflects the elements that comprise experiment and their interrelations. My intent, to repeat, is not to seek strategies in an *ad hoc* fashion following Mach and Franklin. That is, to refer once again to Gooding, Pinch and Schaffer's well phrased remark that, in Franklin's view "there are epistemological rules which can be applied straightforwardly in the field to separate the wheat of a genuine result from the chaff of error" (Gooding et al. 1989, pp. 22–23). The objective is not to list such rules in an eclectic way, but rather to construct a constrained scheme of "the chaff of error" that reflects the structure of experiment as an instrument of inquiry designed to secure knowledge.

In the spirit of the metaphorical language of Bacon and following his idols of the theatre, I suggest to discern four kinds of idol that beset experiment: idols of the *script*, the *stage*, the *spectator* and the *moral*. The image of theatrical play constitutes a convenient and useful metaphorical setting for experiment since, like a play enacted on stage, an experiment is the result of an activity that has truly "a show" at its centre (Cantor 1989, pp. 173–176). In an experiment, nature is made, if you will, to display a show on a stage conceived and designed in some script. The show is observed and registered by a human or automated spectator and, finally, interpretation is proposed with a view to providing a moral – that is, the outcome of the experiment as knowledge of the physical world.

Error is a multifarious epistemological phenomenon. It is an expression of divergence whose mark is discrepancy – a discrepancy which emerges from a procedure of evaluation against a chosen standard. The nature of this discrepancy, the reason for its occurrence, how to treat it and what can be learnt from it once it has been perceived and comprehended, constitute the vast subject of the problem of error. Each of the four different idols depicts different kinds of cause of discrepancy that may arise at different stages of the process that makes an experiment.

Experiments proceed essentially in two stages: *preparation* and *test*. In the preparation stage the experimenter sets up the initial conditions of the apparatus and the system within which the experiment is designed to evolve – this is the theoretical and the material framework of experiment. Once the experimenter sets the framework, the experiment may commence its runs: the testing – the evolution of the system within the designed framework. I should underline that I use the term "test" in a very loose sense: an experiment is not necessarily a test of some theory. In fact, many experiments (e.g., in physics) have to do with determining some constant of a certain material or a system. However, the dichotomy between these two distinct stages: the *preparation* and the *test*, is crucial in the sense that experiment always exhibits the evolution of a prepared system. (For further analysis see Hon 1998a, §6.)

Constituting a typology of sources of error, the idols reflect the roles that faulty elements would play in the overall structure of experiment. It may be seen immediately that the idols of the script and of the stage are associated with the *preparation*, whereas the idols of the spectator and of the moral pertain to the *testing*. In this way the idols cover all possible faults in terms of the different contexts in which sources of error may crop up in experiment. The claim then is that possible sources of error arranged as they are in four different idols, illuminate the structure of experiment.

A distinct characteristic of the proposed taxonomy is its focus on the source rather than on the resultant error. By concentrating on the definitions of different classes of source of error, the typology illuminates from a negative perspective the elements which are involved in experiment and their inter-relations. Thus:

- An incorrect or ill-suited background theory (e.g., the application of Stokes' law to the very tiny and irregular, jagged metal dust particles in Ehrenhaft's alleged discovery of subelectrons (Hon 1989b)) – an idol of the script, is different from
- Assuming erroneously that certain physical conditions prevail in the set-up (e.g., technical difficulties in establishing and continually maintaining in a systematic fashion the physical conditions required for the determination of the Hall effect: a metal specimen kept in very high temperatures and subject to a strong magnetic field) – an idol of the stage.
- Physical, physiological and psychological elements interfering with the depiction of the displayed phenomenon or with the reading of a measuring device (e.g., Blondlot's auto-suggestive perception of N rays (Nye 1980)) – an idol of the spectator, is different from
- Conferring an erroneous interpretation on experimental results (e.g., Franck and Hertz's interpretation that the first critical potential they measured was an ionisation potential (Hon 1989a)) an idol of the moral.

Which way we look at them, errors – that is, experimental errors – would be covered, I submit, by one of the four idols. (For an elaboration of the account of the four idols as classes of experimental error together with historical illustrations see Hon 1989b.)

An important feature of the typology is that it characterises "the script" – the conceptual, theoretical guiding lines of apparatus and instruments, that is, the background theories – as analytically distinct from "the moral": theories that provide the basis for the interpretation of the outcome of experiment. This distinction is logically crucial since it keeps apart the theories that constitute the conceptual framework of experiment and the theories that render the outcome of experiment meaningful. One of the crucial features of the modern method of experimentation, namely, procedures of correction and reduction of data, was recognised at the outset by Galileo. The experimenter should be, as Galileo demands, a good accountant:

Just as the computer who wants his calculations to deal with sugar, silk, and wool must discount the boxes, bales, and other packings, so the mathematical scientist (*filosofo geometra*), when he wants to recognize in the concrete the effects which he has proved in the abstract, must deduct the material hindrances, and if he is able to do so, I assure you that things are in no less agreement than arithmetical computations. The errors, then, lie not in the abstractness or concreteness, not in geometry or physics, but in a calculator who does not know how to make a true accounting (Galileo 1974, pp. 207–208).

Clearly, to conduct successfully this true accounting the experimenter would need to resort to a theory. This theory should be provided by "the script" and not by "the moral", lest the argument would be circular.

Duhem's insightful logical analysis of the correction procedure of systematic error is rightly based on theories that belong to "the script" and not to those that belong to "the moral" of experiment. Duhem observes that a physical experiment is not merely the observation of a group of facts produced under some controlled constraints. If it were so, it would have been absurd to bring in corrections,

for it would be ridiculous to tell an observer who had looked attentively, carefully, and minutely: "What you have seen is not what you should have seen; permit me to make some calculations which will teach you what you should have observed" (Duhem 1974, p. 156).

Following Duhem, observations in experiment have to be capable of translation into a symbolic language, e.g., an equation, and it is physical theories that provide the required rules of translation. The experimenter has constantly to compare, to continue Duhem's line of argumentation, two objects: on the one hand, the real, concrete object which is being physically manipulated – the apparatus, and on the other hand the abstract, symbolic object upon which one reasons (*ibid.*, p. 156). This crucial comparative activity in experimentation, which allows for the introduction of necessary correction terms, depends entirely on "the script". By contrast, the theories that provide the basis for interpretation, that is, "the moral", are brought as it were from without; they are not involved in the process of correcting systematic errors. They are however crucial for correcting errors of interpretation. However, this analytical purism of separating the script from the moral is not strictly adhered to in the laboratory. In the actual practice of experimentation one encounters frequently the toing and froing between the script and the moral in an attempt to stabilise the result. As philosophers we should caution the practicing experimenter of this shoddy logic.

I am now in a position to look critically at Hacking's typology. Hacking, it may be recalled, has grouped experimental elements into three classes: "ideas, things, and marks" (Hacking 1992, p. 44). As I have indicated, my proposed scheme of idols that beset experiment reflects, albeit negatively, Hacking's typology. The scheme of idols diverges however from the typology which Hacking has proposed on two important points. Roughly, "ideas" correspond to "idols of the script", "things" to "idols of the stage" and finally "marks" relate to elements of "idols of the spectator". There remains the class of "idols of the moral" which Hacking's typology appears not to cover; or, alternatively, in his typology "ideas" cover both the background and the outcome of experiment without distinguishing between these two sets of elements. I agree with Hacking that flexibility and interplay of elements are crucial to the stability of experimental results, and so one may cover the fourth set of idols, "idols of the moral", by "ideas". This is, as I have pointed out, a realistic view of experimental practice since "the script" – "ideas" in Hacking's terms – often informs the interpretations of experimental results.

Nevertheless, I do hold strongly that for analytical, logical reasons there should be a clear separation between "the script" and "the moral". Hacking's taxonomy eliminates the crucial difference between these two sets of idols. Again, the "script" consists of theories that are presupposed to govern and shape the experiment – both the working of the apparatus and the application of instruments. The experimenter does not put these theories to the test; they are presupposed at the preparation stage for the purpose of setting up the initial conditions of the experiment and therefore considered correct. These theories provide the framework for the execution of experiment. By contrast, theories that belong to the "idols of the moral" are being tested and may be dispensed with, replaced or rejected and indeed proved false without affecting at all the overall experiment, its argument and the body of its accumulated data.

A fine historical example is the Franck-Hertz experiment. It required just a change of interpretation to render the experiment worthy of the Nobel Prize. Franck and Hertz interpreted their measurement of the initial critical velocity of slow electrons colliding with gas molecules as corresponding to the energy required to initiate the ionization process. Thus, the very first experiment that demonstrated directly and graphically the existence of quantum energy levels in an atom – a curve exhibiting distinct peaks that indicated a stepwise transfer of energy within the atom – was interpreted not as a quantum but as a classical phenomenon. Only in light of a suggestion by Bohr in 1915 did Franck and Hertz reinterpret their experimental results. They resisted Bohr's suggestion for a while, but once they accepted it, namely, that the critical velocity of the accelerated electrons indicated excitation and not ionization of the bombarded atoms, they recognized their experimental results as strong evidence for the existence of atomic energy levels. The experiment thus contributed to the acceptance of Bohr's atomic theory. This was acknowledged by the Nobel committee, who in 1925 declared that Franck and Hertz had demonstrated the existence of energy levels of the type called for by Bohr's theory of the atom. Franck and Hertz received the Nobel prize in physics, as the citation reads, "for their discovery of the laws governing the impact of an electron upon an atom" (Hon 1989a and forthcoming). Clearly, the moral came as it were from without and it was not part of the script.

Furthermore, the alternative of grouping together the "spectator" and the "moral" under Hacking's class of "marks" should also be objected to. Again, the sources of error and procedures of correction that take place in reading data are distinct from analysing, reducing and interpreting the data and rendering them an experimental result. Thus, from the negative perspective, that is, from the perspective of error, it is instructive to split Hacking's "marks" into two different, distinct classes – "spectator" and "moral". The reader may recall that these two idols comprise the second stage of experiment – the test.

CONCLUDING REMARKS

Against the background of collapse and decline of Scholastic epistemology, a breakdown that led to the proliferation of often conflicting views of knowledge, Bacon conceived of a science in which one seeks "to discover the powers and actions of bodies, and their laws limned in matter. Hence this science", according to Bacon, "takes its origin not only from the nature of the mind but from the nature of things". Bacon developed a new logic, which he had designed in order "to dissect nature truly" (Bacon 2000, pp. 219–220 (II, lii); Solomon 1998, p. xv). This new logic should vouch, in Bacon's view, for the true "Interpretation of Nature" (Martin 1992, p. 147). It consists essentially of two moves. The first, as Bacon put it, is the "expurgation of the intellect to qualify it for dealing with truth" (quoted by Martin, *ibid.*), and the move to follow is "the display of a manner of demonstration for natural philosophy superior to traditional logic" (*ibid.*). Bacon developed the scheme of idols to facilitate the first move; the second move proceeds by founding philosophy on natural and experimental history – the furnishing of the material of knowledge itself (*ibid.*, pp. 146–147).

My proposed scheme of the idols of experiment takes its cue from this Baconian two-tier approach to the true way of interpreting nature. However, the point of my scheme is not epistemological but rather methodological – it is here that the analogy to Bacon's approach ends. The proposed scheme carries the critical, Baconian program over to experimentation itself.

The scheme focuses on the different kinds of possible sources of error that may crop up in experiment. In that sense, the scheme reflects the normative aspect of experiment: the practice of seeking to minimize, if not eliminate altogether, experimental errors. However, once the typology is set up, it may be seen that the different kinds of source of error present four different contexts, which together make experiment. In other words, the four idols: the script, the stage, the spectator and the moral, cover all possible sources of error, each idol characterizing a class of sources of error which arise in the same context, that is, discrepancies of similar origin. The constraints imposed by the scheme with its clear delineation of the classes, provide a comprehensive overview of experiment from a negative perspective that does not depend on open lists. It is hoped that studies of the relations between the elements that comprise the idols could provide an insight into the epistemological underpinnings of experimentation. By transcending the list, the set of idols of experiment provides us with both a normative and a comprehensive, conceptual view of experimentation.

Acknowledgments

I am grateful to the Alexander von Humboldt Foundation for facilitating the completion of this study and to my host at the Max Planck Institute for the History of Science, Jürgen Renn, for the generous hospitality. I wish to thank further Raffaella Campaner and Gereon Wolters for their remarks and comments and Maria Carla Galavotti for the invitation.

Department of Philosophy, University of Haifa
Haifa, Israel

REFERENCES

Bacon, F. [1620] 1859. *Novum Organum*. Translated by A. Johnson. London: Bell & Daldy Fleet Street.

Bacon, F. [1620] 1960. *The New Organon*. Edited with an introduction by F. H. Anderson. Englewood Cliffs, New Jersey: Prentice Hall.

Bacon, F. [1620] 1989. *The Works of Franics Bacon*. Faksimile-neudruck der Ausgabe von Spedding, Ellis und Heath, London 1857–1874, in vierzehn Bänden, erster Band. Stuttgart: Frommann-Holzboog Verlag.

Bacon, F. [1620] 2000. *The New Organon*. Edited by L. Jardine and M. Silverthorne. Cambridge: Cambridge University Press.

Buchwald, J. Z. 1993. "Design for Experimenting". In P. Horwich (ed.), *World Changes*. Cambridge: MIT Press, 1983, pp. 169–206.

Cantor, G. 1989. "The Rhetoric of Experiment". In Gooding et al. (eds.), 1989, pp. 159–180.

Coquillette, D. R. 1992. *Francis Bacon*. Stanford, California: Stanford University Press.

Darrigol, O. 1999. "Baconian Bees in the Electromagnetic Fields: Experimenter-Theorists in Nineteenth-Century Electrodynamics". *Studies in History and Philosophy of Modern Physics* 30: 307–345.

Duhem, P. [1906] 1974. *The Aim and Structure of Physical Theory*. New York: Atheneum.

Franklin, A. 1986. *The Neglect of Experiment*. Cambridge: Cambridge University Press.

Franklin, A. 1989. "The Epistemology of Experiment". In Gooding et al. (eds.), 1989, pp. 437–460.

Franklin, A. 1990. *Experiment, Right or Wrong*. Cambridge: Cambridge University Press.

Galilei, G. 1974. *Dialogue Concerning the Two Chief World Systems*. Translated by S. Drake. Berkeley, Los Angeles, London: University of California Press.

Gooding, D., Pinch, T. and Schaffer, S. (eds.) 1989. *The Uses of Experiment*. Cambridge: Cambridge University Press.

Hacking, I., 1984. "Experimentation and Scientific Realism". In Leplin (ed.), 1984, pp. 154–172.

Hacking, I. 1992. "The Self-Vindication of the Laboratory Sciences". In A. Pickering (ed.), *Science as Practice and Culture*. Chicago and London: the University of Chicago Press, 1992, pp. 29–64.

Harré, R. 1983. *Great Scientific Experiments*. Oxford and New York: Oxford University Press.

Heidelberger, M. and Steinle, F. (eds.) 1998. *Experimental Essays – Versuche zum Experiment*. Baden-Baden: Nomos Verlagsgesellschaft.

Hintikka, J. 1988. "What is the Logic of Experimental Inquiry?". *Synthese* 74: 173–190.

Hon, G. 1989a. "Franck and Hertz versus Townsend: A Study of Two Types of Experimental Error". *Historical Studies in the Physical and Biological Sciences* 20: 79–106.

Hon, G. 1989b. "Towards a Typology of Experimental Errors: an Epistemological View". *Studies in History and Philosophy of Science* 20: 469–504.

Hon, G. 1995. "Going Wrong: To Make a Mistake, to Fall into an Error". *The Review of Metaphysics* 49: 3–20.

Hon, G. 1998a. "'If This Be Error': Probing Experiment With Error". In Heidelberger and Steinle 1998, pp. 227–248.

Hon, G. 1998b. "Exploiting Errors". *Studies in History and Philosophy of Science* 29: 465–479.

Hon, G. Forthcoming. "From Propagation to Structure: The Experimental Technique of Bombardment as a Contributing Factor to the Emerging Quantum Physics". *Physics in Perspective*.

Jardine, L. 1974. *Francis Bacon: Discovery and the Art of Discourse*, Cambridge: Cambridge University Press.

Kuhn, S. T. 1970. *The Structure of Scientific Revolutions.* Second edition, enlarged. Chicago: the University of Chicago Press.

Leplin, J. (ed.) 1984. *Scientific Realism.* Berkeley, Los Angeles, London: University of California Press.

Mach, E. [1905] 1976. *Knowledge and Error.* Dordrecht: Reidel.

Martin, J. 1992. *Francis Bacon, the State, and the Reform of Natural Philosophy.* Cambridge and New York: Cambridge University Press.

Mayo, D. 1996. *Error and the Growth of Experimental Knowledge.* Chicago and London: Chicago University Press.

Nye. M. J. 1980. "N-Rays: An Episode in the History and Psychology of Science". *Historical Studies in the Physical Sciences* 11: 125–156.

Pickering, A. 1989. "Living in the Material World: on Realism and Experimental Practice". In Gooding et al. (eds.), 1989, pp. 275–297.

Radder, H. 1998. "Issues for a Well-Developed Philosophy of Scientific Experimentation". In Heidelberger and Steinle (eds.), 1998, pp. 392–404.

Solomon, J. R. 1998. *Objectivity in the Making: Francis Bacon and the Politics of Inquiry.* Baltimore and London: Johns Hopkins University Press.

Whitehead, A. N. 1929. *Science and the Modern World.* Cambridge: Cambridge University Press.

RAFFAELLA CAMPANER

AN ATTEMPT AT A PHILOSOPHY OF EXPERIMENTAL ERROR
A COMMENT ON GIORA HON

Giora Hon's paper aims at presenting a philosophical analysis of experiments and their role in the acquisition of scientific knowledge. In an attempt to answer the question "What is a scientific experiment?", the author suggests what he himself defines as a "negative way" to experimentation: in order to uncover the crucial features of experiments, we shall focus on experimental errors, and, more precisely, on their sources.

Hon maintains that accounts of experiments already presented in the literature, such as those by Ian Hacking and Allan Franklin, are inadequate, insofar as they fail to elaborate a coherent and convincing philosophical view of experiment. Instead of looking for some "principle of organisation of kinds of experiments according to their goals" (this volume, p. 266), as the other authors have been doing, Hon puts forward a restricted classification meant to organise errors according to their sources. He maintains that errors are related to the following:
1) An incorrect or ill-suited background theory;
2) Assuming erroneously that certain physical conditions prevail in the set-up;
3) Physical, physiological and psychological elements interfering with the depiction of the displayed phenomenon or with the reading of a measuring device;
4) Conferring an erroneous interpretation on experimental results.
"Which way we look at them" – Hon claims – experimental errors will fall under one of these four "idols" (this volume, p. 278).

These categories reflect, I believe, a conception of experiment as having a strong theoretical component. As suggested by the literature on the topic over the last fifteen years (*cf.*, for example, Pickering, Galison and Gooding, as well as Hacking and Franklin, already mentioned), it is necessary to take into account a much wider range of aspects having to do with scientific activity and experimental practice to elaborate an adequate philosophy of experiment From the list 1) – 4), there emerges a peculiar, almost exclusive attention for theoretical aspects of knowledge: Hon's concern focuses on background *theory*, *assumptions* about set-ups and *interpretations* of results. The only category which seems to leave wider scope for practical aspects is number 3), although it, too, lays more emphasis on *depiction* of the displayed phenomena and *reading* of devices more than, for example, their manipulation. The list mirrors a rather "biased" conception of possible sources of errors if the target is to identify an overall, well-developed epistemology of the latter. A number of actual, practical skills and, so to speak, "practicalities" do not constitute a secondary, peripheral or accessory aspect, but play a pre-eminent role in the performance of a large number of experiments. Non-verbal or pre-linguistic skills and mastery of experimental apparatus, techniques and procedures have been more and more emphasized by recent attempts to reconstruct the peculiar features of scientific research[1]. To refer to some episodes in the history of science analysed in detail in the pertinent literature, Gooding, for example, describes the experiments performed by the French physicist J. B. Biot around 1820, exploring the interaction between electric currents and magnetized needles:

> Biot reports that when the wire was brought close to a horizontally suspended magnetic needle, there was an immediate deviation of the needle [...] But the possibility of observing anything but chaotic needle behaviour depends on skilful manipulation of the wire, and this takes some time to acquire. [...] As we shall see with Faraday and Morpurgo, would-be observers have to *do* quite a bit in order to see anything at all. [...] Scientists engage nature in the fine structure of their experiments. That is where they gain the practical mastery of a phenomenal domain that enables them to develop the linguistic resources and the demonstrative experiments that they use to establish facts about nature (Gooding 1990, p. 133).

And discussing Faraday's investigation of electromagnetism, Gooding highlights that:

> Recent repetition of these experiments has shown the difficulty of seeing what Faraday recorded he saw, even after considerable practice. [...] skilful interaction with the phenomenal world is needed as well as a concept of what might be elicited. When Biot and Faraday arranged their operations and the outcomes as images or instruments they embodied their experience and associated observational skills which had been *impossible* to communicate in verbal and material representations that were *easy* to communicate (Gooding 1990, p. 134; p. 137).

Experimental activity involves a good deal of manipulation of the entities and phenomena investigated and may require some highly sophisticated skills (let us think, for example, of those necessary for chemical syntheses, or of performances of microinjections in cell cultures in molecular biology), as well as skills in elaborating visualizations and in reading of visual images. Hon criticises Franklin's list of strategies "that provide reasonable belief in the validity of an experimental result" (Franklin 1990, p. 103) for being *ad hoc*, and "neither exclusive nor exhaustive" (this volume, p. 267), not inspired by any general guiding principle. Hon's conceptual scheme seems, though, to be based on an overgeneral, or perhaps partial, typology of sources of error, which runs the risk of failing to account for observational and procedural abilities. Practical aspects, not strictly theoretical, logical, or linguistic ones, seem to be particularly important if, as it is the case here, an epistemology of *experiment* is the final target of the whole inquiry. Hon's analysis is meant to shed light specifically on the crucial features of experiments. "Practicalities" and skills ought therefore to be given a very specific place among the possible sources of *experimental* errors: it is necessary to ensure that Hon takes into consideration all relevant features to make it illuminating specifically for a theory of experiment, and not simply for any general theory of knowledge[2]. The author himself remarks that capturing the enormous variety of variables involved in experimentation is an extremely arduous and puzzling task. Experiments consist in:

a play of operations in a field of activity, which I call the experimenter's space. The place of the experiment is not so much a physical location [...] as a set of intersecting spaces where different skills are exercised (Pickering 1992, p. 75).

If practical, concrete elements are maintained to be such an essential component of experiments, it seems reasonable to acknowledge their primary role also within an attempt to identify sources of experimental errors.

Hon faces the task of elaborating a philosophical map of the complex array of heterogeneous elements experiments consist in.

The shaping of experimental systems is a contingent process. It is embedded in instruments, apparatus, technical procedures, materials at hand, and model objects, on the one hand, and it is closely linked to local crafts, research traditions, and wider epistemic as well as practical interests on the other. The decisive question is how these particular segments get articulated, how they condense to a structure that finally develops a dynamics that was not inherent in these parts per se, and therefore serves as a crystallisation point for unprecedent knowledge (Hagner and Rheinberger, p. 363).

Is it enough to present a categorisation of errors by source to capture such a complex, intertwined set of elements? At least two issues seem to be at stake: on the one hand, it is to be established whether Hon's typology covers all the possible sources of error; on the other hand, whether an analysis of errors as such can accomplish the challenging enterprise of giving a satisfactory insight on experimentation as a scientific activity. Is it possible – for example – to elaborate a good philosophy of experiment without considering at all the social side of experimental practice, its economical components, human intentionality, plans and goals, or even the role of "common sense" in deciding when to consider an experiment concluded?[3] Although these cannot be strictly regarded as "sources of error", a certain social, public and economic dimension should perhaps be given some space. In many cases, dozens or even hundreds of scientists combine a diverse range of resources in a collaborative effort to perform a single, massively sophisticated experiment (*cf.*, for example, modern light-energy and quantum physics). These features are to be taken into account as playing some role in the working of experiments, but it seems unlikely that they can be satisfactorily represented in terms of error source analysis.

Hon's explicit concern is with what he calls the "methodological cluster of obstacles to the construction of a philosophy of experiment" (this volume, p. 260). When addressing such a cluster,

> we are concerned with the transition from the myriad of strategies, methods, procedures, conceptions, styles and so on, to some general, cohesive and coherent view of experiment as a method of extracting knowledge from nature (this volume, p. 264).

It is debatable whether one can formulate an account of experimental error to solve solely the issues raised by the methodological cluster of problems, without also dealing with those raised by the epistemological cluster, namely the transition from material processes to propositional knowledge. Hon views experiments as arguments:

> An experiment – I claim- can be cast into a formal argument whose propositions instantiate partly states of affairs of material systems and partly inference schemata and some lawful, causal connections. In other words, an experiment implies an argument the premises of which are assumed to correspond to the states of the physical systems involved, e.g. the initial conditions of some material systems and their evolution in time. These premises warrant the argument's conclusion (Hon 1998, p. 235).

It might, however, also be necessary to give some more consideration to the move from material procedures to propositional knowledge, from the performance of instrumental devices and their manipulation to their translation in accounts of phenomena. When trying to identify as completely as possible the sources of error, we need to reconstruct the whole performance of the experiment, that is the process of its occurring, or, in other words, its development. While *historical* studies on experiments have recently been flourishing, a *philosophy* of experiments – Hon highlights – has yet to be advanced. Even if we espouse the author's point of view, we will still need an account of the *dynamic* process of which the experiment consists. Scientific practice has an intrinsic real-time structure[4]. If we are content with merely a classification of possible sources of errors, we might get simply a static image, a "photograph" of how experiments *have* worked, or, rather, of how they failed to work, instead of reaching an understanding of effective inner *workings* of experiments[5]. Some reference to the temporal dimension of the experimental activity seems to be particularly important, especially given that the target of Hon's own paper is to "develop a *historically informed* philosophy of experiment" (this volume, p. 259, italics added).

Following Hon's proposal, to provide a satisfactory answer to the original question "What is a scientific experiment?", the question to be raised is then: "What is an experimental error?". If the experimenters have insufficient practical skills or do not possess sufficient manual dexterity with tools and procedures, the theory and the storage of knowledge behind given techniques cannot be of much use. The experimenter may not possess the necessary practical competences. These often cannot be conveyed verbally, but require lengthy practice to be mastered. Not only the apparatus, but also the experimenter may not be working properly. Is a lack of abilities of this sort to be considered an error, or rather as an "oversight", or some sort of "miscalculation", or a still different kind of "fault"?

In another paper, Hon draws a line between the concept of "error" and that of "mistake"[6]. In Hon's perspective, errors are associated with unavoidable ignorance; they come about when one applies techniques to novel phenomena and is therefore groping, so to speak, in the dark. They occur because an exploration of a *terra incognita* is taking place. Mistakes, in contrast, are associated with avoidable ignorance. They occur while we are walking on a *terra firma* and could be avoided since checking procedures are known and available (*cf.* Hon 1995a, p. 6). In this respect, "material aspects" and practical skills should hence be considered particularly important precisely in the light of the innermost, distinctive features of errors. In experimental enquiry, which tries to breach the borders of acquired knowledge and gropes its way in a vanguard position, practical abilities and manipulative techniques have an extremely important discovery role: encounters with bits of the world not anticipated by any theoretical knowledge often occur through material procedures.

In order to turn a classification of errors into an efficient means toward an epistemology of experiment, the analysis needs to include some understanding of how a number of distinct elements happen to come together into that special source of knowledge an experiment is. The problem comes down to whether a fixed classification of sources of error can on its own provide a satisfactory insight in such a complicated concrete process or activity as experimentation. A classification of errors might run the risk of being considered a kind of *a posteriori* operation: only after errors have been clearly identified and attributed to some causes, is it possible to define their sources. The individuation of sources of errors might be suspected to already presuppose a specific, possibly biased, conception of experiments, their structure and their functioning, whereas the real challenge here is to understand what the added value of a negative route towards a philosophy of experiment is with respect to the "standard", "positive" way. We should, in other words, question whether an analysis of errors constitutes a viable access to the essential functioning of experiments, or, rather, whether an epistemology of experimental errors cannot but presuppose an already quite refined view of what experiments are like. The immediacy of this risk is suggested by a different notion of "error" given by Hon. Error is "an expression of divergence whose mark is discrepancy – a discrepancy which emerges from a procedure of evaluation against a chosen standard" (this volume, p. 277). This seems to hint at the existence of some already established standard or datum point in the definition of error, which would not be present, in this case, as science is proceeding towards what is largely a *terra incognita*.

Hon's attempt to build a new philosophy of experiment in terms of errors and their sources no doubt represents an original approach in the literature on the topic. Especially because of its originality, this shifting of perspective is worthwhile of further development, in order to avoid falling victim of some other "idols" and to shed more light on the complex web of elements that makes an experiment: "knowledge can arise only when there is a possibility of being wrong" (Hon 1995, p. 15), or a possibility of lacking material, manual skills and acting incorrectly.

Department of Philosophy, University of Bologna
Bologna, Italy

NOTES

[1] Commenting on Gooding's work, Thomas Nickles, for example, recalls: "Given the highly formal treatment of reasoning by many philosophers, even to speak of experimental reasoning already threatens to impose an overly verbal, rule-based, indeed theoretical, perspective and to ignore the skilled practice and judgmental behaviour which characterise experimental work. While philosophers and other students of science have long debated the theory-ladenness of observation, Gooding point out what we might call the technique- or skill-ladenness of observation [...] He reminds us that at the frontier experimentalists are, in some respects, novices rather than experts, and he is thereby able to backlight the surprisingly large gaps between the initial detection of observation novelty, its eventual cognitive organisation in the work of an individual, and its later articulation as a finished scientific communication" (Nickles 1988, p. 300).

[2] See, for example, the following reflections on the topic: "...grounding rational lines of inquiry in lucky discoveries of improvement in apparatus seems embarrassing to experimenters, who might like to be granted powers of thought, and who might also crave an image of scientific rationality. Therefore, it is not all that frequent that an experimental paper freely admits that a breakthrough occurred when someone tried some 'sticky tape', 'waste plastic material that happened to be at hand' or 'a new kind of oil' to doctor a balky piece of equipment, but such accidents occur. So, there's a bias against sticky tape in the original accounts, and then again in philosophical reflections. In my opinion, we have to work against the temptation to produce smooth symmetric theories of experimentation. Let me come back to Allan Franklin. [...] The only real representation of experiment [...] in his first book is the glorious photo of a mess of a laboratory on the dust jacket [...] Philosophers still need to get sticky tape on their fingers. In short, we need to get down and get dirty before we will have an appropriate understanding of experimentation" (Ackermann 1990, p. 456); and: "The more abstract, theory-based conception of knowledge familiar from earlier socio-historical studies is gradually turning into a more particularistic conception of the material sites, artefacts and techniques of 'knowledge production'. The focus is more intensive and 'internal' [...], as the aim is to identify the pragmatic strategies and informal judgments made at the worksite when researchers sort through 'messy' arrays of data and decide whether equipment is working properly" (Lynch 1990, p. 476).

[3] *Cf.*, for example, Pickering 1995a, especially pp. 17–23.

[4] On the essential temporal dimension of scientific practice, *cf.* Pickering 1995a and 1995b.

[5] As Hon acknowledges, even if "we have no choice but to analyze experiment *in vitro*", we should "keep a wide eye on its features as an activity *in vivo*" (this volume, p. 261).

[6] See Hon 1995a, especially pp. 6–7.

REFERENCES

Ackermann R. 1990. "Allan Franklin, Right or Wrong". In A. Fine, M. Forbes and L. Wessels (eds.), PSA 1990, vol. 2, pp. 451–457.

Buchwald J. (ed.) 1995. *Scientific Practice*. Chicago and London: The University of Chicago Press.

Buchwald J. 1998. "Issues for the History of Experimentation". In M. Heidelberger and F. Steinle (eds.), 1998, pp. 374–391.

Franklin A. 1989. "The Epistemology of Experiment". In D. Gooding, T. Pinch and S. Schaffer (eds.), 1989, pp. 437–460.

Franklin A. 1990. *Experiment, Right or Wrong*. Cambridge: Cambridge University Press.

Galison P. and Assmus A. 1989. "Artificial Cloud, Real Particles". In D. Gooding, T. Pinch and S. Schaffer (eds.), 1989, pp. 225-274.

Galison, P. 1998. "Context and Constraints". In J. Buchwald (ed.), 1995, pp. 13–41.

Gooding, D. 1989. "'Magnetic Curves' and the Magnetic Field: Experimentation and Representation in the History of a Theory". In D. Gooding, T. Pinch and S. Schaffer (eds.), 1989, pp. 183–223.

Gooding, D., Pinch T. and Schaffer S. (eds.) 1989. *The Uses of Experiment*. Cambridge: Cambridge University Press.

Gooding D. 1990. "Theory and Observation: the Experimental Nexus". *International Studies in the Philosophy of Science* 4: 131–148.

Hagner M. and Rheinberger H.J. 1998 "Experimental Systems, Objects of Investigation and Spaces of Representation". In M. Heidelberger and F. Steinle (eds.), 1998, pp. 355–373.

Heidelberg, M. and Steinle, F. (eds.) 1998. *Experimental Essays – Versuche zum Experiment*. Badeb-Baden: Nomos Verlagsgesellschaft.

Hon, G. 1989a. "Franck and Hertz versus Townsend: a Study of Two Types of Experimental Error". *Historical Studies in the Physical and Biological Sciences* 20: 79–106.

Hon, G. 1989b. "Towards a Typology of Experimental Errors: An Epistemological View". *Studies in History and Philosophy of Science* 20: 469–504.

Hon, G. 1995a. "Going Wrong: To Make a Mistake, To Fall Into an Error". *The Review of Metaphysics* 49: 3–20.

Hon, G. 1995b. "Is the Identification of Experimental Error Contextually Dependent? The Case of Kaufmann's Experiment and Its Varied Reception". In J. Buchwald (ed.), 1995, pp. 170–223.

Hon, G. 1998. "If This Be Error: Probing Experiment With Error". In M. Heidelberger and F. Steinle (eds.), 1998, pp. 227–248.

Hon, G. "An Attempt at a Philosophy of Experiment". This volume, pp. 259–584.

Lynch, M. 1990. "Allan Franklin's Trascendental Physics". In A. Fine, M. Forbes and L. Wessels (eds.), PSA 1990, vol.2, pp .471–485.

Nickles T. 1988. "Justification and Experiment". In D. Gooding, T. Pinch, S. Schaffer (eds.), 1989, pp. 299–333.

Pickering, A. (ed.) 1992. *Science as Practice and Culture*. Chicago and London: The University of Chicago Press.

Pickering, A. 1995a. *The Mangle of Practice*. Chicago: The University of Chicago Press.

Pickering, A. 1995b. "Beyond Constraint: the Temporality of Practice and the Historicity of Knowledge". In J. Buchwald (ed.), 1995, pp. 42–55.

Radder, H. 1998. "Issues for a Well-Developed Philosophy of Scientific Experimentation". In M. Heidelberger and F. Steinle (eds.), 1998, pp. 392–405.

GEREON WOLTERS

O HAPPY ERROR
A COMMENT ON GIORA HON

In Catholic Easter Vigil liturgy at a certain point the joy about the resurrection of Christ finds its solemn expression in the paradoxical declamation: *"O felix culpa!"* – "O happy fault!"[1] – That means that mankind ought to be happy to have been laden with the guilt of original sin and other sins, because only such guilt made possible our salvation through the death and the resurrection of Jesus Christ.[2]

In his very elegant paper Giora Hon pursues the same strategy as the fathers of the church, when they coined the happy fault paradox: "O happy experimental error", he seems to be declaiming, "thou shows us the truth about experiment!"[3]

As Hon maintains, this truth about experiment should consist in "a historically informed philosophy of experiment" (p. 260). He distinguishes two approaches that students of the philosophy of experimentation have developed so far: (a) the *epistemological* approach, which claims to bridge the gap between the "material process, which is the very essence of experiment, and propositional knowledge – the very essence of scientific knowledge" (p. 263); (b) the *methodological* approach, which pertains to the "level of manipulation of matter - the very essence of physical experiment" (p. 264). Hon is dissatisfied with the work done to date in both approaches and wants himself to offer an improvement to the methodological approach. This improvement consists basically in his suggestion that unconstrained lists of "epistemological strategies" (p. 266) that are pursued in experiments should be replaced by a comprehensive typology of possible error sources. There are, according to him, exactly four possible source-kinds of experimental error and they positively enable at the same time – o happy fault! – "four contexts which together make experiment" (p. 282).

I would like to question or challenge Hon's view in three respects:

(1) I would like to question in a more general way the approach to truth by way of error.

(2) I am doubtful about Hon's conviction that his typology of experimental error and the resulting "comprehensive overview of experiment from a negative perspective" (p. 282) covers everything that may be rightly designated as experiment. In other words, his typology seems to me to be too narrow.

(3) My skepticism about the completeness of his typology of experiment entails some doubt on Hon's dismissal of methodological approaches like those of Mach and Allan Franklin. Their approaches to experiment allegedly consist of open lists of epistemological strategies that one ought to pursue in order to achieve successful experiments. Such lists, in Hon's view, however, do not provide constraints, and without constraints there cannot be anything like genuine "classification", or "generalization" (p. 266).

I turn now to my first line of criticism. I am of the opinion that Hon's concept of experiment, and consequently both his typology of experimental error and the resulting four basic features of successful experimentation are too narrow for the purpose of exhaustively classifying everything that may be rightly called experiment.

What is an experiment, according to Hon? He cites two different characterizations of the experimental activity. The first – and this is explicitly his own – considers "experiment a philosophical system that aims at furnishing knowledge claims about the world" (p. 262). Here he is obviously thinking in an epistemological context. The second characterization of experiment, which is quoted in the context of the methodological approach, regards experiment "as a method of extracting knowledge from nature" (p. 264). It is not clear to me to what extent Hon himself shares also this second characterization. But I take it, that in a rough way he accepts it.

My thesis now is that there are experiments as exemplifications of methods of extracting knowledge from nature that do not fit into Hon's typology.

This typology consists of four classes that elegantly correspond to Francis Bacon's four sources of error. Hon accordingly calls his four sources of experimental error "idols of the script", "idols of the stage", "idols of the spectator", and "idols of the moral". The "idol of the script" consists in assuming an ill-suited background theory. This means, positively, that the first stage of experiment invokes a background theory, or background theories, respectively. The "idol of the stage" originates from wrong assumptions about the prevailing physical conditions of the material setup of the experiment. The "idol of the spectator" is due to "physical, physiological and psychological elements interfering with the depiction of the displayed phenomenon or with the reading of the measuring device" (p. 278). This means, positively, that correctly registering the outcomes of an experiment is an essential component of a comprehensive theory of experimentation. Fourthly and finally, "the idol of the moral" consists in erroneous interpretations of experimental results, which, in turn, means, positively, that the *interpretation* of experimental results is the final component of a comprehensive concept of experiment.

In addition Hon calls the first two positive components of experimentation – i.e. background theory and physical setup – "preparation" whereas reading the results and interpreting them form a second component in performing experiments, which he calls "test" (p.277). "Preparation" and "test" seem to introduce a temporal ordering, two fundamental stages, to the components of the theory of experimentation.

However, there is, in my view, an important new class of experiments that does not fit well into Hon's four components' scheme. I am referring here to a recent article titled "Equipping scientists for the new biology" by three biologists in the journal *Nature Biotechnology* (Vol. 18, April 2000).[4] This one-page paper mostly deals with science policy, or more exactly with the funding of what the authors call "discovery science." In their conception discovery science is opposed to "hypothesis-driven science." Hypothesis-driven science is roughly science as we know it: you somehow generate a hypothesis that subsequently is submitted to tests. Discovery science, on the other hand, is characterized by what the authors claim to be a "new research method." The prototype of applying this new method is the Human Genome Project (HUGO). They also assert that "discovery science requires large-scale facilities for genome-wide analyses, including DNA sequencing, gene expression measurements, and proteomics." More generally, I would like to characterize discovery science as the collecting and analyzing of gigantic masses of data, in order to find characteristic patterns. Discovery science is, as the authors say, a "technology-driven approach to biology and the biomedical sciences." One could call this approach in more traditional terms also "experimental natural history."[5] The authors of the paper rightly remark the following: "Discovery science ... enumerates the elements of a system irrespective of any hypothesis on how the system functions."

Now here it seems to me that discovery science, first of all, is experimental, in the sense assumed by Hon, in that it "extracts knowledge from nature". In large computer runs discovery science detects patterns that cry out, as it were, for interpretation. What is missing here, however, seems to be the background theory that is the first ingredient of stage one, i.e. "preparation", of Hon's typology. There is no background theory in discovery science, at least not in the sense that one finds it in hypothesis-driven science. The other components of Hon's typology, however, do seem to apply also in the case of discovery science. In this context one should note that Hon is cautious enough not to take the word "test" too strictly, when he says that "an experiment is not necessarily a test of some theory" (p. 277).

My second line of criticism has to do with Hon's dismissal of such somehow "rhapsodic" conceptions of experiment that consist in giving open lists of "epistemological strategies" of experiment or similar devices, as have been provided by Mach and others. Against such open lists of strategies Hon states that without constraints they are *ad hoc* and thus somehow unphilosophical (p. 266). I do not believe that in such lists there is no constraint in the sense of an "overall guiding principle" (p. 267). In my view there is such a principle and it is *success*. By choosing success as a constraint for lists of epistemological experimental strategies one achieves in my opinion two goals. On the one hand, one becomes more flexible: one can easily include basic changes in the overall conception of science itself. For this "discovery science" seems to be an example. Thus, one avoids creating a theory that would become the target of Buchwald's warning, quoted by Hon, that "axiomatics and definitions are the logical mausoleums of physics" (p. 260). On the other hand, by pursuing the success strategy, one achieves – to the delight of the philosopher – a unifying perspective of everything that might be included in the list of epistemic experimental strategies.

My third and last point of criticism concerns Hon's approach for arriving at truth by way of a typology of experimental error. To be sure, one can and should learn from errors, in order to avoid them the next time. But the very expression "the next time" points to a problem. You can learn from error only if those situations in which you have fallen into error are basically of the same type as the ones you have to newly cope with. You are, however, at a loss, when fundamentally new situations occur. So, in a general way, Hon's approach of arriving at a true concept of experiment by way of experimental error does not seem to leave sufficient room for experiments that are of a basically new type. We do not know what the future of science will bring forth. And we should be open for surprises.

I have always found the happy-fault formula rather awkward, logically and theologically. In this commentary I have outlined my difficulties with its analogical transfer to the philosophy of science. In short, I suggest replacing Hon's happy-error approach to experiment by a happy-success approach.

Department of Philosophy, University of Konstanz,
Konstanz, Germany

NOTES

[1] The verse in which "*O felix culpa*" occurs is part of the hymn *Exultet jam angelica turba caelorum.* The full verse is: "O felix culpa, quae talem ac tantum meruit habere redemptorem!" – "O happy fault which we received as its reward so great and good a redeemer!". – The *felix culpa* formula seems to originate from a sermon of St. Augustin. From there it made its way via church fathers like Leo the Great to the *Summa Theologiae* of St. Thomas Aquinas (s.th. 3,1,3 ad 3).

[2] When searching the Internet for "felix culpa" I found besides a German rock band of that name a book by Tom Peters with the title *O Felix Culpa...O Happy Fault: How Bad Guys Keep Good Guys Going.* This title seems to express nicely most of the theological content of the *felix culpa*-formula.

[3] I was very proud to have found this analogy between the early Christian theology of salvation and Hon's approach, but – alas! – Hon himself had used it already years before (see Hon, G. 1991. "A Critical Note on J.S. Mill's Classification of Fallacies". *British Journal for the Philosophy of Science* 42: 263–268, p. 264.).

[4] I would like to thank Eric Kubli (Zurich) for directing my attention to this paper.

[5] I owe this very fitting denomination to a conversation with Michael Friedman (Stanford).

COLIN HOWSON

BAYESIAN EVIDENCE

BAYESIANISM = LOGIC?

In his seminal essay "Truth and Probability" F.P. Ramsey advanced an idea which seemed novel but had in fact a long historical pedigree, that the probability axioms are consistency constraints of a logical character[1]. The idea that epistemic probability is an extension of deductive logic was already a very old one, going right back to the beginnings of the mathematical theory in the seventeenth century. Leibniz in the *Nouveaux Essais* and elsewhere said so explicitly, and the idea runs like a thread, at times more visible, at times less, through the subsequent development of the epistemic view of probability. Strangely enough, however, Ramsey did more than anyone else to deflect it out a logical path by choosing to embed his discussion not within the theory of logic as it was then being transmuted into its modern form largely at the hands of Gödel and Tarski, but within a theory of *utility*. His achievement was remarkable, inaugurating the current paradigm for decision theory, but it put epistemic probability in a setting where its logical character is effectively obscured. Ramsey's talk of consistent preferences is more loose folk-usage than anything to do with logical consistency; indeed, most people have preferred to see Ramsey's, Savage's etc., axioms as very general rationality constraints, and rationality is not logic. Indeed, I believe that the idea of consistency is conceptually quite distinct from whatever principles underly prudent behaviour. Whether a set of beliefs is consistent should not depend, for example, on whether you have transitive preferences. For this reason I believe that Ramsey and those who have followed him in developing the theory of subjective probability as a subtheory of a general theory of utility thereby removed it from its proper conceptual environment, i.e. logic. In the next few pages I shall try to bring it home, so to speak, by developing a view of the axioms of epistemic probability as authentic consistency constraints, by which I mean that their relation to the concept of deductive consistency is both clear and close.

To that end, I propose to return to the older custom of using the agent's estimate of the "true" odds as the measure of uncertainty[2], where these odds are the agent's informal estimate of the chances favouring the proposition, call it A, to those favouring -A. Perhaps surprisingly, no precise meaning has to be given to these "chances". What they are and whether they exist in any determinate sense at all is irrelevant (which is just as well in view of the almost wholly negative results of investigations from James Bernoulli to Carnap): it is sufficient for what follows that the agent expresses in their odds an opinion of what he/she *believes* their relative magnitude. Indeed, all the subsequent work will be done by just two properties of these odds: they are positive ratios and, as far as the uncertainty of A is concerned, they are supposed to give no advantage to either side of any bet.

The odds scale was quickly realised as not a good one on which to measure uncertainty, as it is infinite in the positive direction and unbalanced in that even odds are located very near one end and infinitely far from the other, and for this reason the measure adopted to measure uncertainty became instead the normalised odds, i.e. the quantity $p = o/(o+1)$ obtained from the odds o. p is called the agent's *fair betting quotient*. From this we obtain the reverse identity $o = p/(1-p)$, where $0 \le p \le 1$. If the agent's fair betting quotient is p and the payoffs in a bet are Q if A is true and -R if not then the bet is fair, according to the agent, just in case $R/Q = p/(1-p)$, that is,

$$Qp - R(1-p) = 0 \qquad (1)$$

i.e. *just in case something that looks formally like an expected gain is zero.*

That a person's uncertainty can be numerically measured in this particularly simple way has come in for vigorous objections from the utility camp. Mostly they are variations on Savage's claim that judgments of null advantage cannot be divorced from considerations of how the corresponding potential gains and losses are valued:

to ask which of two "equal" betters has the advantage is to ask which of them has the preferable alternative. (1954, p. 63)

This is the crux of the difference between the modern view, for which Savage has been such a powerful advocate[3], that it is impossible in principle to develop a theory of personal probability independently of a theory of utility, and the quite different older tradition based on the zero-expected-gain rule, which has numbered such distinguished people as James Bernoulli, Condorcet, Laplace (who defines "advantage" explicitly as expected gain (1820, p. 170)), Carnap (who also defines a fair-betting quotient as one which has zero expected gain (1950, p. 170)) and more recently Hellman (1997).

Savage concedes the existence of this other, older view in a remark which rather weakens the force of his earlier claim:

Perhaps I distort history somewhat in insisting that early problems were framed in terms of choice among bets, for many, if not most, of them were framed in terms of equity, that is, they asked which of two players, if either, would have the advantage in a hypothetical bet.

The "Perhaps" is surely disingenuous: those early problems were indeed "framed in terms of equity", but equity of advantage, characterised in terms of expected gain, not expected utility, a concept which arrived later. Savage does provide some arguments against the expected gain criterion, though these are surprisingly weak, deriving what force they have from a persistent conflation of prudence and fairness. They are also not original, repeating in the main objections made much earlier by Daniel Bernoulli, like this one for example:

Suppose a pauper happens to acquire a lottery ticket by which he may with equal probability win either nothing or 20,000 ducats. Will he have to evaluate the worth of the ticket as 10,000 ducats; and would he be acting foolishly, if he sold it for 9,000 ducats?

Savage, who quotes this approvingly, follows with his own example:

Thus a prudent merchant may insure his ship against loss at sea, though he understands perfectly well that he is thereby increasing the insurance company's expected wealth, and to the same extent decreasing his own. *Such behaviour is in flagrant violation of the principle of mathematical expectation.* (1954, pp. 92–93; my italics.)

And finally, of course, there is the St Petersburg Problem, made into a sort of experimentum crucis by Daniel Bernoulli, and widely supposed as a consequence of his classic discussion to deliver the death-blow to the expected-gain criterion. That is how Savage tells it too. The expected gain criterion tells us that one dollar is the fair price to pay for the promise of receiving 2^n dollars if the first head in a sequence of tosses of a fair coin occurs at the nth toss, for each of n = 1,2,3, ... If you were to agree to all such contracts for n up to 10,000, say, you would almost certainly lose a large sum of money for very little in return, which seems, and is, foolish. If you were to agree to all of them you would be faced with the prospect of staking an infinite sum, which is not only foolish but impossible in principle. So the idea of fairness as zero expected gain seems to lead to absurdity.

The reply to all these supposed counter-examples is simply that a fair game is not one which it is necessarily prudent to join, especially if you are not wealthy. To accept any substantial number of the St Petersburg bets, for example, would be certainly imprudent for most people, but they are not unfair bets. Indeed, as Condorcet pointed out[4], in the context of repeatable uncorrelated trials with constant p there is a good chance, tending to 1, that their fairness will actually be *observable:* by Chebychev's inequality the zero expectation will tend to be reflected in (will converge in probability to) a corresponding long-run zero gain. The conceptual divorce between preference and expected gain is made complete by noting that "gain" can be characterised entirely neutrally in terms of payoffs in any medium one cares to consider. As Bertrand was the first to point out, they can in principal be entirely notional; he considered hydrogen molecules as one possibility. The stake certainly does not have to have value in any conventional sense, and could in principle be anything as long as it is sufficiently divisible: "sand or manure if you prefer, or better an ideal, continuous fluid." (Hellman 1997, p. 195)

I think we can conclude that Savage's claim that subjective probability can be developed only within a theory of utility is without foundation, despite the almost unanimous support it has received. Let us move on. Suppose the bet with stake $S = R+Q$ is fair. Its payoff table can be represented in the convenient form that de Finetti made familiar in his classic 1937 paper:

A

T $S(1-p)$

F $-pS$

Where I_A is the indicator function of A, the bet can therefore be expressed as a random quantity $S(I_A-p)$; this will be useful later. Besides ordinary odds there are also *conditional odds*, in bets on a proposition A which require the truth of some proposition B for the bet to go ahead. The betting quotient in such a conditional bet is called a conditional betting quotient. A conditional bet on A given B with stake S and conditional betting quotient p clearly has the form $I_B S(I_A-p)$.

Let us now think of what it means for these assignments to be *consistent*. There is already at hand a well-known notion of consistency for assignments of numbers to compounds involving number-variables, and this is *equation-consistency*, or *solvability*. A set of equations is consistent (solvable) if at least one single-valued assignment of values to its variables satisfies it. Although this aspect is not mentioned in the usual logic texts, deductive consistency itself is really nothing but solvability in this sense. This may seem surprising, because consistency is usually seen as a property of sets of sentences. However, it is not difficult to see why we can equivalently regard it as a property of truth-value assignments. Firstly, note that according to the classical Tarskian truth-definition for a first or higher-order language conjunctions, disjunctions and negations are homomorphically mapped onto a Boolean algebra of two truth-values, {T,F}, or {1,0} or however they are to be signified. Now consider any attribution of truth-values to some set Σ of sentences of L, i.e. any function from Σ to truth-values. Call any such assignment CONSISTENT if it is capable of being extended to a single-valued function from the entire set of sentences of L to truth-values which satisfies those homomorphism constraints. The theory of "signed" semantic tableaux or trees is a syntax adapted to such a way of looking at, and testing for, CONSISTENT assignments. ("Signing" a tableau just means appending Ts and Fs to the constituent sentences. The classic treatment is Smullyan 1968, pp. 15–30; a simplified account is in Howson 1997.) Here is a very simple example:

$$A\ T$$
$$A \rightarrow B\ T$$
$$B\ F$$

The tree rule for [A→B T] is the binary branching

$$/ \quad \backslash$$
$$F\ A \quad B\ T$$

Appending the branches beneath the initial signed sentences results in a *closed tree*, i.e. one on each of whose branches occurs a sentence to which is attached both a T and an F. A soundness and completeness theorem for trees (Howson 1997a, pp. 107–111) tells us that any such tree closes if and only if the initial assignment of values to the three sentences A, A→B and B is inCONSISTENT, i.e. unsolvable over L subject to the constraints of the general truth-definition.

To see that CONSISTENCY and consistency are essentially the same concept note that an assignment of truth-values to a set Σ of sentences is CONSISTENT just in case the set obtained from Σ by negating each sentence in Σ assigned F is consistent in the standard (semantic) sense. In algebraic treatments of logic the identity becomes more apparent: to show that a set of propositional formulas is consistent is to show that as a system of simultaneous Boolean polynomial equations (equated to 1, the maximal element of the Boolean algebra) it has a solution over the propositional variables. Thus in deductive logic (semantic) consistency can be equivalently defined in the equational sense of a truth-value assignment being solvable, i.e. extendable to a valuation over all sentences of L satisfying the general rules governing truth-valuations, and we can call such an extension a *model* of the initial assignment.

The variables evaluated in terms of betting quotients are traditionally propositions and not sentences, but this is no great formal dissimilarity since propositions can be regarded simply as equivalence classes of sentences, with an obvious rule for distributing the probability to the constituent sentences. By analogy with deductive CONSISTENCY we can say that an assignment of fair betting quotients is consistent just in case it can be solved in a analogous sense, by being extendable to a single-valued assignment to all the propositions in the domain of discourse, or *language* L (*cf.* Paris 1994, p. 6), subject to suitable constraints analogous to the Tarskian ones in the deductive case. What should those constraints be? We are talking about *fair* betting quotients (in the agent's estimation at any rate), and just as the Tarskian conditions characterise truth-in-general so the constraints here should be those formally characterising the *purely general* content of the notion of fairness: call them (F). The odds expressing the strength of your beliefs about A are fair (relative to that estimation) if they give no calculable advantage to either side. What this implies in the way of completely general constraints seem to amount to:

(a) If p is the fair betting quotient on A, and A is a logical truth, then p =1; if A is a logical falsehood p = 0.

(b) Fair bets are invariant under change of sign of stake.

The reason for (a) is not difficult to see. If A is a logical truth and p is less than 1 then in the bet $S(I_A-p)$ with betting quotient p, I_A is identically 1 and so the bet reduces to the positive quantity $S(1-p)$ received come what may. Hence the bet is not fair since one side has a manifest advantage. Similar reasoning shows that if A is a logical falsehood then p must be 0. As to (b), changing the sign of the stake reverses the bet, and according to (F) the fair betting quotient confers no advantage to either side, from which (b) follows immediately.

We have not finished yet. Suppose we are given information $\Im(M)$ describing a repeated sampling procedure whose data are generated by a specified statistical model M, and that x describes some data so generated whose probability according to M is $P_M(A)$. Thus according to $\Im(M)$ the objective chance, or chance-density, of an x-outcome being generated (i.e. the chance that X = x, where X is the random variable one of whose values is x) is $P_M(A)$. The criterion of fairness would naturally seem to require that

$$fbq(x) = P_M(A),$$

where "fbq" stands for "fair betting quotient". Three cautionary remarks are in order however. Firstly, the equation above as it stands will not do, since M appears on the right-hand side and not the left. Secondly, (F) is supposed to be non-domain-specific. That more or less tells us that $\Im(M)$ should function as an instance of *variable* information on which the left-hand side is conditioned. Thus we should have

(c) $\quad fbq(x|\Im(M)) = P_M(x),$

(c) is what used to be called, BL^5, the principle of direct probability. But, and now I come to the third cautionary remark, the justification of (c) is based on very little other than a possible mere homonymy. Why should (c) hold (Lewis makes it hold by mere fiat (1981, p. 269))? Before that question can be answered a further constraint under (F) needs to be considered.

This condition is a natural closure condition saying that your views of what betting quotients are fair must, as betting quotients, respect the structural relations between bets. It is well-known by professional bookmakers that certain combinations of bets amount to a bet on some other event, inducing a corresponding relationship between the betting quotients. For example, if A and B are mutually inconsistent then simultaneous bets at the same stake are extensionally the same as a bet on AvB with that stake, and if p and q are the betting quotients on A and B respectively, we easily see that $S(I_A-p)+S(I_B-q) = S(I_{AvB}-r)$ if and only if $r = p+q$. Add to this that if each of a set of bets gives zero advantage then the net advantage of anybody accepting all of them should also be zero, and we arrive at the condition:

(Closure) *If a finite or denumerable sum of fair bets is equivalent to a bet on a proposition B with betting quotient q then q is the fair betting quotient on B.*

In fact, we can generalise the condition to say that the sum of n fair bets, where it is defined[6], should itself be fair. Closure obviously follows immediately from this.

I said that Closure, or its generalisation, is a natural condition, and so it is. Closure may not be a *provable* thesis, but then neither are the axioms of set theory, for example, yet that does not stop them being so compelling that they are given their axiomatic status. And there is a pertinent analogy from deductive logic. We are all familiar with the fact that a multiple conjunction is true if all the conjuncts separately are true. But that "fact" is a consequence of defining (binary) conjunction in the standard truth-functional way, and definitions involving terms in preformal use are really just disguised postulates. Before this particular "definition" succeeded so completely in acquiring canonical status it would have been open to a sceptical critic to point out that one cannot *prove* in any non-question-begging way that a conjunction is true if all the conjuncts are true separately, and maybe even advance "counterexamples": for example, "She had a baby" may be true, and "She got married" may be true, but ordinary usage does not pronounce "She had a baby and she got married" necessarily true in consequence. At any rate, Closure seems so fundamentally constitutive of the ordinary notion of a fair game I believe that we are entitled to adopt it as a postulate to be satisfied by any formal explication. And, of course, identifying the informal expectation condition (1) with zero expected value within the fully developed mathematical theory, we have as a theorem that, since expectation is a linear functional, all expectations, zero or not, add over sums of random variables. Moreover, this result is maintained over all ways of describing the sample space, i.e. over all measurable refinements and coarsenings.

Now we can return to the discussion of (c). Consider a large number of independent samplings, at each one of which you place a bet on $X = x$ with the same stake and what you believe to be fair betting quotient r. In common with most working natural scientists, and even some philosophers, I shall take objective chance to be characterised as a long run-frequency distribution over the outcome-space (e.g. the variant of von Mises's theory given in Howson and Urbach 1993, Chapter 13). A frequency characterisation of P_M is of great importance, since given it, and only given it, we know that after sufficiently many trials you will make a nonzero gain/loss which predictably remains nonzero if r is not equal to $P_M(x)$. By the generalisation of closure above, the sum of those bets is fair to you, and clearly you can escape contradiction only by setting r equal to $P_M(x)$.

We shall suppose from now on that the propositions in question are members of B(L), the Borel field generated by the atomic quantifier-free sentences of some first order language L (Gaifman 1964, p. 4). We can now make use of the following arithmetical facts:

(i) $-S(I_A-p) = S(I_{\neg A}-(1-p))$.

(ii) Suppose {Ai} is a denumerable family of propositions in B(L) where $A_i \& A_j = \bot$ for $i \neq j$, that p_i are corresponding betting quotients, and that Σp_i exists (this last condition will be seen later to be satisfied). Then:

$\Sigma S(I_{Ai} - p_i) = S(I_{VAi} - \Sigma p_i)$.

(iii) If p,q>0 then there are nonzero numbers S,T,W such that:

$S(I_{A\&B} - p) + (-T)(I_B - q) = I_B W(I_A - p/q)$ (T/S must be equal to p/q).

The right hand side is clearly a conditional bet on A given B with stake W and betting quotient p/q.

Closure tells us that if the betting quotients on the left hand side are fair then so are those on the right. The way the betting quotients on the left combine to give those on the right is, of course, just the way the probability calculus tells us that probabilities combine over compound propositions and for conditional probabilities.

Now for the central definition. Let Q be an assignment of personal fair betting quotients to a subset X of B(L). By analogy with the deductive case, we shall say that Q is *consistent* if it can be extended to a single-valued function on all the propositions of L satisfying suitable constraints, in this case the general conditions of fairness including closure. It is now a short step to the following:

Theorem: An assignment Q of fair betting quotients (including conditional fair betting quotients) is consistent if and only if Q satisfies the constraints of the countably-additive probability calculus together with the principle of direct probability.

The proof is straightforward. Necessity is obvious in the light of (i)–(iii) above, and sufficiency follows from a simple argument from expected values (for details see Howson 2000, pp. 130–132)

I pointed out that there is a soundness and completeness theorem for first order logic which establishes an extensional equivalence between a semantic notion of consistency, as a solvable truth-value assignment, and a syntactic notion, as the openness of a tree from the initial assignment. In the theorem above we have an analogous soundness and completeness theorem for a quantitative logic of uncertainty, establishing an extensional equivalence between a semantic notion of consistency, i.e. having a model, and a syntactic one, deductive consistency with the probability axioms when the probability functor P signifies the fair betting quotients in Q. The Bayesian theory is an authentic *logic*.

The logical perspective proves to be a very enlightening one, dissolving what were, and to a great extent still are, regarded as difficult, sometimes intractable and certainly always controversial issues. The first, associated with the traditional form of Dutch Book arguments, is why invulnerability to an assured negative gain, rather than invulnerability to an assured non-positive gain, should be the condition for rational betting behaviour. The former issues in (given the other, most unrealistic, condition that the agent be willing to bet impartially on or against at some characteristic betting quotient) the usual finitely additive probability axioms, the latter those plus the condition that the probability function be strictly positive. It can now be seen that the apparent necessity to make an arbitrary choice is nothing more than an artefact of the Dutch Book set-up.

Second is the problem of unsharp probabilities. The fact that people generally cannot produce odds which in their minds exactly equilibrate advantage, whereas classical Bayesianism deals in point-valued probabilities, has been thought to be a problem by many, a constructive consequence of which has been the development of a theory of unsharp probabilities, so-called upper and lower probabilities. I believe that a deductive parallel is highly informative here. In the way of looking at the Bayesian theory which I have been suggesting the deductive analogue of probability values is of course truth-values. Yet the model of correct deductive reasoning provided by formal deductive logic, i.e. first order logic and its various modal etc. extensions, is invariably provided with with sharp truth-values, despite the fact that the reality being modelled, natural language, typically does not have sharp values, as the many varieties of Sorites demonstrate. This is not thought generally to detract from whatever explanatory virtues the model possesses. These are generally thought to be great: hence the significance given the classic theorems of Gödel, Löwenheim, Skolem, Church, Tarski, Cohen etc. Similarly the explanatory power of the sharp-valued Bayesian model, or models, is also correctly thought to be considerable, exhibited in the various stability and convergence theorems, and in the remarkable and far-reaching properties of the deceptively simple Bayes formula. I think enough has been said to justify the continued use of "unrealistic" sharp personal probabilities.

And some methodological implications

Those observations bring us to the topic of applications. The Bayesian theory is known primarily as, and was invented specifically to be, a theory of inductive inference from evidence. The idea that the theory is a species of logic fits well with the traditional view of it as an inductive logic. Let us see what methodological consequences flow from this view. Firstly, we should note that any alleged "principle" not derivable from the probability axioms is not generally valid. Several such principles often have been canvassed: the Principle of Indifference, the Maximum Entropy principle and its variant the Principle of Minimum Information, invariance principles, and the Principle(s) of Conditionalisation. The list is not complete but these are the major items. They are not independent: the Principle of Indifference is a consequence of the Maximum Entropy Principle relative to the null constraint set; the only form of the maximum Entropy Principle to be consistent for continuous distributions is the Principle of Minimum Information; and the two main principles of conditionalisation, Bayesian and Jeffrey conditionalisation, are consequences of the Principle of Minimum Information. And there are additional interconnections. These various principles have been extensively discussed and there is now something of a consensus that all except Bayesian conditionalisation are controversial, and if appropriate to some contexts certainly not of general validity. The exception, conditionalisation, is still widely regarded as a core Bayesian principle of universal validity. I shall claim that it is not. Indeed, I shall show that it is demonstrably not, and the logical perspective explains very clearly why not.

Conditionalisation

Recall that Bayesian conditionalisation is the rule that if your current belief function is the probability function P, and you learn that a proposition A is true, but no more, then you should update P to a new probability function Q according to the rule

$$Q(\,.\,) = P(\,.\,|A) = P(\,.\,\&A)/P(A) \qquad (2)$$

It is well-known that violating conditionalisation can result in a Dutch Book (Teller, 1973, who, however, attributed it to David Lewis). Admittedly this is not a proof that anyone who infringes conditionalisation at the time the new data is acquired (and who is willing to bet at their fair betting odds, etc.) can be Dutch Booked; it is not difficult to see that there can be no Dutch Book against such a person. It is a proof that there exists a Dutch Book against anyone who announces in advance that they will follow an updating rule different from conditionalisation. Essentially the same Dutch Book argument for conditionalisation is also alleged to establish the following identity, called by van Fraassen the "Reflection Principle" (1984):

$$P(B|Q(B) = r) = r \text{ for all } r, 0 \leq r \leq 1 \tag{3}$$

The Lewis-Teller Dutch Book argument for (2) (and (3)) is widely supposed to show that the only consistent updating strategy is that of conditionalisation. This is not true, and demonstrably not true: on the contrary, *there are circumstances where conditionalisation is an inconsistent strategy*. Here is a simple example. Suppose B is a proposition, e.g. "2 + 2 = 4", of whose truth you are P-certain; i.e. $P(B) =1$. Suppose also that for whatever reason – you believe you may be visited by Descartes's Demon, for example – you think it distinctly P-possible that $Q(B)$ will take some value q less than 1; i.e. $P(Q(B) = q) > 0$. Given the circumstances, it follows by the probability calculus that $P(B|Q(B) = q)=1$. But suppose at the appropriate time you learn $Q(B) = r$ by introspection; then $Q(B) = q$. But if you conditionalise on this information then you must set $Q(B) = (B|Q(B) = q) = 1$. By a nice parallel, though not a surprising one once it is realised that (3) is conditionalisation in disguise (see Howson 1997b, p. 198), the counterexample to conditionalisation is also a counterexample to this, for we know that $P(B|Q(B) = r) = 1$ by the probability calculus alone.

An objection that we can immediately foreclose is that "second-order" probabilities, i.e. probabilities like Q that appear as an argument of the function P, have not been and possibly cannot be formally justified. The objection fails because there is nothing *formally* second-order about Q, nor is there any need to provide a formal justification for allowing $Q(B) = q$ as an argument. The justification already exists in the standard (Kolmogorov) mathematics, for Q is simply a random variable, with *parameter* B. defined in the possibility space generating the propositions/events in the domain of P (it is assumed that these possibilities include information about the agent's future beliefs).

"Dynamic modus ponens"

Consider the following deductive parallel to the "dynamic" rule of conditionalisation. I shall call it "dynamic modus ponens". This is the deductive "updating rule" that if at time t I accept as true A→B (material conditional) and at time t' > t I increase my knowledge stock by just (the truth of) A then I should accept B as true at time t' also. This "dynamic rule" is unsound for just the same reasons the dynamic rule of conditionalization is. Consider the following example. At time t I accept the material conditional A→B. But suppose that A is actually equivalent to the negation of B, and note that in this case A→B is equivalent to B. My acceptance of A→B might even be just a pedantic way of saying I accept B. But at time t+1, for whatever reason, I decide to accept A as true. Clearly, it would result in inconsistency in my accepted beliefs at t+1 if I were to invoke "dynamic modus ponens" and accept B.

Two obvious questions arise. (i) How can there be a Dutch Book argument for an unsound principle? This certainly isn't the case with the "synchronic" probability axioms. (ii) Is there *some* valid principle there? I'll deal with these questions in turn.

(i) The Dutch Book argument for conditionalisation penalises the offender not for inconsistency, but for changing their mind. Take "dynamic modus ponens" first. Suppose I were certain of B, and I choose rather unorthodoxly, but logically quite correctly, to express this certainty by offering to offer or accept very high odds on A→B where A is −B, the bet to be enforced should A be seen to be true. I then come to believe that B is in fact false, and express this by offering to give or accept very large odds on −B. Obviously I can have money taken from me in these circumstances, but that is only because of two peculiar features of the situation: (a) I am always willing to put my money where my mouth is; and (b) I imprudently make enforceable conditional bets in situations where it is conceivable that the discovery of the truth of the conditioning proposition may change my belief function. This is, of course, what happens in the counterexample above: I offer infinite odds on 2+2 = 4 conditional on a conceivable state of affairs coming to pass in which I will doubt that same proposition.

(ii) Can conditionalisation be stated in a suitable way such that so stated it is a sound principle? Yes, easily. A conditional probability function P(. |D) is the restriction of P to the subuniverse D; in other words, for any given C, P(C|D) measures your degree of belief in C on the additional supposition that D is true. As Ramsey pointed out, this does not have the implication that P(C|D) is what your degree of belief in C would be were you to learn D. Actually learning D might, as Ramsey observed, cause you to change your conditional degree of belief in C given D, just as learning A in the "dynamic modus ponens" example positively demands the removal of

A→B from the stock of things you accept as true. The reason that conditionalisation sounds plausible is because *it is implicitly assumed that learning D causes no change in your probability of C conditional on D [truth-value assignment to D→C]*. But then conditionalisation [in both cases] would not only be plausible, but mandatory.

Here is the argument more formally. It is easy to see that a probabilistic rule of the form

$$\frac{Q(A) = 1 \quad Q(B|A) = r}{Q(B) = r} \tag{4}$$

is demonstrably sound, i.e. a consequence of the probability axioms. Substituting P(B|A) for r, we obtain equally sound instances of the principle of conditionalization:

$$\frac{Q(A) = 1 \quad Q(B|A) = P(B|A)}{Q(B) = P(B|A)} \tag{5}$$

Similarly, from the following way of expressing modus ponens

$$\frac{\tau(A) = 1 \quad \tau(A \to B) = r}{\tau(B) = r} \tag{6}$$

where r is either 1 (true) or 0 (false), we obtain the analogue of (4) placing the limits on "dynamic modus ponens":

$$\frac{\tau(A) = 1 \quad \tau(A \to B) = v(A \to B)}{\tau(B) = v(A \to B).}$$

Observe that when r = 1 in the probabilistic rule of conditionalization (4) and in the general form (6) of modus ponens, we seem to have two models for reasoning about conditionals when both the conditional "assertion" and its "antecedent" are accepted (I use scare quotes because we know from Lewis's so-called Triviality Theorem that unless we are prepared to countenance a non-Boolean structure for propositions then we are not dealing with propositions qua elements of a propositional algebra).

What goes here for Bayesian conditionalisation goes, mutatis mutandis, for Jeffrey conditionalisation. The condition of validity is the same as for Bayesian conditionalisation: when an exogenous shift from P(A) to Q(A) occurs on a proposition A (or on the members of a partition) then it is well known that the necessary and sufficient condition for the validity of Jeffrey's rule (i.e. its validity relative to the "synchronic" probability axioms) is that the following equations hold:

$$Q(\,.\,|A) = P(\,.\,|A), Q(\,.\,|\text{-}A) = P(\,.\,|\text{-}A)$$

(Howson and Urbach 1993, pp. 105–110).

It may be objected that conditionalisation, particularly Bayesian conditionalisation, is almost universally regarded as the core principle of Bayesian *methodology*: it is supposed to tell the Bayesian reasoner how to update their belief-function on new evidence. Yet on the view of the Bayesian theory proposed here conditionalisation is not a valid principle! Does it follow that on this view Bayesianism has no methodological significance? It would be funny if restating a principle in such a way that it is generally valid deprives it of methodological application, and indeed this is not so here. Conceding that conditionalisation is valid just in case all conditional probabilities remain unchanged by learning the new evidence in no way restricts the applicability of the Bayesian theory.

Hume's other principle

I believe that Hume's celebrated circularity argument, that any justification for believing "that the future will resemble the past" must explicitly or implicitly assume what it sets out to prove, shows in completely general terms that a sound inductive inference must possess, in addition to whatever observational or experimental data is specified, at least one independent assumption (an inductive assumption) that in effect weights some of the possibilities consistent with the evidence more than others. I shall call this "Hume's Other Principle"[7], and I personally take it to be a profound *logical* discovery, comparable to that of deductive inference itself. If it is correct, then any satisfactory theory of uncertainty must satisfy an important *weakness* constraint (a weakness constraint sounds odd but there it is): it must not tell us what the uncertainties of contingent propositions are. Bayesian probability clearly satisfies this principle: unconditional probability distributions are exogenous to the theory. For so long seen as a problem for the Bayesian theory, *this indeterminacy can now be seen as a natural and inevitable phenomenon*, entirely appropriate in a theory of uncertainty which respects Hume's sceptical argument.

This view of things is entirely in harmony with the logical perspective. By general agreement arrived at long ago, the provable theses of deductive logic are either trivial or else are of the *conditional* form "If such and such statements are true then necessarily so is this". As Ramsey, emphasising the similarly conditional nature of probabilistic inferences, clearly puts it:

This is simply bringing probability into line with ordinary formal logic, which does not criticize premises but merely declares that certain conclusions are the only ones consistent with them. (1931, p. 91)

On the logical interpretation of the Bayesian theory, just as in deductive logic, you only get out as the conclusion of valid reasoning some transformation of what you put in. Bacon notoriously castigated the contemporary version of deductive logic, Aristotelian syllogistic, for what he perceived as a crucial failing: deductive inference does not enlarge the stock of factual knowledge. But Bacon's condemnation has with the passage of centuries become modulated into a recognition of what is now regarded as a if not the fundamental property of logically sound inferences which, as we see, probabilistic reasoning shares, and in virtue of which it is an authentic logic: *sound inference does not beget new content.*

But if this is so, then once again it might seem reasonable to ask how the Bayesian theory can give us anything of methodological significance. And it is a question that certainly has been asked: the standard objection to subjective Bayesianism is precisely that in regarding prior distributions as, in effect, free parameters, it has neither methodological nor explanatory significance. The problem is resolved by recognizing that the theory tells us that to reason soundly in a very wide range of circumstances *is necessarily to reason inductively.* I shall not go into details because the features of the Bayesian theory which do this are well-known. Firstly, there are the convergence-of-opinion theorems, a typical one of which (Halmos 1950, p. 213) states that the consistent agent must assign probability one to the proposition that his/her posterior distribution will convergence to the truth for a class of hypotheses definable in the product space of denumerably many observations. These theorems are the formal analogues of a similar constellation of theorems of statistical mechanics where the evolution of a system is determined only with probability one. Thus Khinchin:

The most important problem of general dynamics is the investigation of the character of the motion of an arbitrary mechanical system on the initial data, or more precisely the determination of such characteristics of the motion which in one sense or another "almost do not depend" on these initial data. (1949, p. 10; "almost" is of course the probabilist's shorthand for "with probability one")

Such theorems are accorded a good deal of explanatory significance in physics, and if they are there then so too should they be here where, if anything, the Bayesian results are stronger: the physics theorems characteristically apply only to one type of measure, Lebesgue measure, while the Bayesian theorems apply to any measure satisfying some usually very weak conditions.

Methodologically, perhaps, these theorems are of more limited interest, since in the first place they are asymptotic, and in the second they tell us only that the consistent agent should regard him/herself as approaching the truth, not that they actually will do so. And indeed they need not do so, and in some cases provably will not, e.g. if they place prior probability 0 on the true hypothesis; Kelly (1996, p. 308) has some more interesting examples. The methodological interest of the Bayesian theory characteristically arises in the context of agreement on some statistical or other model, and shows how the impact of suitably favourable evidence can cause the Bayes factor to become very large, giving a posterior probability close to one to the hypothesis (subject to its having nonzero prior probability). These results can be highly informative. One of the most striking concerns a classic problem of statistical inference: given a normal model, a null hypothesis H_0 placing the mean m at m_0, and an iid sample of size n with sample mean x, what should we infer about H_0? In a remarkable paper (1957), Lindley showed that under any prior distribution concentrated on a finite interval of values $m \neq m_0$, as long as H_0 has some positive prior probability then a value of x significant at any fixed value α, however small α is as long as it is not zero, will for sufficiently large n give H_0 a posterior probability arbitrarily close to 1. Lindley's result shows dramatically how both Neyman–Pearson and Fisher tests of significance greatly overestimate the strength of the evidence against the null hypothesis in large samples, and, more deeply, explains why inferential theories which, like those, disobey the Likelihood Principle[8] will in predictable circumstances fail to evaluate correctly the evidential import of the data (see also Shafer 1982, Howson 2002).

Conclusion

The simplicity of this way of developing personal probability and its laws stands in very great contrast to the utility-based approach, which requires first the development of a general theory of preference over acts. Besides depending on questionable principles like the Sure-Thing principle, and problems about distinguishing acts and states, a theory of what it is rational to do is surely conceptually quite distinct from a theory of consistency between judgments, which is of the species of *logic*. Small wonder a specifically logical focus issues in a theory of personal probability both simpler and more elegant than the alternative.

Department of Philosophy, Logic and Scientific Method, London School of Economics and Political Science
London, United Kingdom

NOTES

[1] That other great twentieth century pioneer, de Finetti, also seemed to be pointing to something like a logical notion of consistency with his concept of the internal coherence of assignments of probabilities.

[2] E.g. Proteus: But now he parted hence, to embark for Milan.
    Speed: Twenty to one, then, he is shipp'd already.
(William Shakespeare, *Two Gentlemen of Verona*. I am grateful to Vittorio Girotto and Michael Gonzalez for bringing this quotation to my attention.).

[3] And in which he has been enthusiastically followed by philosophers; for example Earman "Degrees of belief and utilities have to be elicited in concert" (1992, pp. 43–44). Not so, as we shall see.

[4] "Réflexions sur la règle générale qui prescrit de prendre pour valeur d'un évènement in certain la probabilité de cet évènement, multipliée par la valeur de l'évènement en lui-même."

[5] David Lewis of course redubbed it "The Principal Principle".

[6] The denumerable sum of the St Petersburg bets we considered earlier is not defined, since the stake is infinite.

[7] "Hume's Principle" in the current literature in philosophy of mathematics is the principle that two collections are equinumerous just in case their members can be put in a one-one relation. Before this rediscovery the principle was attributed to Cantor.

[8] The Likelihood Principle states that the evidential import of any data x is determined by the likelihood function $L_x(t) = p(x|t)$, where the members of the hypothesis space are parametrised by t. It follows that theories like Fisher's and Neyman–Pearson's where the 'significance' or otherwise of x involves reference to possible values of the data other than x do not obey the principle. Birnbaum showed that the principle follows from two very plausible assumptions both of which would be acceptable to classical statisticians (Birnbaum 1962; see also Lee 1997 pp. 193–197).

REFERENCES

Bayes, T. 1763. "An Essay Towards Solving a Problem in the Doctrine of Chances". *Philosophical Transactions of the Royal Society,* vol. 53: 370–418.

Birnbaum, A. 1962. "On the Foundations of Statistical Inference". *Journal of the American Statistical Association* 57: 269–306.

Carnap, R. 1950. *Logical Foundations of Probability.* Chicago: Chicago University of Chicago Press.

De Finetti, B. 1937. "Foresight, Its Logical Laws, Its Subjective Sources". Translated and reprinted in H. Kyburg and H. Smokler (eds.), *Studies in Subjective Probability.* New York: Wiley, 1964, pp. 93–159.

Earman, J. 1992. *Bayes or Bust? A Critical Examination of Bayesian Conformation Theory.* Cambridge, Mass.: MIT Press.

Gaifman, H. 1964. "Concerning Measures in First Order Calculi". *Israel Journal of Mathematics*: 1–18.

Halmos, P. 1950. *Measure Theory.* New York: Van Nostrand, Reinhold.

Hellman, G. 1997. "Bayes and Beyond". *Philosophy of Science* 64: 190–210.

Howson, C. 1997a. *Logic With Trees.* London: Routledge.

Howson, C. 1997b. "Bayesian Rules of Updating". *Erkenntnis* 45: 195–208.

Howson, C. 2000. *Hume's Problem: Induction and the Justification of Belief.* Oxford: Clarendon Press.

Howson, C. 2002. "A Logic for Statistics". *Proceedings of the 2001 British Academy Bayes Theorem Conference.* Oxford: Oxford University Press.

Howson, C. and Urbach, P.M. 1993. *Scientific Reasoning: the Bayesian Approach.* Chicago: Open Court, second edition.

Jeffreys, H. 1961. *Theory of Probability.* Oxford: Oxford University Press, third edition.

Kelly, K. 1996. *The Logic of Reliable Inquiry.* Oxford: Oxford University Press.

Khinchin, A.I. 1949. *Mathematical Foundations of Statistical Mechanics.* New York: Dover.

Laplace, P.S. de, 1820. "Essai philosophique sur les probabilités". Page references are to *Philosophical Essay of Probability.* New York: Dover, 1951.

Lee, P.M. 1997. *Bayesian Statistics.* London: Arnold, second edition

Lewis, D. 1981. "A Subjectivist's Guide to Objective Chance". In R.C. Jeffrey (ed.), *Studies in Inductive Logic and Probability.* Berkeley: University of California Press, pp. 263–293.

Lindley, D.V. 1957. "A Statistical Paradox". *Biometrika* 44: 187–192.

Paris, J. 1994. *The Uncertain Reasoner's Companion*. Cambridge: Cambridge University Press.

Ramsey, F.P. 1926. "Truth and Probability". In *The Foundations of Mathematics and Other Logical Essays*. London: Routledge and Kegan Paul, 1931, pp. 156–199.

Savage, L.J. 1954. *The Foundations of Statistics*. New York: Wiley.

Shafer, G. 1982. "Lindley's Paradox". *Journal of the American Statistical Association* 77: 325–334.

Smullyan, R.M. 1968. *First Order Logic*. New York: Dover.

Teller, P. 1973. "Conditionalisation and Observation". *Synthese* 26: 218–258.

van Fraassen, B. van 1984. "Belief and the Will". *Journal of Philosophy* 81: 235–256.

IGOR DOUVEN

ON BAYESIAN LOGIC
COMMENTS ON COLIN HOWSON

Bayesians typically hold that rationality mandates that we conform our degrees of belief to the axioms of the probability calculus. Starting with Ramsey (1926) and de Finetti (1937), they have sought to defend their position by arguing that we are susceptible to "Dutch books", i.e. bets that ensure a negative net pay-off in every possible future, just in case our degrees of belief violate the axioms of probability. It is by now widely recognized that such Dutch book arguments rest on premises that are – to say the least – fairly problematic. For instance, these arguments assume that we can tell an individual's degrees of belief from her willingness to engage in certain bets. More exactly, they assume that we can identify, for any proposition A, an individual's degree of belief in A with her fair betting quotient for A[1], that is, with the highest/lowest price, in units utile, at which she is willing to place/accept a bet on A that pays one utile if A is true (and nothing otherwise). While this assumption may have seemed plausible at the time when operationalism and behaviourism ruled the day (when Ramsey and de Finetti devised their Dutch book arguments), to the modern eye it looks rather suspect. Schick (1986), Bacchus, Kyburg and Thalos (1990) and Waidacher (1997), among others, have pointed to further contentious assumptions underlying Dutch book arguments.

A different approach to justifying the Bayesian theory thus seems more than welcome. Howson (2003) promises to offer just that. His aim is to defend the Bayesian theory as a logic, concerned with consistent reasoning, and not as being part of a broader theory concerned with rational (betting) behaviour. In this note I will argue that, although I can see nothing wrong with thinking about the Bayesian theory as a logic, it is far from clear how its status as a logic can contribute to the theory's justificational status. My main point of critique will be that the fact that Bayesian logic can be proved to be sound and complete, as Howson demonstrates, is by itself of very little significance, contrary to what Howson seems to believe.

According to Howson (2000, p. 127), a theory has logical status exactly if it satisfies three conditions. The first two are that the theory must be about relations between statements and that there must be no restrictions as regards the subject matter of these statements, respectively. The third, and in the context of this note most important condition, says that the theory "should incorporate a semantic notion of consistency which can be shown to be extensionally equivalent to one characterizable purely syntactically" (*ibid.*). That probability theory satisfies the first two conditions is immediate. That it also satisfies the third is not immediate, but Howson shows that it does nonetheless by proving a soundness and completeness theorem for probability theory. More specifically, he shows that, given some language L, an assignment of fair betting quotients to a set of propositions expressible in L has a model (in a particular sense; see below) exactly if that assignment conforms to the probability axioms (Howson 2000, pp. 130ff). This soundness and completeness theorem is, from a purely technical point of view, certainly an important result. However, I fail to see how it can offer a justification of the probability axioms. That probability theory is sound and complete in the sense meant by Howson certainly does not count *against* that theory, but that, in my view, is about all we can conclude.

To explain this, let me first briefly recall that there is an issue of justification not only for inductive logic but also for deductive logic. There exists a welter of rival (deductive) logics, all seemingly purporting to formalize what we might call the laws of truth.[2] The question which of these is the *true* logic has been, and still is, hotly debated in philosophy. For present concerns just notice that to defend a particular logic by claiming that it is sound and complete would be entirely disingenuous. Although not all logics are sound and complete,[3] many of them are.

This is not necessarily to say that the justification of a particular logic as being the correct logic is wholly detached from considerations that have to do with soundness and/or completeness. As will be remembered, to say of a logic that it is sound, or that it is complete, is to say that it is sound/complete *with respect to a given semantics*. To be very precise, then, we should say that a logic is sound with respect to semantics S just in case reasoning according to the rules of that logic is X-preserving, where X is the central semantic concept of S. Similarly, we should say that it is complete with respect to S just in case any sentence that has property X, on the condition that all elements of a certain (possibly empty) collection of sentences have X, can be derived from the sentences in that collection by means of the rules of the logic. And depending on what the central semantic concept is, a soundness cum completeness proof may be of more or less significance (even if, as I suspect, it may never be wholly decisive as a justification of a given logic).

For instance, in semantics for deductive logics, the central semantic concept invariably is called "truth". That does not mean – and given the differences between the various logics, could not mean – that these semantics all simply systematize our ordinary way of thinking about truth and the laws it obeys (and thus, we might say, it does not mean that they are all *really* about truth or, alternatively, that they all are about *real* truth). In fact, semantics have been devised which are entirely at odds with our intuitions regarding truth (for example, standard semantics for so-called paraconsistent logics postulates the existence of impossible worlds in which e.g. contradictions can be true – see Priest 2001, Ch. 9). Now it would clearly seem more telling if a logic is shown sound and complete with respect to a semantics that reflects our pre-analytic conception of truth than if it is shown to be sound and complete with respect to some semantics we find intuitively bizarre. Indeed, I would guess that many who believe that standard first-order logic is the truly correct logic do so at least partly on the basis of the fact that this logic is sound and complete with respect to a semantics that we feel reflects the logical structure of the world, so to speak, and not just that of the actual world, but of every possible world.[4] We have a virtually unshakeable belief that, for any assertoric sentence and any possible world, either it or its negation is true of that world, but never both. We firmly believe that, for every possible world w, if some sentence A is true of w and some other sentence B is also true of w, then $A\&B$ is true of w, and vice versa. We firmly believe that, for every possible world w, if an open formula Ax (with only x free) is made true be every assignment of an inhabitant of w as a referent to x, then $(x)Ax$ is true of w, and vice versa. And so on for the other clauses of "standard" semantics. Exactly because this semantics codifies these – what perhaps with more right than anything else may be called – truisms, that we, or at any rate most of us, regard it as an important fact that first-order logic can be proved to be sound and complete with respect to it.

To put the foregoing in a nutshell, the *mere* fact that a logic is sound and complete is not significant at all. The value, if any, of a soundness cum completeness theorem depends on what semantics is supposed in its proof. With this clarified, we now return to Howson.

Howson's soundness and completeness theorem, as stated above, refers to the notion of a model. That is Howson's "semantic notion of consistency" which he has shown to be equivalent to the syntactic notion of conformity to the probability axioms. What exactly is meant by this notion of a model? Howson (2000, p. 128) defines it thus: an assignment of fair betting quotients to a collection of propositions in some language L has a model exactly if there is a function extending that assignment and defined on *all* propositions in L which satisfies "the general criteria for assigning fair betting quotients" (*ibid.*), that is, roughly, criteria which ensure that fair betting quotients determine fair bets. In its turn, fairness of bets is defined in terms of advantage: a bet on *A* is fair for a given person just in case, relative to the state of knowledge of that person, neither of the parties involved in the bet on *A* has a calculable advantage (*ibid.*).

At this juncture one might start wondering whether Howson's approach is really so different from the Dutch book defences of Bayesianism, given that his fundamental semantic notion is ultimately to be understood in terms of avoidance of a calculable advantage to either side in a bet. For what else is avoidance of a calculable advantage to either side in a bet than avoidance of Dutch book vulnerability of either side? The difference between the two is that, as Howson emphasizes, the word "advantage" in the explication of his semantics is *not* to be understood in terms of value or utility: "The stake certainly does not have to have value in any conventional sense, and could in principle be anything as long as it is sufficiently divisible" (This volume, p. 304). Citing Hellman, he says that it could for instance be sand or manure. And presumably a theory of rational action has little or nothing to say about whether or not you should engage in bets in which nothing of value to you is at stake.

But while it is clear how Howson does *not* want us to understand the notion of advantage, it is considerably less clear how we *are* to understand it. I am not a native speaker of English and so have to be a bit careful here, but it seems to me that we normally do not say a bet gives you a calculable advantage if what you stand to win from it is of no value to you – such as, we may suppose, manure. Suppose, for instance, that your local bookie's fair betting quotient in some proposition exceeds 1. Then in what intelligible sense could that give you an advantage if you are not to bet with your bookie on that proposition at a stake you would really like to cash? If in a bet nothing of value ("in any conventional sense") to either side is at stake, then, I submit, such terms as "advantage" and "disadvantage" simply fail to apply.

I am fully aware that these quick considerations do not warrant the conclusion that Howson's notion of advantage and, concomitantly, the notion of fairness defined by means of it, are wholly obscure. What I think we can conclude from them, however, is that these notions in any event are *not the intuitive or ordinary notions of advantage and fairness*. And this is important. The nature of the notion of fairness that is central to Howson's semantics is immaterial as long as we are concerned with assessing the value of Howson's technical result *qua* technical result. But if that result is to play a role in justifying probability theory, then it does matter what

Howson's "fairness" refers to – just as it matters what semantics is used in a soundness and completeness proof for a given deductive logic if that proof is adduced as a (or as part of a) justification of that logic. At a minimum, we will want to know whether Howson's "fairness" corresponds to a notion that is pre-analytically significant enough to make having a logic formalizing that notion of interest. And for all Howson has said, it is not clear to me what good there is in being able to syntactically characterize fairness of bets in his sense of fairness, let alone how the fact that probability theory is sound and complete relative to a semantics built on that concept of fairness could justify the claim that the logic sets the standards of right reasoning.[5]

To put my point in different terms: it may be arguable that it is a good thing to have a logic that preserves fairness of bets given an ordinary understanding of fairness. It might be said that such a logic can help us to avoid situations in which we could be bilked. To argue in this way for the Bayesian position, however, would not be essentially different from the traditional Dutch book defence. On Howson's understanding of fairness, on the other hand, the justification via his soundness and completeness result of the claim that probability theory sets the bounds on (static) doxastic rationality does seem to be different from the Dutch book defence. However, on that understanding of fairness it is unclear how Howson's technical result could have any justificatory force.

Nothing said in this note excludes that Howson *can* argue that his concept of fairness is significant to the extent that his soundness and completeness theorem constitutes a veritable justification of Bayesianism, nor that he can do so without any reliance on a theory of rational action. But he has not done so yet. And, as I hope to have shown, until then he has not provided Bayesians with an alternative to the traditional Dutch book defence of their position.

Department of Philosophy, Erasmus University
Rotterdam, The Netherlands

NOTES

[1] Or, at any rate, that we can take the latter to reliably indicate the former.
[2] For a clear presentation of a number of prominent rival logics, see Priest (2001). It must be noted that not everybody will agree that these logics are rivals; *cf.* Haack (1978, Ch. 12) on the controversy between logical monists and logical pluralists.

[3] For instance, second-order logic is not complete (given standard second-order semantics). By Howson's lights, it would thus not really be deserving of the name "logic" (as has also been argued by e.g. Quine and Putnam). As Howson also remarks, however, he thinks we should not attach too much weight to the question whether or not a theory has logical status. Indeed, he could have proposed weaker criteria for what counts as a logic than the three necessary and sufficient conditions cited in the text, and have phrased his main claim as saying that what is really important about probability theory is not simply that it is a logic but that it is a logic that is both sound and complete.

[4] Though, famously, some doubts have been raised about whether it reflects the logical structure of the quantum world and of the mathematical realm.

[5] I must admit that I am not entirely sure whether Howson believes his soundness and completeness proof justifies probability theory as the one true logic of partial belief or just as *a* "correct" logic (leaving open the possibility that there are other modes of right reasoning besides those governed by the probability axioms, that is). However, apart from the fact that it would seem rather uninteresting to argue for the weaker claim – to the best of my knowledge no one has ever claimed that probabilistic reasoning is or can be wrong or misleading – it would also not really make a difference for my point of critique. To play a role in the justification of probability theory even in the weaker sense Howson would have to show that his semantics embodies a concept that intuitively is of such significance that it makes a logic that characterizes that concept syntactically worth having.

REFERENCES

Bacchus, F., Kyburg, H. Jr. and Thalos, M. 1990. "Against Conditionalization". *Synthese* 85: 475–506.

de Finetti, B. 1937. "Foresight: Its Logical Laws, Its Subjective Sources". In H. Kyburg, Jr. and H. Smokler (eds.), *Studies in Subjective Probability*. New York: Krieger, 1980 (second ed.), pp. 53–118.

Haack, S. 1978. *Philosophy of Logics*. Cambridge: Cambridge University Press.

Howson, C. 2000. *Hume's Problem*. Oxford: Oxford University Press.

Howson, C. 2003. "Bayesian Evidence". This volume, pp. 301–320.

Priest, G. 2001. *An Introduction to Non-Classical Logic*. Cambridge: Cambridge University Press.

Ramsey, F. P. 1926. "Truth and Probability". In *The Foundations of Mathematics*. London: Routledge, 1931, pp. 237–256.

Schick, F. 1986. "Dutch Bookies and Money Pumps". *Journal of Philosophy* 83: 112–119.

Waidacher, C. 1997 "Hidden Assumptions in the Dutch Book Argument". *Theory and Decision* 43: 293–312.

PAOLO GARBOLINO

ON BAYESIAN INDUCTION
(AND PYTHAGORIC ARITHMETIC)

Colin Howson has called "Hume's Other Principle" the idea that, in order to make sound inferences from observational data, we need "at least one independent assumption (an inductive assumption) that ... weights some of the possibilities consistent with the evidence more than others" (Howson, this volume, p. 315). I agree with Howson that subjectivist Bayesians are constructive skeptics in the sense that they admit, with Hume, that the circularity of any inductive argument can be broken only by an independent premise, and I would like to show that this premise, in the case of inferences from observed frequencies, is de Finetti's *exchangeability* and its generalisation, *partial exchangeability*.

A finite sequence $X_1,...,X_n$ of binary random variables (I consider only binary random variables for sake of simplicity) is exchangeable if and only if its joint distribution is invariant under permutations:

$$P(X_1 = x_1,...,X_n = x_n) = P(X_1 = x_{\sigma(1)},...,X_n = x_{\sigma(n)})$$

Let $S_n = X_1 + ... + X_n$ denote the number of "successes" in the sequence. If my judgmental probability P satisfies such a measure preserving condition, then it can be represented as a convex mixture of hypergeometric distributions where the weights are uniquely determined by P:

$$P = \sum_{r=0}^{n} p_r H_{n,r}$$

Here $p_r = P(S_n = r)$, and $H_{n,r} = P(X_1 = x_1,...,X_n = x_n | S_n = r)$. If the exchangeable sequence is infinite, then the limit of the random quantity defined as the proportion of "successes" in the sequence exists with probability 1, and the variables X_i are mutually independent and identically distributed, given this random quantity.

The simplest extension of exchangeability is *marginal partial exchangeability*, which obtains when a non-exchangeable sequence can be partitioned into exchangeable subsequences. A kind of analogical induction can be partially pursued within this framework (von Plato 1981). Another type of partial exchangeability is relative to Markov chains (*Markov exchangeability*): all the sequences with the same initial value and the same number of state transitions are judged equiprobable, and the mixture is taken to range over the possible transition matrices (Diaconis and Freedman 1980). Representation theorems for the most general type of partial exchangeability have been provided by (Diaconis and Freedman 1984) and (Lauritzen 1988).

From the subjectivist point of view, "Hume's Other Principle" is a *genus*, to use Richard Jeffrey's term (Jeffrey 1988, p. 249), whose species are different exchangeability judgements, each one of them representing in a rigorous way the "independent inductive assumption" we need for the corresponding mode of inductive inference. I shall briefly show how simple exchangeability satisfies Howson's requirements: (i) it enjoys a convergence-of-opinion property, (ii) it is not a logically valid principle, (iii) it does not tell us what the probabilities are, and (iv) it is the sole assumption independent from the axioms of probability calculus we need for the appropriate mode of inductive inference. The first three conditions are obviously satisfied by exchangeable probability distributions but I think it is worthwhile to stress how exchangeability is different from the old-fashioned "symmetry" assumptions which plagued the history of inductive probability (Zabell 1988).

Consider finite exchangeability, so that the hypothesis that the value of the total number of "successes" is R, in a finite population of N individuals, can be verified by any member of the set of all possible sequences of N observations with the same number of "successes". From an abstract point of view we have a group of invariance transformations partitioning the set of all possible sequences of observations into equivalence classes *modulo* the number of "successes": the members of the equivalence classes are equivalent representations of the same physical state of affair. This congruence relation is independent of my personal state of information: it is a "global symmetry" that can be fairly supposed to be normative for any belief state. So far, Carnap's *structure descriptions* picked up the "right" symmetry but equivalent sequences of observations are not bound by logic alone to happen with the same probability. This is a further step which is not independent of my personal state of information:

a judgement of equiprobability of the members of the equivalence classes is a "local" symmetry that depends on personal belief states, for it contains empirical information to the effect that the operation of observing the number of "successes" is not performed in such a way to make some particular sequence more likely to occur.

Exchangeability is sufficient, together with the basic probability axioms, to update our state of opinion given the observation of empirical frequencies. We don't need to *postulate* conditionalisation as an independent principle for exchangeable sequences, because the predictive probabilities are independent of P. For a finite exchangeable sequence of total length N, with a total number R of "successes", the probability that the next observation will be a "success", given that in the first n observations we have observed r "successes", is given by:

$$P(X_{n+1} = x | S_n = r) = \frac{(R-r)}{(N-n)}$$

Let's denote by P and Q my probability distributions at time t and t', respectively. If P and Q are both exchangeable, my predictive probabilities, given R and N, will be the same:

$$P(X_{n+1} = x | S_n = r) = Q(X_{n+1} = x | S_n = r)$$

That is, the partition whose members are the equivalence classes *modulo* the number of "successes" is a *sufficient statistics*, and sufficiency is logically equivalent to conditionalisation, when the new information I receive at time t' amounts to know the frequency of "successes" in n observations, i. e. $Q(S_n = r) = 1$. Bayesian statistical models implicitly require "time-invariant exchangeability", so to speak, by partitioning an exchangeable random vector into a vector $(X_1, X_2, ..., X_n)$ observed before the time the decision is to be taken, and another random vector $(X_{n+1}, X_{n+2}, ..., X_{n+m})$ that influences the outcome of the decision.

The general case of partial exchangeability can be dealt with by specifying some sufficient statistics and conditional distributions for the data given the sufficient statistics. Models of this type are also known as *intersubjective models* (Dawid 1982, 1985), (Wechsler 1993). Brian Skyrms has extended what I have called "time-invariant exchangeability" into a theory of subjectivist learning of physical probabilities, the "unknown" parameters in an objectivist reading of the representation theorems (Skyrms 1984, pp. 37–62).

I would make a last point about "the principle of direct probability" and its justification. For subjectivist Bayesians the "principle" is, again, a straightforward logical consequence of exchangeability. In a finite exchangeable sequence of binary random variables all the observations have the same probability of "success":

$$P(X_1 = x_1) = P(X_2 = x_2) = \cdots = P(X_n = x_n) = p$$

It follows immediately that the expected value of the frequency S_n/n of "successes" in the sequence is equal to p. Conversely, if I know the value of the frequency of "successes" in an exchangeable sequence, then it follows that my probability of "success" in any individual observation is equal to this known frequency.

The conclusion is that for probability assignments which satisfy exchangeability it is true that "sound inference does not beget new content" (Howson, this volume, p. 316). Inductive inferences are just a matter of deducing complex probabilistic judgements, according to the axioms and theorems of probability calculus, from premises containing other (possibly more simple) probabilistic judgements. This is the reason why de Finetti said, in the last talk before retirement he gave at the University of Rome in 1976, that he considered the expression "Bayesian induction" to be redundant, as saying "Pythagoric arithmetic" (de Finetti 1989, p. 165). The provocative slogan "Probability does not exist" he put in the preface to *Theory of Probability* (de Finetti 1974) is famous enough. I think that the real philosophical import of de Finetti's work might be epitomised by a paraphrase of his slogan: if by "inductive logic" we mean a set of rules characterising valid inductive arguments, either in addition or in the place of the rules of probability theory, then we may as well say that "Inductive logic does not exist".

Scuola Normale Superiore
Pisa, Italy

REFERENCES

Dawid, A. P. 1982. "Intersubjective Statistical Models". In G. Koch and F. Spizzichino (eds.), *Exchangeability in Probability and Statistics*. Amsterdam: New Holland, pp. 217–232.

Dawid, A. P. 1985. "Probability, Symmetry and Frequency". *British Journal for the Philosophy of Science* 36: 107–128.

De Finetti, B. 1974. *Theory of Probability*. Chichester: Wiley, vol. I.

De Finetti, B. 1989. "La Probabilità: Guardarsi dalle Contraffazioni!" In B. de Finetti, *La Logica dell'Incerto*. Milano: Il Saggiatore, pp. 149–188.

Diaconis, P. and Freedman, D. A. 1980. "De Finetti's Theorem for Markov Chains". *Annals of Probability* 8: 115–130.

Diaconis, P. and Freedman, D. A. 1984. "Partial Exchangeability and Sufficiency". In J. K. Ghosh and J. Roy (eds.), *Statistics: Applications and New Directions*. Calcutta: Indian Statistical Institute, pp. 205–236.

Howson, C. 2003. "Bayesian Evidence". This volume, pp. 301-320.

Jeffrey, R. C. 1988. "Conditioning, Kinematics, and Exchangeability". In B. Skyrms and W. L. Harper (eds.), *Causation, Chance and Credence*. Dordrecht: Kluwer, vol. I, pp. 221–255.

Lauritzen, S. L. 1988. *Extremal Families and Systems of Sufficient Statistics*. Berlin: Springer.

Skyrms B. 1984. *Pragmatics and Empiricism*. New Haven: Yale University Press.

von Plato J. 1981. "On Partial Exchangeability as a Generalization of Symmetry Principles". *Erkenntnis* 16: 53–59.

Wechsler S. 1993. "Exchangeability and Predictivism". *Erkenntnis* 38: 343–350.

Zabell, S. 1988. "Symmetry and its Discontents". In B. Skyrms and W. L. Harper (eds.), *Causation, Chance and Credence*. Dordrecht: Kluwer, vol. I, pp. 155–190.

ILKKA NIINILUOTO

PROBABILITY AND LOGIC
COMMENTS ON COLIN HOWSON

In his stimulating paper "Bayesian Evidence" Colin Howson advances the thesis that the "proper conceptual environment" of probability is logic. He argues that F. P. Ramsey and L. J. Savage "effectively obscured" the logical character of epistemic probability by putting it in the setting of decision theory and utility theory. Instead, Bayesian probability should be analysed in terms of expected gains in fair betting: probability axioms then appear as formal constraints that are analogous to the logical principles of consistency.

Howson also makes some interesting remarks concerning Bayesian inferences. His examples against the principle of conditionalisation involve cases where we become uncertain of a mathematical truth or learn something that logically contradicts our earlier beliefs. I agree that such cases cannot be treated by simple conditionalisation models, but rather belong to the domain of belief revision (see Gärdenfors 1988). For this reason, I shall concentrate my remarks on the first part of Howson's paper, viz. his discussion on the relations of probability and logic.

Varieties of probability logic

According to Howson, Ramsey treated probability axioms as "consistency constraints of a logical character", but then strangely deflected from this logical path "by choosing to embed his discussion not within the theory of logic as it was then being transmuted into its modern form largely at the hands of Gödel and Tarski". I think this remark invites us to put Ramsey's seminal paper "Truth and Probability" (1926) into a historical perspective. In this way, we can also briefly survey various attempts to correlate logic and probability.

First, as a student of Bertrand Russell, Ramsey himself was an expert on logic who did important work on type theory.

Secondly, there was an important British tradition in logic and probability. Classical works in logic such as Augustus De Morgan's *Formal Logic* (1847) and George Boole's *The Laws of Thought* (1854) devoted a lot of attention to probability. Works in scientific method such as J. S. Mill's *A System of Logic* (1843) and Stanley Jevons' *The Principles of Science* (1874) discussed probability and induction. John Venn, who pioneered the frequency interpretation of probability in the *The Logic of Chance* (1866), distinguished "conceptualist" and "materialist" treatments of logic in *The Principles of Empirical and Inductive Logic* (1889). The former deals with consistency among human ideas, while the latter is interested in the truth and falsity of judgements (see Niiniluoto 1988).

This theme seems to form the background to Ramsey's article. He starts by asserting that "the Theory of Probability is taken as a branch of logic, the logic of partial belief and inconclusive argument" (Ramsey 1931, p. 157). His main results on betting ratios prove that "the laws of probability are laws of consistency, an extension to partial beliefs of formal logic" (*ibid.*, p. 182). But, in addition to the Logic of Consistency, there is the Logic of Truth. Ramsey's paper, which is a landmark in the foundations of Bayesianism, thus ends with a section which is inspired by a rival approach to probabilistic inference, viz. C. S. Peirce's frequentist reliabilism.

Thirdly, the new logic – Frege's system, Russell's type theory, Hilbert's proof theory, Löwenheim's, Skolem's, and Tarski's model theory – had nothing to do with probability. This was no accident, as probability was systematically excluded from the pure core of logic (*cf.* Niiniluoto 1988). To see why this was the case, let us recall the famous dictum of G. W. Leibniz in 1678: *probabilitas est gradus possibilitatis,* i.e., probability is "degree of possibility". Hence, probabilistic statements such as "it is probable that p" and "it is probable to the degree that p" are *modal* expressions – comparable to "it is possible that p" and "it is necessary that p". However, at the same time the Aristotelian tradition of modal syllogistic had fallen into "neglect, if not contempt" (to quote Thomas Reid's assessment in 1774). Immanuel Kant made a distinction between problematic, assertoric, and apodeictic judgements, but he argued that these modalities qualify only our epistemological relation to a judgement, not its content. This was the received view in the psychological logic of the 19th century Germany, and it was repeated by Gottlob Frege when he excluded modality from his *Begriffsschrift* in 1879. Russell also denied that modalities could be assigned to propositions. The first attempt to introduce modality as an operator in the syntax of logic was made by C. I. Lewis in 1912. The semantical treatment of probability had to wait until the 1960s for the first applications of possible worlds semantics.

Fourthly, some Continental logical work existed that was relevant to the study of probability, but it was based on the idea that probabilities are assigned to propositional functions, i.e., open formulas with a free object or time variable. Jan Łukasiewicz gave a precise account of the finite frequency interpretation of probability along these lines in *Die logische Grundlagen der Wahrscheinlichkeitsrechnung* in 1913. Ramsey did not refer to Łukasiewicz, but the latter's main results were known to him (see Ramsey 1931, p. 158). Eino Kaila published his *Wahrscheinlichkeitslogik* in 1926, and Hans Reichenbach gave a full account of his frequentist probability logic in 1932 (see Reichenbach 1949). The Polish logicians, among them Janina Hosiasson, refused to acknowledge Reichenbach's probability logic as an extension of classical logic (*cf.* Niiniluoto 1998). Alfred Tarski pointed out that probability is intensional or non-truth-functional and concluded that it thereby differs also from many-valued logic. New impetus for probability logic arose in 1963, when Jerzy Łos applied model theory to assign probabilities to first-order formulas with quantification (see Fenstad 1968).

The status of probability logic

Ramsey rejected the logical interpretation of probability, as developed by J. M. Keynes in 1921 in his theory of unique probabilistic relations between propositions. Ramsey would also have rejected Rudolf Carnap's inductive logic of the 1940s, with its hope to establish that quantitative probability statements are analytically true. As a Bayesian, Howson is of course not advocating a return to logical probability as partial entailment. He is not claiming (with Carnap) that *probability values* are determined by logical considerations. Rather, he suggests that the *axioms of probability* are a sort of logical consistency principles. Howson gives an elegant formulation of the important result that fair betting quotients satisfy probability axioms. But I think this can be seen as a theorem of pure mathematics, and it does not yet show the "logical" character of probability axioms.

Howson regards it as significant that the condition of fairness can be expressed in terms of relations between judgements. He protests against Ramsey and Savage, who embedded personal probabilities in a more complex theory involving preferences, actions, decisions, and utilities. I do not share this concern with him. If we are flexible in using the term "logic", then we can have studies with titles such as "logic of action" and "preference logic". Monetary bets constitute a special case of the more general theory of utility, and the situation does not essentially change if the stakes are given in, e.g., sand or fluid. The important point is not the medium of payoffs but – as Ramsey argued – the attempt to measure beliefs by our willingness to act upon them. In order to link the abstract notion of a betting quotient to degrees of belief, some reference to human action and decision is indispensable.

After giving his justification of probability axioms, Ramsey asserts that "the theory of probability is in fact a generalization of formal logic", but still "one of the most important aspects of formal logic is destroyed": "the calculus of objective partial belief cannot be immediately interpreted as a body of objective tautology" (Ramsey 1931, pp. 181–187).

Savage (1954, p. 57) states that the mathematical theory of probability enables "the person using it to detect inconsistencies in his own or envisaged behavior". If a person has detected such an inconsistency, he will remove it.

Kyburg and Smokler (1964, p. 12) warn that the Bayesian notion of "coherence" (i.e., avoidance of a Dutch Book) should not be confused with the logical concept of "consistency". Howson presents an original argument to show that coherence is indeed a consistency requirement. He first extends the ordinary logical notion of consistency from sets of propositions to truth value assignments. He then defines an assignment of betting quotients to a set of propositions to be *consistent* if it can be extended to all propositions so that the constraint of fairness is satisfied.

This argument is beautiful, but I find it puzzling. The analogy between truth values and betting quotients fails for the reason that the probability axioms as external constraints are different from the principles of logic, since probability values are not truth-functional. Further, this analogy points in two directions, neither of which is acceptable to a Bayesian. One is Reichenbachian probability logic, where truth equals probability one. The other is inductive logic: a set of statements is logically consistent if and only if some truth value assignment makes each statement true; this is equivalent to the existence of a Carnapian state description which implies all the statements. Consistency is thus equivalent to logical possibility. But if probability is degree of possibility, as Leibniz asserted, then the next step is to determine probabilities by calculating how many state descriptions satisfy a statement: more consistent, more probable. But this leads us towards the logical interpretation of probability – not to the direction where the axioms of probability have a logical justification.

Department of Philosophy, University of Helsinki
Helsinki, Finland

REFERENCES

Carnap, R. 1950. *The Logical Foundations of Probability*. Chicago: The University of Chicago Press.

Fenstad, J. E. 1968. "The Structure of Logical Probabilities". *Synthese* 18: 1–27.

Gärdenfors, P. 1988. *Knowledge in Flux: Modeling the Dynamics of Epistemic States*. Cambridge, Mass.: The MIT Press.

Kyburg, H. E., Jr., and Smokler, H. (eds.) 1964. *Studies in Subjective Probability*. New York: John Wiley.

Niiniluoto, I. 1988. "From Possibility to Probability: British Discussions on Modality in the Nineteenth Century". In S. Knuuttila (ed.), *Modern Modalities*. Dordrecht: D. Reidel, pp.275–309.

Niiniluoto, I. 1998. "Induction and Probability in the Lvov-Warsaw School". In K. Kijania-Placek and J. Wolenski (eds.), *The Lvov-Warsaw School and Contemporary Philosophy*. Dordrecht: Kluwer, pp. 323–335.

Ramsey, F. P. 1931. "Truth and Probability". In *The Foundations of Mathematics and Other Logical Essays*. London: Kegan Paul, Trench, Trubner & Co., pp. 156–199. Reprinted in H.E. Kyburg and H. Smokler (1964), pp. 61–92.

Reichenbach, H. 1949. *The Theory of Probability*. Berkeley: The University of California Press.

Savage, L. J. 1954. *The Foundations of Statistics*. New York: John Wiley.

INDEX OF NAMES

Aaronson, D., 125
Aaronson, M., 125
Abelson, R., 123
Ackermann, R., 291
Adrian, E.D., 4
Althusser, L., 190, 192, 193
Ampère, A.-M., 171
Anderson, J.R., 129
Arbib, M.A., 121
Arbuthnot, J., 107
Aristotle, 2, 105, 209–211, 221–223, 229, 237–238, 251–252, 270, 273
Aspect, A., 215, 224, 239, 240
Aspray, W., 115, 116
Atkinson, D., 224, 227–235, 237–241, 246–247, 251–256
Augustin, 299

Baars, B.J., 121
Babbage, C., 113–116, 118–119, 128, 131
Bacchus, F., 321
Bachelard, G., 164, 181, 189–192
Bacon, F., 96, 99, 181, 199, 262, 270–277, 281–282, 296, 316
Baltas, A., 88
Barthes, R., 193
Bartlett, F., 120, 126
Batens, D., 196
Bayes, T., 107, 111, 310, 317
Beck, A., 4
Becker, O., 63
Bell, J., 32, 34, 58, 60, 214–215, 219, 224, 239–240, 247, 253–255
Berg, C.C., 63, 64
Berger, H., 4
Berkeley, G., 14, 15, 29
Bernoulli, D., 71, 79, 303
Bernoulli, J., 302
Bertrand, J.-L.-F., 304
Berzelius, J., 172–173
Bicchieri, C., 80
Biot, J.B., 286

Birdsall, T.G., 104
Birnbaum, A., 318
Birnbaum, M.H., 105, 111
Birolini, A., 96
Blackwell, R.J., 130
Blondlot, R., 278
Boerhaave, H., 161
Bohm, D., 58, 214–215, 219, 239, 247, 253–255
Bohr, N., 214–216, 219, 239, 247, 253, 280–281
Boole, G., 132, 334
Borsboom, D., 245
Bouwmeester, D., 39
Bradshaw, G.L., 101
Braithwaite, R., 99
Brehmer, B., 108
Brown, J.R., 212, 244, 246, 250
Brown, W., 103
Brunswik, E., 107–108, 110
Buchwald, J., 260–261, 264, 298

Campaner, R., 282
Canguilhem, G., 191–192
Cantor, G., 277, 318
Carmichael, L., 4
Carnap, R., 95, 130, 302, 328, 335
Carnot, L., 114
Carrier, M., 255–256
Castellan, N.J., 126
Caton, R., 4
Cavaillès, J., 191–192
Cavallo, T., 3
Caverni, J.P., 47
Christie, M., 182
Church, A., 310
Clauser, J.F., 34
Cohen, L.J., 112
Cohen, R.S., 310
Colebrooke, H., 113
Coles, M.G., 4
Conant, J., 196
Condillac, S.B. de, 113

Condorcet, J.M.N., 113, 302–303
Cooley, J.W., 11
Copernicus, N., 129
Coquillette, D.R., 275
Cronbach, L., 108

d'Alembert, J.B., 113
Dalton, J., 173
Darrigol, O., 264–265
Darwin, C., 129
Dashiell, J.F., 108
Daston, L., 113–114, 131
Davis, H., 4
Dawid, A.P., 329
Dawid, R., 256
de Barros, J.A., 30, 36, 38–39, 57–60
de Finetti, B., 304, 318, 321, 323, 327, 330
Deleuze, G., 193
De Morgan, A., 333
Derrida, J., 190, 193, 195
Descartes, R., 261–262, 312
Descombes, V., 196
de Saussure, F., 190–193
de Staël, G.N. Madame de, 131
Diaconis, P., 328
Dosse, F., 196
Dotherty, M.E., 110
Draaisma, D., 155–156
Du Bois-Reymond, E., 3
Dufay, C., 171
Duhem, P., 182, 206, 259, 263, 276, 279
Duncker, K., 120–121

Earman, J., 318
Edwards, A.W.F., 96
Egidi, M., 80
Ehrenhaft, F., 278
Einstein, A., 57, 129–130, 154–155, 213–217, 219, 224, 239, 243–244, 247, 253–254, 256
Einthoven, V., 4

Epelboim, J., 12, 14, 22–23

Faraday, M., 100, 171, 265, 286
Fechner, G. T., 100, 103, 107, 126
Fenstad, J.E., 335
Feynman, R., 131
Fine, A., 34, 57–58, 61
Fischer, K-H., 63
Fisher, R.A., 96, 105, 107, 111–112, 317
Fludd, R., 155
Fodor, J., 128
Foucault, M., 190, 193
Fouraker, L.E., 65
Fourier, J.B., 8–9, 12, 19
Franck, J., 278, 280–281
Franklin, A., 266–269, 276, 285, 287, 296
Freedman, D.A., 328
Frege, G., 334
Friedman, M., 299

Gaifman, H., 308
Galanter, E., 120, 124
Galavotti, M.C., 282
Galilei, G., 209–212, 218–223, 227–234, 236–238, 251–252, 254, 279
Galison, P., 24, 130–131, 175, 181, 195, 285
Galton, F., 100, 108
Galvani, L., 2–3
Garceau, L., 4
Gärdenfors, P., 333
Gardner, H., 103, 110
Gauss, C.F., 113
Gavin, E.A., 105
Gendler, T.S., 212, 219–220, 229, 238, 244, 246, 251–252
Gibbs, E., 4
Gibbs, F.A., 4
Gigerenzer, G., 99, 102, 104, 106–108, 111–112, 115, 126–127, 139, 141, 145–150, 153-157

Gilboa, I., 96
Girotto, V., 47, 54, 318
Glas, E., 249, 256
Gleick, J., 131
Gödel, K., 301, 310, 333
Goldstein, D.G., 112, 115, 127
Gonzalez, M., 318
Gonzalez, W., 80
Gooding, D., 267–268, 276, 285–286, 291
Goodman, N., 176, 182, 196
Gosset, W.S., 107
Green, B., 123
Greenberger, D.M., 1, 31, 35, 245
Greif, A., 96
Gruber, H., 110
Grupsmith, E., 125
Guilford, J.P., 103
Güth, W., 66

Haack, S., 325
Hacking, I., 111–112, 130–131, 159, 162, 183, 196, 261, 267–270, 280–281, 285
Hagner, M., 287
Halmos, P., 316
Hammond, K.R., 108
Han, B., 5–6, 12, 14, 22–23
Hanson, N.R., 100, 207
Harré, R., 265–266
Harsanyi, J., 71, 79
Häselbarth, V., 63–64, 79
Haugeland, J., 196
Heidelberger, M., 181, 259
Heisenberg, W.K., 214
Hellman, G., 302, 304, 324, 326
Hempel, C.G., 130
Hendry, D.F., 93
Herbart, J.F., 103
Hertz, H., 278, 280–281
Hicks, J., 96
Hilbert, D., 334
Hilgard, E., 108
Hintikka, J., 260

Hirschman, A., 87
Holland, J.H., 80
Holland, P.W., 31
Holmes, F.L., 182
Holton, G., 129–130
Holyoak, K.J., 80
Hon, G., 260, 262, 264, 269, 276–278, 281, 285–291, 295–299
Horne, M.A., 1, 31, 35, 255
Hosiasson, J., 335
Howard, D., 256
Howson, C., 143–144, 149, 305, 308–309, 312, 315, 317–318, 322–328, 330, 333, 335–336
Hull, C.L., 129
Hume, D., 14–15, 29, 315, 318, 327–328
Humphreys, P., 245, 256

Irvine, A.D., 244

Jacobson, R., 193
Jaffe, A., 249
James, W., 120
Jammer, M., 239
Janis, A.J., 246, 254
Jardine, L., 183, 271–272
Jasper, H.H., 4
Jeffrey, R., 311, 314–315, 328
Jevons, S., 334
Johnson-Laird, P.N., 47–49, 52, 54, 128
Joyce, C.R.B., 108

Kahneman, D., 53–54, 111
Kaila, E., 335
Kalish, G.K., 65
Kant, I., 75, 334
Kauffman, S., 96
Kaufman, L., 5
Kekulé, von S., 100
Kelley, H.H., 102, 105–106
Kelly, K., 317
Kelvin, Lord W., 4

Kendall, M., 107
Kepler, J., 156
Keynes, J.M., 96, 335
Kita, S., 128
Klein, U., 174, 180–182, 187–191, 194–196, 199–201, 203–207
Koehler, D.K., 50
Köhler, E., 69
Köhler, W., 120–121, 126
Kolmogorov, A.N., 215, 312
Koyré, A., 191–192
Kubli, E., 299
Kuhn, T., 130–131, 191, 196, 205–207, 245, 256, 260–261
Kühne, U., 243, 252, 256
Kuipers, T., 218
Külpe, O., 123
Kyburg, H., 321, 336

Lacan, J., 192–193
Lakatos, I., 243–245, 249–251, 256
Langley, P., 101, 121
Laplace, P.S. de, 109, 132, 302
Latour, B., 175–176, 181–182
Lauritzen, S.L., 328
Lavoisier, A-L., 182, 200
Laymon, R., 244
Leary, D.E., 102, 107, 108, 130
Lee, P.M., 318
Legendre, A., 114
Legrenzi, M., 47, 54
Legrenzi, P., 47, 54
Leibniz, G.W. von, 99, 301, 334, 336
Lennox, W.G., 4
Lenoir, T., 100, 130
Levelt, W.J.M., 127–128
Levinson, S., 128
Lévi-Strauss, C., 190, 192–193
Lewis, C.I., 307, 311–312, 334
Lindley, D.V., 317
Linsley, D., 4
Locke, J., 14–15
Lopes, L.L., 105, 112
Łos, J., 335

Lovie, A.D., 103
Löwenheim, L., 310, 334
Lu, Z.L., 5, 6, 12, 14, 22–23
Luce, D., 109, 126
Łukasiewicz, J., 335
Lynch, M., 291

Mach, E., 212, 243, 246, 248–249, 252, 263, 266–268, 277, 296, 298
Macieszyna, S.J., 4
March, J.G., 80
Marris, R., 80
Martin, J., 281
Marx, K., 192–193
Matteucci, C., 3
Matthews, B.H.C., 4
Mayo, D., 260
McArthur, L.A., 105
McCabe, K. A., 80
McCorduck, P., 117, 119, 132
McCulloch, W., 115–116
McLaughlin, R., 142
Mellenbergh, G.J., 245
Melton, A.W., 107
Mermin, N.D., 35
Metropolis, N., 131
Michaela, I.L., 105
Michelson, A.A., 129
Michotte, A., 105
Mill, J.S., 334
Miller, G.A., 120, 123–124
Millikan, R.A., 129, 254
Milnor, J. W., 65
Mises, von R., 308
Murdock, B.B., 104
Murray, D.J., 102, 104, 106, 111
Mynatt, C.R., 110

Nash, J. F., 65, 71, 79
Nering, F. D., 65
Newell, A., 117–125, 128, 155
Newton, I., 100, 154–155, 211, 217, 220–222, 244

Neyman, J., 102–104, 107, 111–112, 126, 141, 148–150, 317
Nickles, T., 100, 156, 291
Niiniluoto, I., 334–335
Nisbett, R. E., 80
Norman, D.A., 105
Norton, J.D., 212, 229, 244, 256
Nye, M.J., 278

Oas, G., 30
Ockenfels, A., 76
Oppenheim, A.V., 8
Ørsted, H.C., 243

Papert, S., 116
Papineau, D., 144
Paris, J., 306
Pavlov, I.P., 126
Pearson, E. S., 102–104, 107, 111–112, 126, 141, 148–150, 317
Pearson, K., 107–108
Peijnenburg, J., 224
Peirce, C.S., 176, 334
Penfield, W., 4
Piaget, J., 105, 126
Pickering, A., 181–182, 263–264, 267–268, 285, 287
Pinch, T., 267, 276
Pitts, W., 115–116
Plato, 118, 155, 272
Podolsky, B., 213, 219, 239
Poincaré, H., 100
Poni, C., 96
Popper, K., 99–100, 114, 121, 130, 143, 157, 228, 241
Pribram, K.H., 120, 124
Priest, G., 323, 325
Prony, de G.R., 114, 118
Putnam, H., 326

Quine, W.V.O., 182, 206–207, 227, 232, 326
Quinn, F., 249

Radder, H., 259
Ramsey, F.P., 301, 313, 316, 321, 323, 333–336
Rassenti, S.J., 80
Rausch, E., 65
Ravelhofer, B., 95
Raymond, P., 196
Rédei, M., 240
Redi, F., 2
Reichenbach, H., 99–100, 107, 129–130, 130, 154–157, 240, 335
Reid, T., 334
Reinfeld, M., 63
Renn, J., 282
Rescher, N., 74–75, 80, 243
Rheinberger, H-J., 175–176, 181–182, 195, 201, 287
Ricciardi, F.M., 65
Rocke, A.J., 182
Rosen, N., 213, 219, 239
Rosenbaum, P.R., 31
Rosenfeld, L., 239
Roth, A., 71, 73, 79
Rubin, G.S., 25
Rubinstein, A., 92, 95
Rucci, A.J., 107
Rugg, M.D., 4
Russell, B., 117, 119, 132, 333–334

Salviati, 209, 210, 222, 239
Sauermann, H., 63–65, 71, 79
Savage, L.J., 301–304, 333, 335–336
Scazzieri, R., 87, 96
Schafer, R.W., 8
Schaffer, S., 114, 182, 267, 276
Schelling, T.C., 96
Schick, F., 321
Schmeidler, D., 96
Schmittberger, R., 66
Schuster, K.G., 63
Schwarze, B., 66
Searle, J., 128
Secord, J.A., 183
Selten, R., 71–80, 85, 88, 96

Sen, A., 87, 95
Shafer, G., 317
Shannon, C.E., 132
Shapin, S., 182
Shaw, J.C., 119, 124
Shepard, R., 123
Shimony, A., 35
Siegel, S., 65
Simon, H., 65, 71, 75, 77, 80, 86, 101, 117–128, 155
Simplicio 210, 222–223, 239
Skinner, Q., 126, 196
Skolem, T., 310, 334
Skyrms, B., 329
Smith, A., 114
Smith, L.D., 129
Smith, V., 65, 80
Smokler, H., 336
Smullyan, R.M., 305
Solomon, J.R., 281
Spary, E.C., 183
Spiegel, M., 149
Spinoza, B., 193
Spranzi, M., 236
Stahl, G., 141–142
Steinle, F., 181, 259
Sterling, T.D., 107
Stevin, S., 248
Stewart, T.R., 109
Stirpe, F., 96
Stokes, G., 278
Stöltzner, M., 254, 256
Stove, D., 142
Stroup, A., 182
Stump, D.J., 195
Suppes, P., 5–6, 12–14, 22–23, 25–26, 29–31, 33, 36, 38–39, 57–60, 89, 95
Swets, J.A., 102, 104, 120
Swijtink, Z.G., 106
Szabó, Á., 243
Szabó, L., 57, 61

Tanner, W.P., 102–104, 120, 148, 150
Tarski, A., 301, 310, 333–335

Teigen, K.H., 112
Teller, P., 311–312
Thagard, P.R., 80
Thalos, M., 321
Thirring, W., 256
Thomas Aquinas, 299
Thomson, G.H., 103
Thorndike, R.F., 108
Thurston, W., 246
Thurstone, L.L., 109–110
Tietz, R., 63, 66, 68
Titchener, E.B., 103
Tolman, E., 120, 129
Toulmin, S., 102, 130
Tukey, J.W., 11
Turano, K., 25
Turing, A., 115–117, 155
Tversky, A., 50, 53–54, 111
Tweney, R.D., 107, 110

Urbach, P., 143–144, 149, 308, 315

Valeriani, L., 87
van Bendegem, J.P., 196
van Fraassen, B., 205–206, 312
van Heerden, J., 245
Venn, J., 34
Viale, R., 80
Volta, A., 3
von Neumann, J., 115–117, 155, 214, 253–255
von Plato, J., 328

Waidacher, C., 3 21
Walpole, H., 91
Weber, E.H., 103
Weber, H.-J., 64, 68,
Weinberg, S., 2 48
Wertheimer, M., 121
Wechsler, S., 329
Whitehead, A.N., 117, 119, 132, 261
Whittaker, E.T., 2
Wickelgreen, W.A., 105

Williamson, S., 5
Witten, E., 217
Witt-Hansen, J., 243
Wolters, G., 282
Wong, D.K., 13, 25–26, 29
Woolgar, S., 181–182
Wundt, W., 64, 123, 126

Zabell, S., 328
Zanotti, M., 31, 33
Zeilinger, A., 1, 31, 35, 255
Zhou, L., 92
Zytkow, J.M., 101

Boston Studies in the Philosophy of Science

Editor: Robert S. Cohen, *Boston University*

1. M.W. Wartofsky (ed.): *Proceedings of the Boston Colloquium for the Philosophy of Science, 1961/1962.* [Synthese Library 6] 1963   ISBN 90-277-0021-4
2. R.S. Cohen and M.W. Wartofsky (eds.): *Proceedings of the Boston Colloquium for the Philosophy of Science, 1962/1964.* In Honor of P. Frank. [Synthese Library 10] 1965   ISBN 90-277-9004-0
3. R.S. Cohen and M.W. Wartofsky (eds.): *Proceedings of the Boston Colloquium for the Philosophy of Science, 1964/1966.* In Memory of Norwood Russell Hanson. [Synthese Library 14] 1967   ISBN 90-277-0013-3
4. R.S. Cohen and M.W. Wartofsky (eds.): *Proceedings of the Boston Colloquium for the Philosophy of Science, 1966/1968.* [Synthese Library 18] 1969   ISBN 90-277-0014-1
5. R.S. Cohen and M.W. Wartofsky (eds.): *Proceedings of the Boston Colloquium for the Philosophy of Science, 1966/1968.* [Synthese Library 19] 1969   ISBN 90-277-0015-X
6. R.S. Cohen and R.J. Seeger (eds.): *Ernst Mach, Physicist and Philosopher.* [Synthese Library 27] 1970   ISBN 90-277-0016-8
7. M. Čapek: *Bergson and Modern Physics.* A Reinterpretation and Re-evaluation. [Synthese Library 37] 1971   ISBN 90-277-0186-5
8. R.C. Buck and R.S. Cohen (eds.): *PSA 1970.* Proceedings of the 2nd Biennial Meeting of the Philosophy and Science Association (Boston, Fall 1970). In Memory of Rudolf Carnap. [Synthese Library 39] 1971   ISBN 90-277-0187-3; Pb 90-277-0309-4
9. A.A. Zinov'ev: *Foundations of the Logical Theory of Scientific Knowledge (Complex Logic).* Translated from Russian. Revised and enlarged English Edition, with an Appendix by G.A. Smirnov, E.A. Sidorenko, A.M. Fedina and L.A. Bobrova. [Synthese Library 46] 1973   ISBN 90-277-0193-8; Pb 90-277-0324-8
10. L. Tondl: *Scientific Procedures.* A Contribution Concerning the Methodological Problems of Scientific Concepts and Scientific Explanation. Translated from Czech. [Synthese Library 47] 1973   ISBN 90-277-0147-4; Pb 90-277-0323-X
11. R.J. Seeger and R.S. Cohen (eds.): *Philosophical Foundations of Science.* Proceedings of Section L, 1969, American Association for the Advancement of Science. [Synthese Library 58] 1974   ISBN 90-277-0390-6; Pb 90-277-0376-0
12. A. Grünbaum: *Philosophical Problems of Space and Times.* 2nd enlarged ed. [Synthese Library 55] 1973   ISBN 90-277-0357-4; Pb 90-277-0358-2
13. R.S. Cohen and M.W. Wartofsky (eds.): *Logical and Epistemological Studies in Contemporary Physics.* Proceedings of the Boston Colloquium for the Philosophy of Science, 1969/72, Part I. [Synthese Library 59] 1974   ISBN 90-277-0391-4; Pb 90-277-0377-9
14. R.S. Cohen and M.W. Wartofsky (eds.): *Methodological and Historical Essays in the Natural and Social Sciences.* Proceedings of the Boston Colloquium for the Philosophy of Science, 1969/72, Part II. [Synthese Library 60] 1974   ISBN 90-277-0392-2; Pb 90-277-0378-7
15. R.S. Cohen, J.J. Stachel and M.W. Wartofsky (eds.): *For Dirk Struik.* Scientific, Historical and Political Essays in Honor of Dirk J. Struik. [Synthese Library 61] 1974   ISBN 90-277-0393-0; Pb 90-277-0379-5
16. N. Geschwind: *Selected Papers on Language and the Brains.* [Synthese Library 68] 1974   ISBN 90-277-0262-4; Pb 90-277-0263-2
17. B.G. Kuznetsov: *Reason and Being.* Translated from Russian. Edited by C.R. Fawcett and R.S. Cohen. 1987   ISBN 90-277-2181-5

Boston Studies in the Philosophy of Science

18. P. Mittelstaedt: *Philosophical Problems of Modern Physics*. Translated from the revised 4th German edition by W. Riemer and edited by R.S. Cohen. [Synthese Library 95] 1976
ISBN 90-277-0285-3; Pb 90-277-0506-2
19. H. Mehlberg: *Time, Causality, and the Quantum Theory*. Studies in the Philosophy of Science. Vol. I: *Essay on the Causal Theory of Time*. Vol. II: *Time in a Quantized Universe*. Translated from French. Edited by R.S. Cohen. 1980   Vol. I: ISBN 90-277-0721-9; Pb 90-277-1074-0
Vol. II: ISBN 90-277-1075-9; Pb 90-277-1076-7
20. K.F. Schaffner and R.S. Cohen (eds.): *PSA 1972*. Proceedings of the 3rd Biennial Meeting of the Philosophy of Science Association (Lansing, Michigan, Fall 1972). [Synthese Library 64] 1974
ISBN 90-277-0408-2; Pb 90-277-0409-0
21. R.S. Cohen and J.J. Stachel (eds.): *Selected Papers of Léon Rosenfeld*. [Synthese Library 100] 1979
ISBN 90-277-0651-4; Pb 90-277-0652-2
22. M. Čapek (ed.): *The Concepts of Space and Time*. Their Structure and Their Development. [Synthese Library 74] 1976   ISBN 90-277-0355-8; Pb 90-277-0375-2
23. M. Grene: *The Understanding of Nature*. Essays in the Philosophy of Biology. [Synthese Library 66] 1974   ISBN 90-277-0462-7; Pb 90-277-0463-5
24. D. Ihde: *Technics and Praxis*. A Philosophy of Technology. [Synthese Library 130] 1979
ISBN 90-277-0953-X; Pb 90-277-0954-8
25. J. Hintikka and U. Remes: *The Method of Analysis*. Its Geometrical Origin and Its General Significance. [Synthese Library 75] 1974   ISBN 90-277-0532-1; Pb 90-277-0543-7
26. J.E. Murdoch and E.D. Sylla (eds.): *The Cultural Context of Medieval Learning*. Proceedings of the First International Colloquium on Philosophy, Science, and Theology in the Middle Ages, 1973. [Synthese Library 76] 1975   ISBN 90-277-0560-7; Pb 90-277-0587-9
27. M. Grene and E. Mendelsohn (eds.): *Topics in the Philosophy of Biology*. [Synthese Library 84] 1976   ISBN 90-277-0595-X; Pb 90-277-0596-8
28. J. Agassi: *Science in Flux*. [Synthese Library 80] 1975
ISBN 90-277-0584-4; Pb 90-277-0612-3
29. J.J. Wiatr (ed.): *Polish Essays in the Methodology of the Social Sciences*. [Synthese Library 131] 1979   ISBN 90-277-0723-5; Pb 90-277-0956-4
30. P. Janich: *Protophysics of Time*. Constructive Foundation and History of Time Measurement. Translated from German. 1985   ISBN 90-277-0724-3
31. R.S. Cohen and M.W. Wartofsky (eds.): *Language, Logic, and Method*. 1983
ISBN 90-277-0725-1
32. R.S. Cohen, C.A. Hooker, A.C. Michalos and J.W. van Evra (eds.): *PSA 1974*. Proceedings of the 4th Biennial Meeting of the Philosophy of Science Association. [Synthese Library 101] 1976   ISBN 90-277-0647-6; Pb 90-277-0648-4
33. G. Holton and W.A. Blanpied (eds.): *Science and Its Public*. The Changing Relationship. [Synthese Library 96] 1976   ISBN 90-277-0657-3; Pb 90-277-0658-1
34. M.D. Grmek, R.S. Cohen and G. Cimino (eds.): *On Scientific Discovery*. The 1977 Erice Lectures. 1981   ISBN 90-277-1122-4; Pb 90-277-1123-2
35. S. Amsterdamski: *Between Experience and Metaphysics*. Philosophical Problems of the Evolution of Science. Translated from Polish. [Synthese Library 77] 1975
ISBN 90-277-0568-2; Pb 90-277-0580-1
36. M. Marković and G. Petrović (eds.): *Praxis*. Yugoslav Essays in the Philosophy and Methodology of the Social Sciences. [Synthese Library 134] 1979
ISBN 90-277-0727-8; Pb 90-277-0968-8

Boston Studies in the Philosophy of Science

37. H. von Helmholtz: *Epistemological Writings.* The Paul Hertz / Moritz Schlick Centenary Edition of 1921. Translated from German by M.F. Lowe. Edited with an Introduction and Bibliography by R.S. Cohen and Y. Elkana. [Synthese Library 79] 1977
ISBN 90-277-0290-X; Pb 90-277-0582-8
38. R.M. Martin: *Pragmatics, Truth and Language.* 1979
ISBN 90-277-0992-0; Pb 90-277-0993-9
39. R.S. Cohen, P.K. Feyerabend and M.W. Wartofsky (eds.): *Essays in Memory of Imre Lakatos.* [Synthese Library 99] 1976   ISBN 90-277-0654-9; Pb 90-277-0655-7
40. Not published.
41. Not published.
42. H.R. Maturana and F.J. Varela: *Autopoiesis and Cognition.* The Realization of the Living. With a Preface to "Autopoiesis' by S. Beer. 1980   ISBN 90-277-1015-5; Pb 90-277-1016-3
43. A. Kasher (ed.): *Language in Focus: Foundations, Methods and Systems.* Essays in Memory of Yehoshua Bar-Hillel. [Synthese Library 89] 1976
ISBN 90-277-0644-1; Pb 90-277-0645-X
44. T.D. Thao: *Investigations into the Origin of Language and Consciousness.* 1984
ISBN 90-277-0827-4
45. F.G.-I. Nagasaka (ed.): *Japanese Studies in the Philosophy of Science.* 1997
ISBN 0-7923-4781-1
46. P.L. Kapitza: *Experiment, Theory, Practice.* Articles and Addresses. Edited by R.S. Cohen. 1980   ISBN 90-277-1061-9; Pb 90-277-1062-7
47. M.L. Dalla Chiara (ed.): *Italian Studies in the Philosophy of Science.* 1981
ISBN 90-277-0735-9; Pb 90-277-1073-2
48. M.W. Wartofsky: *Models.* Representation and the Scientific Understanding. [Synthese Library 129] 1979   ISBN 90-277-0736-7; Pb 90-277-0947-5
49. T.D. Thao: *Phenomenology and Dialectical Materialism.* Edited by R.S. Cohen. 1986
ISBN 90-277-0737-5
50. Y. Fried and J. Agassi: *Paranoia.* A Study in Diagnosis. [Synthese Library 102] 1976
ISBN 90-277-0704-9; Pb 90-277-0705-7
51. K.H. Wolff: *Surrender and Cath.* Experience and Inquiry Today. [Synthese Library 105] 1976
ISBN 90-277-0758-8; Pb 90-277-0765-0
52. K. Kosík: *Dialectics of the Concrete.* A Study on Problems of Man and World. 1976
ISBN 90-277-0761-8; Pb 90-277-0764-2
53. N. Goodman: *The Structure of Appearance.* [Synthese Library 107] 1977
ISBN 90-277-0773-1; Pb 90-277-0774-X
54. H.A. Simon: *Models of Discovery* and Other Topics in the Methods of Science. [Synthese Library 114] 1977   ISBN 90-277-0812-6; Pb 90-277-0858-4
55. M. Lazerowitz: *The Language of Philosophy.* Freud and Wittgenstein. [Synthese Library 117] 1977   ISBN 90-277-0826-6; Pb 90-277-0862-2
56. T. Nickles (ed.): *Scientific Discovery, Logic, and Rationality.* 1980
ISBN 90-277-1069-4; Pb 90-277-1070-8
57. J. Margolis: *Persons and Mind.* The Prospects of Nonreductive Materialism. [Synthese Library 121] 1978   ISBN 90-277-0854-1; Pb 90-277-0863-0
58. G. Radnitzky and G. Andersson (eds.): *Progress and Rationality in Science.* [Synthese Library 125] 1978   ISBN 90-277-0921-1; Pb 90-277-0922-X
59. G. Radnitzky and G. Andersson (eds.): *The Structure and Development of Science.* [Synthese Library 136] 1979   ISBN 90-277-0994-7; Pb 90-277-0995-5

Boston Studies in the Philosophy of Science

60. T. Nickles (ed.): *Scientific Discovery.* Case Studies. 1980
    ISBN 90-277-1092-9; Pb 90-277-1093-7
61. M.A. Finocchiaro: *Galileo and the Art of Reasoning.* Rhetorical Foundation of Logic and Scientific Method. 1980   ISBN 90-277-1094-5; Pb 90-277-1095-3
62. W.A. Wallace: *Prelude to Galileo.* Essays on Medieval and 16th-Century Sources of Galileo's Thought. 1981   ISBN 90-277-1215-8; Pb 90-277-1216-6
63. F. Rapp: *Analytical Philosophy of Technology.* Translated from German. 1981
    ISBN 90-277-1221-2; Pb 90-277-1222-0
64. R.S. Cohen and M.W. Wartofsky (eds.): *Hegel and the Sciences.* 1984   ISBN 90-277-0726-X
65. J. Agassi: *Science and Society.* Studies in the Sociology of Science. 1981
    ISBN 90-277-1244-1; Pb 90-277-1245-X
66. L. Tondl: *Problems of Semantics.* A Contribution to the Analysis of the Language of Science. Translated from Czech. 1981   ISBN 90-277-0148-2; Pb 90-277-0316-7
67. J. Agassi and R.S. Cohen (eds.): *Scientific Philosophy Today.* Essays in Honor of Mario Bunge. 1982   ISBN 90-277-1262-X; Pb 90-277-1263-8
68. W. Krajewski (ed.): *Polish Essays in the Philosophy of the Natural Sciences.* Translated from Polish and edited by R.S. Cohen and C.R. Fawcett. 1982
    ISBN 90-277-1286-7; Pb 90-277-1287-5
69. J.H. Fetzer: *Scientific Knowledge.* Causation, Explanation and Corroboration. 1981
    ISBN 90-277-1335-9; Pb 90-277-1336-7
70. S. Grossberg: *Studies of Mind and Brain.* Neural Principles of Learning, Perception, Development, Cognition, and Motor Control. 1982   ISBN 90-277-1359-6; Pb 90-277-1360-X
71. R.S. Cohen and M.W. Wartofsky (eds.): *Epistemology, Methodology, and the Social Sciences.* 1983.   ISBN 90-277-1454-1
72. K. Berka: *Measurement.* Its Concepts, Theories and Problems. Translated from Czech. 1983
    ISBN 90-277-1416-9
73. G.L. Pandit: *The Structure and Growth of Scientific Knowledge.* A Study in the Methodology of Epistemic Appraisal. 1983   ISBN 90-277-1434-7
74. A.A. Zinov'ev: *Logical Physics.* Translated from Russian. Edited by R.S. Cohen. 1983
    [*see also* Volume 9]   ISBN 90-277-0734-0
75. G-G. Granger: *Formal Thought and the Sciences of Man.* Translated from French. With and Introduction by A. Rosenberg. 1983   ISBN 90-277-1524-6
76. R.S. Cohen and L. Laudan (eds.): *Physics, Philosophy and Psychoanalysis.* Essays in Honor of Adolf Grünbaum. 1983   ISBN 90-277-1533-5
77. G. Böhme, W. van den Daele, R. Hohlfeld, W. Krohn and W. Schäfer: *Finalization in Science.* The Social Orientation of Scientific Progress. Translated from German. Edited by W. Schäfer. 1983   ISBN 90-277-1549-1
78. D. Shapere: *Reason and the Search for Knowledge.* Investigations in the Philosophy of Science. 1984   ISBN 90-277-1551-3; Pb 90-277-1641-2
79. G. Andersson (ed.): *Rationality in Science and Politics.* Translated from German. 1984
    ISBN 90-277-1575-0; Pb 90-277-1953-5
80. P.T. Durbin and F. Rapp (eds.): *Philosophy and Technology.* [*Also* Philosophy and Technology Series, Vol. 1] 1983   ISBN 90-277-1576-9
81. M. Marković: *Dialectical Theory of Meaning.* Translated from Serbo-Croat. 1984
    ISBN 90-277-1596-3
82. R.S. Cohen and M.W. Wartofsky (eds.): *Physical Sciences and History of Physics.* 1984.
    ISBN 90-277-1615-3

Boston Studies in the Philosophy of Science

83. É. Meyerson: *The Relativistic Deduction*. Epistemological Implications of the Theory of Relativity. Translated from French. With a Review by Albert Einstein and an Introduction by Milič Čapek. 1985 ISBN 90-277-1699-4
84. R.S. Cohen and M.W. Wartofsky (eds.): *Methodology, Metaphysics and the History of Science*. In Memory of Benjamin Nelson. 1984 ISBN 90-277-1711-7
85. G. Tamás: *The Logic of Categories*. Translated from Hungarian. Edited by R.S. Cohen. 1986
 ISBN 90-277-1742-7
86. S.L. de C. Fernandes: *Foundations of Objective Knowledge*. The Relations of Popper's Theory of Knowledge to That of Kant. 1985 ISBN 90-277-1809-1
87. R.S. Cohen and T. Schnelle (eds.): *Cognition and Fact*. Materials on Ludwik Fleck. 1986
 ISBN 90-277-1902-0
88. G. Freudenthal: *Atom and Individual in the Age of Newton*. On the Genesis of the Mechanistic World View. Translated from German. 1986 ISBN 90-277-1905-5
89. A. Donagan, A.N. Perovich Jr and M.V. Wedin (eds.): *Human Nature and Natural Knowledge*. Essays presented to Marjorie Grene on the Occasion of Her 75th Birthday. 1986
 ISBN 90-277-1974-8
90. C. Mitcham and A. Hunning (eds.): *Philosophy and Technology II*. Information Technology and Computers in Theory and Practice. [*Also* Philosophy and Technology Series, Vol. 2] 1986
 ISBN 90-277-1975-6
91. M. Grene and D. Nails (eds.): *Spinoza and the Sciences*. 1986 ISBN 90-277-1976-4
92. S.P. Turner: *The Search for a Methodology of Social Science*. Durkheim, Weber, and the 19th-Century Problem of Cause, Probability, and Action. 1986. ISBN 90-277-2067-3
93. I.C. Jarvie: *Thinking about Society*. Theory and Practice. 1986 ISBN 90-277-2068-1
94. E. Ullmann-Margalit (ed.): *The Kaleidoscope of Science*. The Israel Colloquium: Studies in History, Philosophy, and Sociology of Science, Vol. 1. 1986
 ISBN 90-277-2158-0; Pb 90-277-2159-9
95. E. Ullmann-Margalit (ed.): *The Prism of Science*. The Israel Colloquium: Studies in History, Philosophy, and Sociology of Science, Vol. 2. 1986
 ISBN 90-277-2160-2; Pb 90-277-2161-0
96. G. Márkus: *Language and Production*. A Critique of the Paradigms. Translated from French. 1986 ISBN 90-277-2169-6
97. F. Amrine, F.J. Zucker and H. Wheeler (eds.): *Goethe and the Sciences: A Reappraisal*. 1987
 ISBN 90-277-2265-X; Pb 90-277-2400-8
98. J.C. Pitt and M. Pera (eds.): *Rational Changes in Science*. Essays on Scientific Reasoning. Translated from Italian. 1987 ISBN 90-277-2417-2
99. O. Costa de Beauregard: *Time, the Physical Magnitude*. 1987 ISBN 90-277-2444-X
100. A. Shimony and D. Nails (eds.): *Naturalistic Epistemology*. A Symposium of Two Decades. 1987 ISBN 90-277-2337-0
101. N. Rotenstreich: *Time and Meaning in History*. 1987 ISBN 90-277-2467-9
102. D.B. Zilberman: *The Birth of Meaning in Hindu Thought*. Edited by R.S. Cohen. 1988
 ISBN 90-277-2497-0
103. T.F. Glick (ed.): *The Comparative Reception of Relativity*. 1987 ISBN 90-277-2498-9
104. Z. Harris, M. Gottfried, T. Ryckman, P. Mattick Jr, A. Daladier, T.N. Harris and S. Harris: *The Form of Information in Science*. Analysis of an Immunology Sublanguage. With a Preface by Hilary Putnam. 1989 ISBN 90-277-2516-0
105. F. Burwick (ed.): *Approaches to Organic Form*. Permutations in Science and Culture. 1987
 ISBN 90-277-2541-1

Boston Studies in the Philosophy of Science

106. M. Almási: *The Philosophy of Appearances*. Translated from Hungarian. 1989
 ISBN 90-277-2150-5
107. S. Hook, W.L. O'Neill and R. O'Toole (eds.): *Philosophy, History and Social Action*. Essays in Honor of Lewis Feuer. With an Autobiographical Essay by L. Feuer. 1988
 ISBN 90-277-2644-2
108. I. Hronszky, M. Fehér and B. Dajka: *Scientific Knowledge Socialized*. Selected Proceedings of the 5th Joint International Conference on the History and Philosophy of Science organized by the IUHPS (Veszprém, Hungary, 1984). 1988 ISBN 90-277-2284-6
109. P. Tillers and E.D. Green (eds.): *Probability and Inference in the Law of Evidence*. The Uses and Limits of Bayesianism. 1988 ISBN 90-277-2689-2
110. E. Ullmann-Margalit (ed.): *Science in Reflection*. The Israel Colloquium: Studies in History, Philosophy, and Sociology of Science, Vol. 3. 1988
 ISBN 90-277-2712-0; Pb 90-277-2713-9
111. K. Gavroglu, Y. Goudaroulis and P. Nicolacopoulos (eds.): *Imre Lakatos and Theories of Scientific Change*. 1989 ISBN 90-277-2766-X
112. B. Glassner and J.D. Moreno (eds.): *The Qualitative-Quantitative Distinction in the Social Sciences*. 1989 ISBN 90-277-2829-1
113. K. Arens: *Structures of Knowing*. Psychologies of the 19th Century. 1989
 ISBN 0-7923-0009-2
114. A. Janik: *Style, Politics and the Future of Philosophy*. 1989 ISBN 0-7923-0056-4
115. F. Amrine (ed.): *Literature and Science as Modes of Expression*. With an Introduction by S. Weininger. 1989 ISBN 0-7923-0133-1
116. J.R. Brown and J. Mittelstrass (eds.): *An Intimate Relation*. Studies in the History and Philosophy of Science. Presented to Robert E. Butts on His 60th Birthday. 1989
 ISBN 0-7923-0169-2
117. F. D'Agostino and I.C. Jarvie (eds.): *Freedom and Rationality*. Essays in Honor of John Watkins. 1989 ISBN 0-7923-0264-8
118. D. Zolo: *Reflexive Epistemology*. The Philosophical Legacy of Otto Neurath. 1989
 ISBN 0-7923-0320-2
119. M. Kearn, B.S. Philips and R.S. Cohen (eds.): *Georg Simmel and Contemporary Sociology*. 1989 ISBN 0-7923-0407-1
120. T.H. Levere and W.R. Shea (eds.): *Nature, Experiment and the Science*. Essays on Galileo and the Nature of Science. In Honour of Stillman Drake. 1989 ISBN 0-7923-0420-9
121. P. Nicolacopoulos (ed.): *Greek Studies in the Philosophy and History of Science*. 1990
 ISBN 0-7923-0717-8
122. R. Cooke and D. Costantini (eds.): *Statistics in Science*. The Foundations of Statistical Methods in Biology, Physics and Economics. 1990 ISBN 0-7923-0797-6
123. P. Duhem: *The Origins of Statics*. Translated from French by G.F. Leneaux, V.N. Vagliente and G.H. Wagner. With an Introduction by S.L. Jaki. 1991 ISBN 0-7923-0898-0
124. H. Kamerlingh Onnes: *Through Measurement to Knowledge*. The Selected Papers, 1853-1926. Edited and with an Introduction by K. Gavroglu and Y. Goudaroulis. 1991
 ISBN 0-7923-0825-5
125. M. Čapek: *The New Aspects of Time: Its Continuity and Novelties*. Selected Papers in the Philosophy of Science. 1991 ISBN 0-7923-0911-1
126. S. Unguru (ed.): *Physics, Cosmology and Astronomy, 1300–1700*. Tension and Accommodation. 1991 ISBN 0-7923-1022-5

Boston Studies in the Philosophy of Science

127. Z. Bechler: *Newton's Physics on the Conceptual Structure of the Scientific Revolution.* 1991
ISBN 0-7923-1054-3
128. É. Meyerson: *Explanation in the Sciences.* Translated from French by M-A. Siple and D.A. Siple. 1991
ISBN 0-7923-1129-9
129. A.I. Tauber (ed.): *Organism and the Origins of Self.* 1991   ISBN 0-7923-1185-X
130. F.J. Varela and J-P. Dupuy (eds.): *Understanding Origins.* Contemporary Views on the Origin of Life, Mind and Society. 1992
ISBN 0-7923-1251-1
131. G.L. Pandit: *Methodological Variance.* Essays in Epistemological Ontology and the Methodology of Science. 1991
ISBN 0-7923-1263-5
132. G. Munévar (ed.): *Beyond Reason.* Essays on the Philosophy of Paul Feyerabend. 1991
ISBN 0-7923-1272-4
133. T.E. Uebel (ed.): *Rediscovering the Forgotten Vienna Circle.* Austrian Studies on Otto Neurath and the Vienna Circle. Partly translated from German. 1991   ISBN 0-7923-1276-7
134. W.R. Woodward and R.S. Cohen (eds.): *World Views and Scientific Discipline Formation.* Science Studies in the [former] German Democratic Republic. Partly translated from German by W.R. Woodward. 1991
ISBN 0-7923-1286-4
135. P. Zambelli: *The Speculum Astronomiae and Its Enigma.* Astrology, Theology and Science in Albertus Magnus and His Contemporaries. 1992
ISBN 0-7923-1380-1
136. P. Petitjean, C. Jami and A.M. Moulin (eds.): *Science and Empires.* Historical Studies about Scientific Development and European Expansion.
ISBN 0-7923-1518-9
137. W.A. Wallace: *Galileo's Logic of Discovery and Proof.* The Background, Content, and Use of His Appropriated Treatises on Aristotle's *Posterior Analytics.* 1992   ISBN 0-7923-1577-4
138. W.A. Wallace: *Galileo's Logical Treatises.* A Translation, with Notes and Commentary, of His Appropriated Latin Questions on Aristotle's *Posterior Analytics.* 1992   ISBN 0-7923-1578-2
Set (137 + 138) ISBN 0-7923-1579-0
139. M.J. Nye, J.L. Richards and R.H. Stuewer (eds.): *The Invention of Physical Science.* Intersections of Mathematics, Theology and Natural Philosophy since the Seventeenth Century. Essays in Honor of Erwin N. Hiebert. 1992
ISBN 0-7923-1753-X
140. G. Corsi, M.L. dalla Chiara and G.C. Ghirardi (eds.): *Bridging the Gap: Philosophy, Mathematics and Physics.* Lectures on the Foundations of Science. 1992   ISBN 0-7923-1761-0
141. C.-H. Lin and D. Fu (eds.): *Philosophy and Conceptual History of Science in Taiwan.* 1992
ISBN 0-7923-1766-1
142. S. Sarkar (ed.): *The Founders of Evolutionary Genetics.* A Centenary Reappraisal. 1992
ISBN 0-7923-1777-7
143. J. Blackmore (ed.): *Ernst Mach – A Deeper Look.* Documents and New Perspectives. 1992
ISBN 0-7923-1853-6
144. P. Kroes and M. Bakker (eds.): *Technological Development and Science in the Industrial Age.* New Perspectives on the Science–Technology Relationship. 1992   ISBN 0-7923-1898-6
145. S. Amsterdamski: *Between History and Method.* Disputes about the Rationality of Science. 1992
ISBN 0-7923-1941-9
146. E. Ullmann-Margalit (ed.): *The Scientific Enterprise.* The Bar-Hillel Colloquium: Studies in History, Philosophy, and Sociology of Science, Volume 4. 1992   ISBN 0-7923-1992-3
147. L. Embree (ed.): *Metaarchaeology.* Reflections by Archaeologists and Philosophers. 1992
ISBN 0-7923-2023-9
148. S. French and H. Kamminga (eds.): *Correspondence, Invariance and Heuristics.* Essays in Honour of Heinz Post. 1993
ISBN 0-7923-2085-9
149. M. Bunzl: *The Context of Explanation.* 1993
ISBN 0-7923-2153-7

Boston Studies in the Philosophy of Science

150. I.B. Cohen (ed.): *The Natural Sciences and the Social Sciences.* Some Critical and Historical Perspectives. 1994 ISBN 0-7923-2223-1
151. K. Gavroglu, Y. Christianidis and E. Nicolaidis (eds.): *Trends in the Historiography of Science.* 1994 ISBN 0-7923-2255-X
152. S. Poggi and M. Bossi (eds.): *Romanticism in Science.* Science in Europe, 1790–1840. 1994 ISBN 0-7923-2336-X
153. J. Faye and H.J. Folse (eds.): *Niels Bohr and Contemporary Philosophy.* 1994 ISBN 0-7923-2378-5
154. C.C. Gould and R.S. Cohen (eds.): *Artifacts, Representations, and Social Practice.* Essays for Marx W. Wartofsky. 1994 ISBN 0-7923-2481-1
155. R.E. Butts: *Historical Pragmatics.* Philosophical Essays. 1993 ISBN 0-7923-2498-6
156. R. Rashed: *The Development of Arabic Mathematics: Between Arithmetic and Algebra.* Translated from French by A.F.W. Armstrong. 1994 ISBN 0-7923-2565-6
157. I. Szumilewicz-Lachman (ed.): *Zygmunt Zawirski: His Life and Work.* With Selected Writings on Time, Logic and the Methodology of Science. Translations by Feliks Lachman. Ed. by R.S. Cohen, with the assistance of B. Bergo. 1994 ISBN 0-7923-2566-4
158. S.N. Haq: *Names, Natures and Things.* The Alchemist Jābir ibn Hayyān and His *Kitāb al-Ahjār* (Book of Stones). 1994 ISBN 0-7923-2587-7
159. P. Plaass: *Kant's Theory of Natural Science.* Translation, Analytic Introduction and Commentary by Alfred E. and Maria G. Miller. 1994 ISBN 0-7923-2750-0
160. J. Misiek (ed.): *The Problem of Rationality in Science and its Philosophy.* On Popper vs. Polanyi. The Polish Conferences 1988–89. 1995 ISBN 0-7923-2925-2
161. I.C. Jarvie and N. Laor (eds.): *Critical Rationalism, Metaphysics and Science.* Essays for Joseph Agassi, Volume I. 1995 ISBN 0-7923-2960-0
162. I.C. Jarvie and N. Laor (eds.): *Critical Rationalism, the Social Sciences and the Humanities.* Essays for Joseph Agassi, Volume II. 1995 ISBN 0-7923-2961-9
    Set (161–162) ISBN 0-7923-2962-7
163. K. Gavroglu, J. Stachel and M.W. Wartofsky (eds.): *Physics, Philosophy, and the Scientific Community.* Essays in the Philosophy and History of the Natural Sciences and Mathematics. In Honor of Robert S. Cohen. 1995 ISBN 0-7923-2988-0
164. K. Gavroglu, J. Stachel and M.W. Wartofsky (eds.): *Science, Politics and Social Practice.* Essays on Marxism and Science, Philosophy of Culture and the Social Sciences. In Honor of Robert S. Cohen. 1995 ISBN 0-7923-2989-9
165. K. Gavroglu, J. Stachel and M.W. Wartofsky (eds.): *Science, Mind and Art.* Essays on Science and the Humanistic Understanding in Art, Epistemology, Religion and Ethics. Essays in Honor of Robert S. Cohen. 1995 ISBN 0-7923-2990-2
    Set (163–165) ISBN 0-7923-2991-0
166. K.H. Wolff: *Transformation in the Writing.* A Case of Surrender-and-Catch. 1995 ISBN 0-7923-3178-8
167. A.J. Kox and D.M. Siegel (eds.): *No Truth Except in the Details.* Essays in Honor of Martin J. Klein. 1995 ISBN 0-7923-3195-8
168. J. Blackmore: *Ludwig Boltzmann, His Later Life and Philosophy, 1900–1906.* Book One: A Documentary History. 1995 ISBN 0-7923-3231-8
169. R.S. Cohen, R. Hilpinen and R. Qiu (eds.): *Realism and Anti-Realism in the Philosophy of Science.* Beijing International Conference, 1992. 1996 ISBN 0-7923-3233-4
170. I. Kuçuradi and R.S. Cohen (eds.): *The Concept of Knowledge.* The Ankara Seminar. 1995 ISBN 0-7923-3241-5

Boston Studies in the Philosophy of Science

171. M.A. Grodin (ed.): *Meta Medical Ethics*: The Philosophical Foundations of Bioethics. 1995
     ISBN 0-7923-3344-6
172. S. Ramirez and R.S. Cohen (eds.): *Mexican Studies in the History and Philosophy of Science.* 1995
     ISBN 0-7923-3462-0
173. C. Dilworth: *The Metaphysics of Science.* An Account of Modern Science in Terms of Principles, Laws and Theories. 1995
     ISBN 0-7923-3693-3
174. J. Blackmore: *Ludwig Boltzmann, His Later Life and Philosophy, 1900–1906* Book Two: The Philosopher. 1995
     ISBN 0-7923-3464-7
175. P. Damerow: *Abstraction and Representation.* Essays on the Cultural Evolution of Thinking. 1996
     ISBN 0-7923-3816-2
176. M.S. Macrakis: *Scarcity's Ways: The Origins of Capital.* A Critical Essay on Thermodynamics, Statistical Mechanics and Economics. 1997
     ISBN 0-7923-4760-9
177. M. Marion and R.S. Cohen (eds.): *Québec Studies in the Philosophy of Science.* Part I: Logic, Mathematics, Physics and History of Science. Essays in Honor of Hugues Leblanc. 1995
     ISBN 0-7923-3559-7
178. M. Marion and R.S. Cohen (eds.): *Québec Studies in the Philosophy of Science.* Part II: Biology, Psychology, Cognitive Science and Economics. Essays in Honor of Hugues Leblanc. 1996
     ISBN 0-7923-3560-0
     Set (177–178) ISBN 0-7923-3561-9
179. Fan Dainian and R.S. Cohen (eds.): *Chinese Studies in the History and Philosophy of Science and Technology.* 1996
     ISBN 0-7923-3463-9
180. P. Forman and J.M. Sánchez-Ron (eds.): *National Military Establishments and the Advancement of Science and Technology.* Studies in 20th Century History. 1996
     ISBN 0-7923-3541-4
181. E.J. Post: *Quantum Reprogramming.* Ensembles and Single Systems: A Two-Tier Approach to Quantum Mechanics. 1995
     ISBN 0-7923-3565-1
182. A.I. Tauber (ed.): *The Elusive Synthesis: Aesthetics and Science.* 1996   ISBN 0-7923-3904-5
183. S. Sarkar (ed.): *The Philosophy and History of Molecular Biology: New Perspectives.* 1996
     ISBN 0-7923-3947-9
184. J.T. Cushing, A. Fine and S. Goldstein (eds.): *Bohmian Mechanics and Quantum Theory: An Appraisal.* 1996
     ISBN 0-7923-4028-0
185. K. Michalski: *Logic and Time.* An Essay on Husserl's Theory of Meaning. 1996
     ISBN 0-7923-4082-5
186. G. Munévar (ed.): *Spanish Studies in the Philosophy of Science.* 1996   ISBN 0-7923-4147-3
187. G. Schubring (ed.): *Hermann Günther Graßmann (1809–1877): Visionary Mathematician, Scientist and Neohumanist Scholar.* Papers from a Sesquicentennial Conference. 1996
     ISBN 0-7923-4261-5
188. M. Bitbol: *Schrödinger's Philosophy of Quantum Mechanics.* 1996   ISBN 0-7923-4266-6
189. J. Faye, U. Scheffler and M. Urchs (eds.): *Perspectives on Time.* 1997   ISBN 0-7923-4330-1
190. K. Lehrer and J.C. Marek (eds.): *Austrian Philosophy Past and Present.* Essays in Honor of Rudolf Haller. 1996
     ISBN 0-7923-4347-6
191. J.L. Lagrange: *Analytical Mechanics.* Translated and edited by Auguste Boissonade and Victor N. Vagliente. Translated from the *Mécanique Analytique, novelle édition* of 1811. 1997
     ISBN 0-7923-4349-2
192. D. Ginev and R.S. Cohen (eds.): *Issues and Images in the Philosophy of Science.* Scientific and Philosophical Essays in Honour of Azarya Polikarov. 1997
     ISBN 0-7923-4444-8

Boston Studies in the Philosophy of Science

193. R.S. Cohen, M. Horne and J. Stachel (eds.): *Experimental Metaphysics*. Quantum Mechanical Studies for Abner Shimony, Volume One. 1997   ISBN 0-7923-4452-9
194. R.S. Cohen, M. Horne and J. Stachel (eds.): *Potentiality, Entanglement and Passion-at-a-Distance*. Quantum Mechanical Studies for Abner Shimony, Volume Two. 1997
ISBN 0-7923-4453-7; Set 0-7923-4454-5
195. R.S. Cohen and A.I. Tauber (eds.): *Philosophies of Nature: The Human Dimension*. 1997
ISBN 0-7923-4579-7
196. M. Otte and M. Panza (eds.): *Analysis and Synthesis in Mathematics*. History and Philosophy. 1997   ISBN 0-7923-4570-3
197. A. Denkel: *The Natural Background of Meaning*. 1999   ISBN 0-7923-5331-5
198. D. Baird, R.I.G. Hughes and A. Nordmann (eds.): *Heinrich Hertz: Classical Physicist, Modern Philosopher*. 1999   ISBN 0-7923-4653-X
199. A. Franklin: *Can That be Right?* Essays on Experiment, Evidence, and Science. 1999
ISBN 0-7923-5464-8
200. D. Raven, W. Krohn and R.S. Cohen (eds.): *The Social Origins of Modern Science*. 2000
ISBN 0-7923-6457-0
201. Reserved
202. Reserved
203. B. Babich and R.S. Cohen (eds.): *Nietzsche, Theories of Knowledge, and Critical Theory*. Nietzsche and the Sciences I. 1999   ISBN 0-7923-5742-6
204. B. Babich and R.S. Cohen (eds.): *Nietzsche, Epistemology, and Philosophy of Science*. Nietzsche and the Science II. 1999   ISBN 0-7923-5743-4
205. R. Hooykaas: *Fact, Faith and Fiction in the Development of Science*. The Gifford Lectures given in the University of St Andrews 1976. 1999   ISBN 0-7923-5774-4
206. M. Fehér, O. Kiss and L. Ropolyi (eds.): *Hermeneutics and Science*. 1999 ISBN 0-7923-5798-1
207. R.M. MacLeod (ed.): *Science and the Pacific War*. Science and Survival in the Pacific, 1939-1945. 1999   ISBN 0-7923-5851-1
208. I. Hanzel: *The Concept of Scientific Law in the Philosophy of Science and Epistemology*. A Study of Theoretical Reason. 1999   ISBN 0-7923-5852-X
209. G. Helm; R.J. Deltete (ed./transl.): *The Historical Development of Energetics*. 1999
ISBN 0-7923-5874-0
210. A. Orenstein and P. Kotatko (eds.): *Knowledge, Language and Logic*. Questions for Quine. 1999   ISBN 0-7923-5986-0
211. R.S. Cohen and H. Levine (eds.): *Maimonides and the Sciences*. 2000   ISBN 0-7923-6053-2
212. H. Gourko, D.I. Williamson and A.I. Tauber (eds.): *The Evolutionary Biology Papers of Elie Metchnikoff*. 2000   ISBN 0-7923-6067-2
213. S. D'Agostino: *A History of the Ideas of Theoretical Physics*. Essays on the Nineteenth and Twentieth Century Physics. 2000   ISBN 0-7923-6094-X
214. S. Lelas: *Science and Modernity*. Toward An Integral Theory of Science. 2000
ISBN 0-7923-6303-5
215. E. Agazzi and M. Pauri (eds.): *The Reality of the Unobservable*. Observability, Unobservability and Their Impact on the Issue of Scientific Realism. 2000   ISBN 0-7923-6311-6
216. P. Hoyningen-Huene and H. Sankey (eds.): *Incommensurability and Related Matters*. 2001 ISBN 0-7923-6989-0
217. A. Nieto-Galan: *Colouring Textiles*. A History of Natural Dyestuffs in Industrial Europe. 2001
ISBN 0-7923-7022-8

Boston Studies in the Philosophy of Science

218. J. Blackmore, R. Itagaki and S. Tanaka (eds.): *Ernst Mach's Vienna 1895–1930.* Or Phenomenalism as Philosophy of Science. 2001   ISBN 0-7923-7122-4
219. R. Vihalemm (ed.): *Estonian Studies in the History and Philosophy of Science.* 2001
     ISBN 0-7923-7189-5
220. W. Lefèvre (ed.): *Between Leibniz, Newton, and Kant.* Philosophy and Science in the Eighteenth Century. 2001   ISBN 0-7923-7198-4
221. T.F. Glick, M.Á. Puig-Samper and R. Ruiz (eds.): *The Reception of Darwinism in the Iberian World.* Spain, Spanish America and Brazil. 2001   ISBN 1-4020-0082-0
222. U. Klein (ed.): *Tools and Modes of Representation in the Laboratory Sciences.* 2001
     ISBN 1-4020-0100-2
223. P. Duhem: *Mixture and Chemical Combination.* And Related Essays. Edited and translated, with an introduction, by Paul Needham. 2002   ISBN 1-4020-0232-7
224. J.C. Boudri: *What was Mechanical about Mechanics.* The Concept of Force Betweem Metaphysics and Mechanics from Newton to Lagrange. 2002   ISBN 1-4020-0233-5
225. B.E. Babich (ed.): *Hermeneutic Philosophy of Science, Van Gogh's Eyes, and God.* Essays in Honor of Patrick A. Heelan, S.J. 2002   ISBN 1-4020-0234-3
226. D. Davies Villemaire: *E.A. Burtt, Historian and Philosopher.* A Study of the Author of The Metaphysical Foundations of Modern Physical Science. 2002   ISBN 1-4020-0428-1
227. L.J. Cohen: *Knowledge and Language.* Selected Essays of L. Jonathan Cohen. Edited and with an introduction by James Logue. 2002   ISBN 1-4020-0474-5
228. G.E. Allen and R.M. MacLeod (eds.): *Science, History and Social Activism: A Tribute to Everett Mendelsohn.* 2002   ISBN 1-4020-0495-0
229. O. Gal: *Meanest Foundations and Nobler Superstructures.* Hooke, Newton and the "Compounding of the Celestiall Motions of the Planetts". 2002   ISBN 1-4020-0732-9
230. R. Nola: *Rescuing Reason.* A Critique of Anti-Rationalist Views of Science and Knowledge. 2003   Hb: ISBN 1-4020-1042-7; Pb ISBN 1-4020-1043-5
231. J. Agassi: *Science and Culture.* 2003   ISBN 1-4020-1156-3
232. M.C. Galavotti (ed.): *Observation and Experiment in the Natural and Social Science.* 2003
     ISBN 1-4020-1251-9

Also of interest:
R.S. Cohen and M.W. Wartofsky (eds.): *A Portrait of Twenty-Five Years Boston Colloquia for the Philosophy of Science, 1960-1985.* 1985   ISBN Pb 90-277-1971-3

Previous volumes are still available.

KLUWER ACADEMIC PUBLISHERS – DORDRECHT / BOSTON / LONDON